T0255009

ACOUSTICS
OF
SMALL ROOMS

ACOUSTICS
OF
SMALL ROOMS

MENDEL KLEINER
JIRI TICHY

CRC Press
Taylor & Francis Group
Boca Raton London New York

CRC Press is an imprint of the
Taylor & Francis Group, an **Informa** business

A SPON BOOK

CRC Press
Taylor & Francis Group
6000 Broken Sound Parkway NW, Suite 300
Boca Raton, FL 33487-2742

© 2014 by Taylor & Francis Group, LLC
CRC Press is an imprint of Taylor & Francis Group, an Informa business

First issued in paperback 2017

No claim to original U.S. Government works
Version Date: 20160322

ISBN 13: 978-1-138-07283-1 (pbk)
ISBN 13: 978-0-415-77930-2 (hbk)

This book contains information obtained from authentic and highly regarded sources. Reasonable efforts have been made to publish reliable data and information, but the author and publisher cannot assume responsibility for the validity of all materials or the consequences of their use. The authors and publishers have attempted to trace the copyright holders of all material reproduced in this publication and apologize to copyright holders if permission to publish in this form has not been obtained. If any copyright material has not been acknowledged please write and let us know so we may rectify in any future reprint.

Except as permitted under U.S. Copyright Law, no part of this book may be reprinted, reproduced, transmitted, or utilized in any form by any electronic, mechanical, or other means, now known or hereafter invented, including photocopying, microfilming, and recording, or in any information storage or retrieval system, without written permission from the publishers.

For permission to photocopy or use material electronically from this work, please access www.copyright.com (http://www.copyright.com/) or contact the Copyright Clearance Center, Inc. (CCC), 222 Rosewood Drive, Danvers, MA 01923, 978-750-8400. CCC is a not-for-profit organization that provides licenses and registration for a variety of users. For organizations that have been granted a photocopy license by the CCC, a separate system of payment has been arranged.

Trademark Notice: Product or corporate names may be trademarks or registered trademarks, and are used only for identification and explanation without intent to infringe.

Library of Congress Cataloging-in-Publication Data

Kleiner, Mendel, 1946-
   Acoustics of small rooms / authors, Mendel Kleiner, Jiri Tichy.
     pages cm
   Includes bibliographical references and index.
   ISBN 978-0-415-77930-2 (hardback)
   1. Acoustical engineering. 2. Rooms--Physiological aspects. I. Tichy, Jiri. II. Title.

TA365.K54 2014
690'.2--dc23
                                                      2014001314

Visit the Taylor & Francis Web site at
http://www.taylorandfrancis.com

and the CRC Press Web site at
http://www.crcpress.com

The authors would like to dedicate this book to their respective wives Marja Missan Kleiner, Dagmar Tichy (deceased), and Iva Apfelbeck-Tichy.

Contents

Preface

WHY A BOOK ON THE ACOUSTICS OF SMALL ROOMS?

In this book, we focus on sound in small rooms that have interior volumes in the range from a few cubic meters to a few hundred cubic meters. Thus, rooms as diverse as car cabins and small lecture rooms, reverberation and anechoic chambers might fit the description. Better and acoustic definitions of small rooms could be rooms in which the individual resonances of the room must be considered at low frequencies, or rooms in which the early reflections by walls, ceiling, and room objects arrive within milliseconds of the direct sound.

The sound reproduction in a room is determined by the room's resonances (often called modes). The resonances are damped by sound absorption in the room, and their decay is noted as reverberation. With many resonances per unit frequency, the subjective response of the room is heard as smooth. In a small room, such as a car compartment, there are only about 10 resonances below 200 Hz, but in a concert hall there will be many more resonances. We can still play back the concert hall recording in the car, but the reproduced sound of the low frequencies of the instruments will sound *tainted* or *colored*.

Also, the way the room affects the time history of voice and music is important subjectively. When the resonances are well damped, the time history for the reflections arriving at the listener will make the difference between a *well-sounding* room and one in which there is coloration or even flutter echo. The way that the room directs the sound from the source, musical instrument, voice, or loudspeaker is essentially determined by the reflecting surfaces of the room and its in-room objects, and their acoustical characteristics. A person, the listener, for example, will absorb and scatter sound, so it is never possible to measure or reproduce exactly the sound simply for the reason that humans move around.

Surround sound, stereo, hi-fi systems are all attempts at producing what one audio equipment manufacturer called "... the closest approach to the original sound." As humans, we are multisensory receptors. We hear, but unavoidable tactile and visual information is added to our listening experience. In addition, we use cognition when judging the quality and characteristics of audio and acoustics. The closest approach to the original sound may not be enough for emotionally equivalent small room reproduction of the sound recorded in a concert hall.

This book has 14 chapters. It starts with the description of sound and its propagation. The physics of sound generation, propagation, diffusion, and absorption are described in Chapter 1 using a mathematical approach. The fundamentals are followed in Chapter 2 by an analysis of the properties of the internal resonances of rooms. For high frequencies, a nonmodal approach called geometrical acoustics can be better used to determine the acoustical conditions for music generation and reproduction. This method is described in Chapter 3. Chapter 4 discusses the principles of sound absorption. Absorption is not sufficient, however, to achieve good listening conditions. Hearing typically prefers temporally and spatially diffuse sound reflections, so specular reflections need to be modified by diffusers, as described in Chapter 5. Chapter 6 introduces the physiology of hearing and describes the fundamental sensitivity properties of hearing. Chapters 7 and 8 are devoted to psychoacoustics, the study of the fundamentals of sound perception considering the time patterns and spatial distribution of room sound reflections reaching the listener. Sound reproduction in small rooms is dealt with in Chapters 9 and 10. Chapter 11 exemplifies typical designs used for control rooms in the recording industry to illustrate the approaches used to optimize room geometry, absorption, and diffusion that can be applied to home listening rooms as well. The acoustics of small rehearsal and practice rooms are discussed in Chapter 12. The last two chapters (Chapters 13 and 14) describe approaches used for the modeling and measurement of room acoustic properties, with an emphasis on small rooms.

Both authors are room acoustics researchers and teach room acoustics to undergraduate as well graduate students. We enjoy classical music, attend live concerts often, and consider ourselves fortunate in having experienced the natural acoustics of a wide range of concert halls. Additionally, we find the science of audio sound reproduction in small rooms fascinating and challenging. We hope that you the reader will come to share our enthusiasm for the study of sound in small rooms.

Acknowledgment

Parts of this book were prepared within the framework program entitled "The Organ as Memory Bank" funded by the Swedish Research Council (Contract-ID 90237101).

Authors

Mendel Kleiner was born in the displaced persons camp at Bergen-Belsen, Germany, in 1946 and came to Sweden as a refugee in 1947. He studied in Sweden and received his MSc in electronic engineering from Chalmers in 1969. He then received his PhD in building acoustics in 1978. Dr. Kleiner became tenured professor of acoustics at Chalmers University of Technology, Gothenburg, Sweden, in 1995. He was also tenured professor of acoustics at the School of Architecture at Rensselaer Polytechnic Institute 2003–2005, director of its Architectural Acoustics Program, and director of doctoral studies in its School of Architecture.

Dr. Kleiner retired from his professorship in 2013 and is now professor emeritus at Chalmers University of Technology, Gothenburg, Sweden. Before his retirement, he taught room acoustics, audio, electroacoustics, and ultrasonics in the Chalmers Master Program on Sound and Vibration in the Division of Applied Acoustics and directed the Chalmers Room Acoustics Group. After retirement, he continues his research in acoustics, as well as teaching and consulting. He has extensive R&D experience in many fields of applied acoustics and worked as a consultant to companies such as Volvo, Electrolux, and Autoliv.

Dr. Kleiner's main research areas include electroacoustics and audio, computer simulation of room acoustics, electroacoustic room acoustics enhancement, room acoustics of auditoria and small rooms, sound and vibration measurement technology, product sound quality, psychoacoustics, musical acoustics, as well as ultrasonics.

He has authored or coauthored more than 55 publications and 110 papers in peer-reviewed journals and conference proceedings, holds two patents, and has given several invited international courses on sound field simulation. He has also authored books on architectural acoustics and electroacoustics and has coauthored a book on the acoustics of churches, mosques, and synagogues. He has been invited to present many keynote lectures and courses on sound field simulation at international conferences on acoustics and noise control. He has also organized an international conference on auralization and been on the organizing committee of many international conferences.

Dr. Kleiner became a fellow of the Acoustical Society of America in 2000 and has been recognized for his pioneering work in audible sound field simulation. He serves on the Audio Engineering Society's Standards Committee on Acoustics and was chair of its Technical Committee on Acoustics and Sound Reinforcement for 20 years. He is also a member of the Institute of Acoustics and of IEEE.

Jiri Tichy was born in 1927 in the Czech Republic. After completing his education in electrical engineering and physics, Dr. Tichy received his PhD in engineering physics from the Czech Technical University in Prague. In 1968, Dr. Tichy was invited to the United States, where he became professor of acoustics at the Pennsylvania State University. In 1965, the university established the Intercollege Graduate Program in Acoustics that was later reorganized as the Graduate Department of Acoustics in the College of Engineering. Dr. Tichy served as the head of this department for 25 years until his retirement. The department has awarded several hundred PhD and MS degrees in acoustics and is a major source of education in airborne and underwater sound, architectural acoustics, noise control, vibrations, and signal processing. The department also provides continuing education in acoustics for students in industry and research organizations.

Dr. Tichy's principal interests are in architectural acoustics, noise control, acoustic intensity technique, active noise control, and virtual reality. He taught courses in architectural and building acoustics, theoretical acoustics, active noise control, and intensity technique at Penn State, Czech Technical University in Prague, and at various US industries.

With his students, Dr. Tichy has published 84 papers in journals and conference proceedings and presented 8 keynote lectures in the United States, Japan, Chile, Brazil, France, Germany, and the Czech Republic. Dr. Tichy has also presented 28 invited papers and 69 contributed papers at various national and international conferences.

In 1999, Dr. Tichy organized the first joint international meeting of the Acoustical Society of America with the European Acoustical Association that was held in Berlin, Germany. This was the largest meeting on acoustics that took place in the previous century. In addition, Dr. Tichy organized and co-organized seven national or international meetings on active control of sound as well as a symposia on sound intensity technique for industry.

Dr. Tichy is strongly involved with several professional societies. His home society is the Acoustical Society of America. He has served or chaired many of its professional committees such as medals and awards, architectural acoustics, noise, publication policy, and others. He was elected a fellow, and in 1994–1995 he became the president of the society. Dr. Tichy was also awarded a gold medal by the Acoustical Society of America for his contributions to acoustics and its education.

Another important society that Dr. Tichy is associated with is the Audio Engineering Society, which is particularly involved in most of the topics covered in this book. Dr. Tichy was also elected as a fellow of this society and continues to serve as a member of the Architectural Acoustics and the Standards Committee.

Dr. Tichy has been involved with other societies, including the Institute of Noise Control Engineering (1986–1987), where he also served as the president. The Czech Acoustical Society elected him as an honorary member of the society. He was also a member of the Acoustical Society of Japan and the New York Academy of Sciences.

In addition to his research activities, Dr. Tichy often consults on a variety of acoustical projects.

Chapter 1

Physics of small room sound fields

1.1 BASIC SOUND FIELD QUANTITIES

A common property of gases, liquids, and solid-state materials is elasticity
[1–3,6]. If some force deflects a certain portion of such a medium, this
deflection propagates away at a speed that depends on the elastic properties
of the medium. The deflection of the molecules results in varying the local
medium density and pressure from its equilibrium level. The total pressure
p_t at certain location is

$$p_t = p_{eq} - p \ \ [\text{Pa}] \tag{1.1}$$

where
  p_{eq} is the equilibrium pressure
  p is called the sound pressure caused by the propagating disturbance

The sound pressure is the most important and useful quantity in acoustics.
The perception of sound is caused by the sound pressure acting on the eardrum. As will be shown in later chapters, the acoustic properties of enclosures depend on the spatial and temporal distribution of sound pressure.
Because the sound pressure is a scalar quantity, it is usually manageable by
simpler equations than equations with vectors.

Sound pressure is measured in pascals [Pa], which has a dimension
[N/m²]. Because of an extremely large range of the ear sensitivity, we are
often representing the sound pressure by a logarithmic scale so that the
pressure is expressed as the sound pressure levels (SPLs) L_p in decibel (dB) by

$$L_p = 20 \log \frac{p_{rms}}{p_0} \ \ [\text{dB}] \tag{1.2}$$

where $p_0 = 2 \cdot 10^{-5}$ Pa is the internationally standardized reference value for
sound waves in the air.

Due to the fluctuations of the sound pressure, the medium density also varies. The total density ρ_t is then equal to

$$\rho_t = \rho_0 + \delta \tag{1.3}$$

where
  ρ_t is the total density
  ρ_0 is the constant equilibrium density
  δ is the density increment due to the sound pressure

Because δ is much smaller than the equilibrium density, ρ_t is very often replaced by the constant ρ_0.

Our primary interest in this book is the sound in the air. For an audible sound, the magnitude of the acoustic quantities is much smaller than the magnitude of their equilibrium values. Therefore, in equations that express their mutual relationship, the higher-order terms will be neglected so that these equations will be linear.

The particles or molecules of the media vibrate with a particle velocity $\vec{u}$ and particle displacement $\vec{s}$. These vectors are related as

$$\vec{u} = \frac{\partial \vec{s}}{\partial t} \quad \text{or} \quad \vec{s} = \int_t \vec{u} \cdot dt \tag{1.4}$$

1.1.1 Speed of sound

The fluctuations in the medium propagate with the speed c that is a vector quantity. The propagation direction of this vector is perpendicular to the wave fronts. These are surfaces of the equal wave phase. The details are discussed later. The sound speed direction denotes the direction of energy propagation. Quite often, only the magnitude of the sound speed is needed in equations where the direction of sound propagation has been defined or the sound speed is related to other general quantities such as temperature.

The speed of sound depends on several nonacoustic quantities such as humidity and temperature. The international ISO 9613 standard provides details for precise calculation of c. One of the significant effects on c has the medium temperature as given for air by

$$c = 343.2 \sqrt{\frac{273.15 + T_{°C}}{293.15}} \text{ m/s} \tag{1.5}$$

where $T_{\circ C}$ is the medium temperature in centigrade. As the earlier equation reveals, the speed of sound for 20°C is 343.2 m/s. The constant 373.15 K is the absolute temperature of 0°C in Kelvin degrees.

Other basic acoustic quantities are linked to energy in the sound field. These are quadratic quantities because they are calculated from a product of two linear quantities. These are energy densities, kinetic and potential, that describe the energy accumulated in sound fields. The propagation of the sound energy is described by the sound intensity. We will defer the definition and use of these quantities to the section on specific sound fields so that we can show their usefulness better.

1.2 WAVE EQUATION FOR FREE WAVES

The basic equation to calculate the spatial and time properties of acoustic fields including the effects of their boundaries is the wave equation. This equation is particularly useful for the analysis and design of sound fields in small spaces, which is the principal subject of this book. The waves are called free because they are initially excited, for instance, by a short sound pulse so that afterward the waves propagate freely.

The radiation and propagation of sound is determined by basic laws of mechanics and thermodynamics. The quantities concerned are sound pressure, density, particle velocity, temperature, as well as other energy-related quantities. The basic three equations can be combined into one equation called the wave equation. This chapter covers the details of these steps, the significance and basic application of these equations particularly relevant to small rooms.

1.2.1 Euler's equation

For the sake of simplicity, we will first consider a 1D sound field in which the waves and the medium particles oscillate in one direction. At this point, we will not consider the generator of the field but will concentrate on the mathematical relationship in the free field. Figure 1.1 shows a very small portion of the sound field of cross section S and length dx.

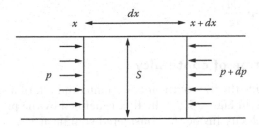

Figure 1.1 Sound pressure acting on a small volume.

The two sides of this small volume, perpendicular to the sound propagation direction are exposed to slightly different forces so that the resulting force F on this volume is equal to their differences

$$S\left[p(x)-p(x+dx)\right] = -S\frac{\partial p}{\partial x}dx = F \tag{1.6}$$

We use Newton's law for force $F = m(du/dt)$, where $m = (\rho + \delta)Sdx$ is the total mass of the considered volume element including the density change due to the sound δ, and change the Lagrange description of acceleration du/dt into an Euler description using the equation

$$\frac{du}{dt} = \frac{\partial u}{\partial t} + \frac{\partial u}{\partial x}\frac{\partial u}{\partial t} \approx \frac{\partial u}{\partial t} \tag{1.7}$$

Because the second term in this equation is much smaller than the first term, we neglect it so that Equation 1.7 becomes linear. Substituting into Equation 1.6 results in

$$-\frac{\partial p}{\partial x} = \rho_t \frac{\partial u}{\partial t} \tag{1.8}$$

This is the important Euler's equation that shows the relationship between the sound pressure and the particle velocity. Its principle use is in the calculation of the particle velocity u from the acoustic pressure p.

The first term in Equation 1.8 is a part of the 3D gradient operator, that is,

$$\text{grad} \equiv \nabla = \frac{\partial}{\partial x}i + \frac{\partial}{\partial y}j + \frac{\partial}{\partial z}k \tag{1.9}$$

Equation 1.8 for a 3D field is

$$\text{grad } p = -\rho_t \frac{\partial \vec{u}}{\partial t} \tag{1.10}$$

1.2.2 Equation of continuity

We will examine the flux of the medium into and out of a small volume element as shown in Figure 1.2. The flux is defined by the product of medium density and velocity inside the considered volume at x in dt and is given by

$$S\rho_t(x)u(x)dt \tag{1.11}$$

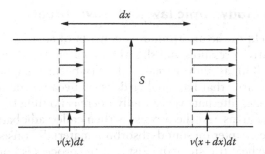

Figure 1.2 Medium flow through a small volume.

The flux going through S at $x + dx$ is

$$S\rho_t\left(x + dx\right)u\left(x + dx\right)dt = S\left[\rho_t\left(x\right) + \frac{\partial \rho_t}{\partial x}dx\right]\left[u\left(x\right) + \frac{\partial u}{\partial x}dx\right]dt$$

$$= S\left[\rho_t u + \frac{\partial\left(\rho_t u\right)}{\partial x}dx + \frac{\partial \rho_t}{\partial x}\frac{\partial u}{\partial x}\left(dx\right)^2\right]dt \qquad (1.12)$$

The last term in the earlier equation is very small and will be neglected so that the medium flow from the considered volume becomes

$$S\left[\rho_t\left(x\right)u\left(x\right) - \rho_t\left(x + dx\right)u\left(x + dx\right)\right]dt = -S\frac{\partial\left(\rho_t u\right)}{\partial x}dxdt \qquad (1.13)$$

The medium flow from the small volume is equal to the change of the matter $(\partial \rho_t/\partial t)dtSdx$ that is formulated by the equation of continuity:

$$\frac{\partial\left(\rho_t u\right)}{\partial x} = -\frac{\partial \rho_t}{\partial t} \qquad (1.14)$$

The particle velocity $\vec{u}$ in the earlier equation is an x component of the velocity vector $\vec{u} = u_x\vec{i} + u_y\vec{j} + u_z\vec{k}$. To proceed with the earlier equation, we will use the *div* operator

$$div\,\vec{u} = \frac{\partial u_x}{\partial x} + \frac{\partial u_y}{\partial y} + \frac{\partial u_z}{\partial z} \qquad (1.15)$$

so that the equation of continuity can be written in the form

$$div\,\vec{u} = -\frac{1}{\rho c^2}\frac{\partial p}{\partial t} \qquad (1.16)$$

1.2.3 Thermodynamic law and wave equation

The sound field in the air around us consists of small volumes with positive and negative pressures and slightly increased and decreased gas densities. The small areas with increased density become slightly warmer than a neighboring area that has smaller density. Because the density and temperature changes alternate very rapidly, there is no time for the heat to flow between these areas and the process is then called adiabatic. However, we will see in the chapter on sound-absorbing materials consisting of very fine fibers that the heat transfer can exist and the process is polytropic.

For an adiabatic process, the relationship between the sound pressure and the density is given by $p = c^2\rho$, where c is a constant equal to the speed of sound. We substitute in Equation 1.12 for ρ in $\partial\rho_t/\partial t$ and obtain for the equation of continuity and Euler's equation:

$$-\rho\frac{\partial u}{\partial x} = \frac{1}{c^2}\frac{\partial p}{\partial t} \tag{1.17}$$

$$-\frac{\partial p}{\partial x} = \rho_t\frac{\partial u}{\partial t} \tag{1.18}$$

These two equations can be combined after deriving the first equation by x and the second by t and making the mixed terms equal. We obtain

$$\frac{\partial^2 p}{\partial x^2} - \frac{1}{c^2}\frac{\partial^2 p}{\partial t^2} = 0 \tag{1.19}$$

This is a 1D wave equation for sound pressure–free waves because there is not a source term. Also, it can be derived for other acoustic quantities such as u and ρ. The wave equation is a fundamental equation that provides the temporal and spatial distributions in bounded and unbounded sound fields.

Equation 1.19 has been derived for 1D waves. However, to solve 3D problems, we replace $\partial^2 p/\partial x^2$ by a 3D Laplace operator Δ for a desired coordinate system. For Cartesian coordinates, the Laplace operator has the form $\Delta(p) = \dfrac{\partial^2 p}{\partial x^2} + \dfrac{\partial^2 p}{\partial y^2} + \dfrac{\partial^2 p}{\partial z^2}$ so that the wave equation is

$$\Delta p - \frac{1}{c^2}\frac{\partial^2 p}{\partial t^2} = 0 \tag{1.20}$$

In cylindrical coordinates (r, φ, z), the wave equation has the form

$$\frac{1}{r}\frac{\partial}{\partial r}\left(r\frac{\partial p}{\partial r}\right) + \frac{1}{r^2}\frac{\partial^2 p}{\partial \varphi^2} + \frac{\partial^2 p}{\partial z^2} - \frac{1}{c^2}\frac{\partial^2 p}{\partial t^2} = 0 \tag{1.21}$$

and in spherical coordinates (r, ϑ, φ), it is given by

$$\frac{1}{r^2}\left(\frac{\partial}{\partial r}\left(r^2\frac{\partial p}{\partial r}\right) + \frac{1}{\sin\vartheta}\frac{\partial}{\partial\vartheta}\left(\frac{1}{\sin\vartheta}\frac{\partial p}{\partial\vartheta}\right) + \frac{1}{\sin^2\vartheta}\frac{\partial^2 p}{\partial\varphi^2}\right) - \frac{1}{c^2}\frac{\partial^2 p}{\partial t^2} = 0 \tag{1.22}$$

1.3 WAVE EQUATION FOR FORCED WAVES

The previous section dealt with a sound field that is generated by some source that acts over a short period so that some sound field is created and then left alone. Our concentration was on the interaction of forces and the propagation of the density fluctuations in the sound field due to the elastic properties of the medium.

This section deals with some of the ways that generate the fluctuations of the medium density externally. In general, these fluctuations can be initiated by external forces, injection of additional masses, and other causes that act on the medium. The source, although it might vary in time, is acting permanently and thus forcing the waves through the medium.

We will concentrate on the kind of sound sources that generate sound in small rooms, such as speech or music. The wave equation was formulated by combining the equation of motion with the equation of continuity. To include the effects of sources, these equations have to be modified. Because of the kind of sources that will be considered, such as loudspeakers, Euler's equation (1.18) does not require any extension. However, the equation of continuity that describes the fluid flux will be amended by a term ρQ that expresses generated mass per unit of volume flow Q. The dimension of Q is [m³/s]. The equation of continuity (1.17), extended and modified, is

$$\rho_0 \cdot \operatorname{div}\vec{u} = -\frac{\partial p}{\partial t} + \rho_0 Q = \frac{1}{c^2}\frac{\partial p}{\partial t} + \rho_0 Q \tag{1.23}$$

We also need Euler's equation for three dimensions:

$$\operatorname{grad} p = -\rho\frac{\partial\vec{u}}{\partial t} \tag{1.24}$$

In order to obtain the wave equation for the sound pressure, we eliminate from the earlier equations $\vec{u}$ and obtain

$$\Delta p - \frac{1}{c^2}\frac{\partial^2 p}{\partial t^2} = -\rho_0 \frac{dQ}{\partial t} \tag{1.25}$$

The right side of the earlier equation for forced waves represents the volume velocity of the source and therefore is applicable on sources that have vibrating surfaces, such as loudspeakers, musical instruments, and vibrating machines. Because the time derivative is applied on the velocity, we see that it is the acceleration that is determining the *strength* of the radiated sound. More details can be found in sections on radiation from sources.

1.4 WAVE EQUATION SOLUTIONS

1.4.1 Plane waves

Because the wave equation is a second-order partial differential equation, the solution is given by two functions with two existing second derivations. The 1D wave equation, derived in the previous section for the sound pressure, is

$$\frac{\partial^2 p}{\partial x^2} - \frac{1}{c^2}\frac{\partial^2 p}{\partial t^2} = 0 \tag{1.26}$$

The solution by the d'Alembert principle is given by any arbitrary function that has the variable $(x - ct)$ or $(x + ct)$. The sound pressure can be formulated as

$$p(x,t) = F(x - ct) + G(x + ct) \tag{1.27}$$

The function F represents a wave traveling in the positive x direction with a velocity c. The direction of propagation is revealed from the following consideration. In order to keep F the same, its argument $(x - ct)$ must be constant. If an observation point is moved to a point $x + \Delta x$, t must be increased in order to keep $(x - ct)$ unchanged, which means that the wave motion repeats at the new point later and the waves travel in the $+x$ direction. Similarly, function G represents a wave that propagates in the $-x$ direction.

In most situations, the sound analyzed by the ear is characterized by the frequency and amplitude of its sound pressure. These waves are called harmonic waves. The sound is a real phenomenon and, therefore, it is mathematically expressed by a real function. However, the mathematical operations linked to wave propagation are simpler and more practical if they are expressed by complex functions. The sound pressure of the plane wave traveling in the +x direction is

$$p(x,t) = p_+ e^{-j(kx-\omega t)} \tag{1.28}$$

where $\omega = 2\pi f$ is the radial frequency, $k = \omega/c = 1/T$ is the wave number, T is the period of sound, and $j = \sqrt{-1}$. The amplitude of the sound wave p_+ is most often a real number but can be complex as well. The particle velocity $\vec{u}$ can be calculated from Euler's equation, which has for harmonic waves the form

$$\vec{u}(x,t) = j\frac{1}{k\rho_0 c}\,\mathrm{grad}(p)\,e^{j\omega t} \tag{1.29}$$

This equation can be simplified by executing the operations so that we obtain

$$u(x) = \frac{1}{\rho_0 c}\,p(x) \tag{1.30}$$

The ratio of the pressure and particle velocity is called specific acoustic impedance. For a plane propagating wave, the earlier equation yields

$$z(x) = \frac{p(x)}{u(x)} \tag{1.31}$$

where $z(x)$ is called characteristic impedance of the medium. The characteristic impedance for 20°C warm air is 414 kg/m²·s. As can be seen from the earlier definition, the characteristic impedance is a measure of the resistance to the wave motion at a given point x.

An observer positioned at a point x in a plane harmonic wave field will notice that the molecules of air will move back and forth in a direction parallel with the x-axis. This motion is in phase at all points that are in a plane, perpendicular to the x-axis, which means that all molecules are, for instance, in zero position at the same time. This plane is called a wave front. As is shown in the next chapters, the wave fronts are very important to visualize the sound fields, the flow of sound energy, the radiation from sound sources, and other matters.

1.4.2 Spherical waves

If the sound is radiated from a source that is so small that we can call it a point source, the wave fronts of the sound waves are concentric spheres.

The sound pressure depends only on one coordinate, which is the distance r from the source. The wave equation (1.22) for the spherical coordinates after removing the dependence on the angle ϑ and φ simplifies into

$$\frac{\partial^2 p}{\partial r^2} + \frac{2}{r}\frac{\partial p}{\partial r} = \frac{1}{c^2}\frac{\partial^2 p}{\partial t^2} \tag{1.32}$$

The d'Alembert general solution for a wave outgoing from the point $r=0$ is

$$p(r,t) = \frac{1}{r}F(r-ct) \tag{1.33}$$

where F is an arbitrary function. Equation 1.34 can be reformulated to a more useful form

$$\frac{\partial^2 (rp)}{\partial r^2} - \frac{1}{c^2}\frac{\partial^2 (rp)}{\partial t^2} = 0 \tag{1.34}$$

This form resembles that of the 1D formulation for plane waves (Equation 1.26) so that the general solution is

$$rp(r,t) = F(ct-r) + G(ct+r) \tag{1.35}$$

which can be shown for $p(r, t)$ to be

$$p(r,t) = \frac{F(ct-r)}{r} + \frac{G(ct+r)}{r} \tag{1.36}$$

The first term represents a wave propagating from point $r=0$ and the second term a wave that travels in opposite direction. As with the plane wave, we will only consider the outgoing harmonic wave so that the sound pressure is

$$p(r,k) = A\frac{e^{-jkr}}{r} \tag{1.37}$$

where A is the wave amplitude. Using Equation 1.29, the particle velocity at a point r is

$$u(r,k) = j\frac{1}{k\rho_0 c}\frac{\partial p(r)}{\partial r} = \frac{A}{j\omega\rho}\left(\frac{jk}{r} + \frac{1}{r^2}\right)e^{-jkr} \tag{1.38}$$

The first term of this equation is real and in phase with the sound pressure. The second term is imaginary and is decreasing with the distance r very rapidly. At very large distance, the sphere has a large radius and over a small patch resembles a plane. If we neglect the second term in the earlier equation, the particle velocity relationship with the sound pressure is the same as with the plane wave, which is

$$u(r) \approx \frac{p(r)}{\rho c} \qquad (1.39)$$

The specific acoustic impedance is

$$z(r,k) = \frac{p(r,k)}{u(r,k)} = \rho_0 c \left(\frac{jkr}{1+jkr} \right) = \rho_0 c \left(\frac{j2\pi r/\lambda}{1+j2\pi r/\lambda} \right) \qquad (1.40)$$

This equation shows that for a large radius, measured in wavelengths, the wave impedance approaches $\rho_0 c$, the impedance of a plane wave, and, therefore, at that point of the field, can be treated as a plane wave, as discussed earlier. Further properties of the spherical waves are discussed in a section on sound sources.

1.5 ACOUSTIC ENERGY DENSITY

The spatial and temporal distribution of energy density is very important to characterize and study the sound fields in enclosures. As discussed earlier, an acoustic wave consists of small fluid volumes that slightly move and change its volume. Therefore, the motion of the mass associated with the volume has a kinetic energy. Due to the fluid's elasticity, the compressions and rarefactions are associated with a potential energy. The instantaneous value of the kinetic energy that the small volume V_0 of the mass $\rho_0 V_0$ carries is

$$w_k(t) = \frac{1}{2} \rho_0 V_0 u^2(t) \qquad (1.41)$$

The potential energy is given by

$$w_p(t) = \frac{V_0}{2\rho_0 c^2} p^2(t) \qquad (1.42)$$

The total energy is the sum of the kinetic and potential energy. We are mostly interested in energy density $w(t)$ that is the energy per unit volume. After dividing the sum of the energies by V_0, we obtain

$$w(t) = w_k(t) + w_p(t) = \frac{1}{2}\rho_0\left(u^2(t) + \frac{p^2(t)}{(\rho_0 c)^2}\right)$$ (1.43)

The energy density $w(t)$ is a function of time and for a general wave, where $p(t)$ and $u(t)$ will have a phase relationship, is time variable.

Because many sound fields can be represented by the sum of plane waves, we will calculate both the instantaneous and time-averaged energy density of a plane progressive harmonic wave. We replace the article velocity $u(t)$ in Equation 1.43 by $p(t)$ using $u(t) = p(t)/\rho_0 c$ and obtain

$$w(t) = \frac{p^2(t)}{(\rho_0 c)^2}$$ (1.44)

The time average of $w(t)$ is given for a harmonic wave with the period T by

$$w = \frac{1}{T}\int_{-T/2}^{T/2} \frac{p^2(t)}{\rho_0 c^2}\, dt = \frac{|p|^2}{2\rho_0 c^2}$$ (1.45)

The quantity $|p|^2$ is the amplitude of the propagating wave that is independent on time and location and, therefore, the energy in the harmonic propagating plane wave is constant.

1.6 SOUND LEVELS

The sound pressure ratio of the weakest sound we can hear and the strongest sound that can cause pain in the ear is extremely high and, depending on frequency, can reach approximately 10^{10}. Therefore, it is practical to express most acoustical quantities on a logarithmic scale as a ratio of the described quantity to a referenced quantity. These have been internationally standardized so that the numerical values of the concerned quantities are numerically convenient. We call these quantities levels and their unit a bell [B] or a decibel [dB]. Because the concerned acoustical quantities are squared, they represent an energy-related quantity rather than a linear quantity.

The SPL was defined in Equation 1.2. Similarly, the sound velocity level is defined as

$$L_u = 20\log\left(\frac{u_{rms}}{u_0}\right) \; [\text{dB}] \tag{1.46}$$

The reference value of the particle velocity for waves in air is usually $u_0 = 5 \cdot 10^{-8}$ m/s, since with this reference value, we obtain for free plane waves in air the same dB values for sound velocity level as for SPL.

The acoustic intensity, which is the sound power per unit area that propagates in a plane wave, is calculated from $I = p \cdot u$ and is expressed as intensity level L_I using a reference level $I_0 = 10^{-12}$ W/m²

$$L_I = 10\log\left(\frac{I}{I_0}\right) \; [\text{dB}] \tag{1.47}$$

The total power P radiated by a source expressed by a power level L_P using the reference value $P_0 = 10^{-12}$ W is

$$L_P = 10\log\left(\frac{P}{P_0}\right) \; [\text{dB}] \tag{1.48}$$

The characteristic impedance of the medium Z_0 is used in many equations. The magnitude of this constant depends on the barometric pressure, the air density, and the temperature. For a defined normal barometric pressure $P_b = 1.01325 \cdot 10^5$ Pa and air density $\rho_0 = 1.292$ kg/m³, the characteristic impedance for temperature 0°C and 20°C is

$$Z_{0,0} = 428.1 \quad \text{and} \quad Z_{0,20} = 413.3 \; \text{N s/m}^3 \tag{1.49}$$

1.7 FREQUENCY ANALYSIS

The sounds around us can be categorized into two distinct groups [4]. One group consists of unwanted and disturbing sounds that we generally call noise. The other group includes all kinds of useful sounds for communication and sounds for pleasure that is in essence music. Most often, we listen to music and communicate in rooms. To analyze the sound fields in enclosures and methods of their optimization requires the knowledge of the temporal and frequency properties of the source signals as well as of the sound fields in the rooms and psychoacoustics of listening.

This chapter deals with an overview of the basic signal processing methods that a reader needs in later sections of the book to understand the analysis and modifications of the room properties to achieve optimum listening conditions. For a more extensive knowledge of signal processing, the reader is advised to study books on signal processing and communication such as Reference 4.

As discussed in the previous sections, the sound is characterized by the fluctuations of sound pressure, medium density, etc., and, therefore, is a function of time. We usually model these functions by a sum of harmonic functions. When we listen to sounds, for instance, from musical instruments, our ear senses the time fluctuations of the sound signal as a periodic signal (assuming that the sound lasts sufficiently long) of certain frequency. The time function of the signal can be transformed into the frequency function by the Fourier transform. This transformation is not only a mathematical modeling, but transformation of the signals from the time into the frequency domain is achieved by our hearing as described in Chapter 6.

For a periodic time signal $f(t)$ that has a finite period, the transformation into the frequency domain is expressed by the Fourier series that has, in an exponential notation, the form

$$f(t) = \sum_{n=-\infty}^{n=\infty} F(n) \, e^{j2\pi n f_0 t} \quad n = \ldots -2,-1,0,1,2,\ldots \tag{1.50}$$

The inverse transformation as well as the spectrum of $f(t)$ is

$$F(n) = \frac{1}{T_0} \int_0^{T_0} f(t) \, e^{-j2\pi n f_0 t} dt \quad n = \ldots -2,-1,0,1,2,\ldots \tag{1.51}$$

where
   f_0 is the fundamental frequency
   n denotes the higher harmonics

The function $f(t)$ is represented by a series of goniometric functions. Equation 1.51 is a transformation of $f(t)$ into its frequency domain representation $F(n)$ that is called a complex line spectrum. $F(n)$ and $f(t)$ are the Fourier transform pairs. The periodic signal is represented by its frequency components $\omega_0, 2\omega_0, 3\omega_0$, etc., with $\omega_0 = 2\pi/T$.

If we have a function with infinitely long period T so that $f(t)$ represents the first cycle (and only one) in the interval $(-\infty < t < \infty)$, the fundamental frequency becomes smaller and smaller as $T \rightarrow \infty$. In this limit, the spectrum becomes continuous. The signal $f(t)$ is now represented by a continuous sum of various components and the two earlier equations become a Fourier transform pair

$$f(t) = \int_{-\infty}^{\infty} F(f)e^{j2\pi ft} \, df \qquad (1.52)$$

and the inverse

$$F(f) = \int_{-\infty}^{\infty} f(t)e^{-j2\pi ft} \, dt \qquad (1.53)$$

The function $F(f)$ is a frequency spectrum of the function $f(t)$, which represents the signal. Its spectrum is usually expressed as

$$F(f) = |F(f)|e^{j\varphi(f)} \qquad (1.54)$$

The complex spectrum consists of the amplitude $|F(f)|$ and the phase $\varphi(f)$ that are both real functions of frequency. The earlier equations represent either a periodical or random signal that never repeats.

Quite often, we are working with steady single frequency tones. Their time function can be represented by a cosine function $\cos(2\pi f_0 t)$, where f_0 is the frequency of sound. The Fourier transform pair is

$$\cos(2\pi f_0 t) \leftrightarrow \frac{1}{2}\left[\delta(f - f_0) + \delta(f + f_0)\right] \qquad (1.55)$$

where δ is the delta function. Figure 1.3 shows both the time function and its line spectrum for the (infinitely long) cosine function consisting of two

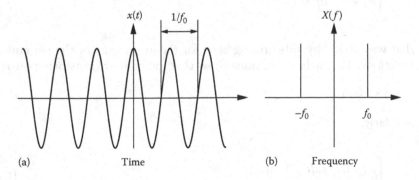

Figure 1.3 Fourier transform pair of a cosine function. (a) Time function and (b) spectrum.

lines symmetrical to the y-axes. The left line represents negative frequency that has often assigned a zero amplitude.

Similarly, the transform of the sin function is

$$\sin(2\pi f_0 t) \leftrightarrow \frac{1}{2j}\left[\delta(f - f_0) - \delta(f + f_0)\right] \tag{1.56}$$

We note that the spectrum magnitude is the same as that of the cosine function.

1.7.1 Impulse function

The delta function is a very important tool of signal processing because the response to the impulse function is linked to spectra. The basic properties of this function are

$$\delta(t) = 0 \quad \text{for } t \neq 0 \quad \text{and} \quad \delta(t) = 1 \quad \text{for } t = 0 \tag{1.57}$$

We will show some often applied Fourier transforms of sound pulses that are often used in the measurements of the room properties. The sound pulse is very short and is usually approximated by a delta function. The use of the δ function can be shown in the following equation:

$$\int_{-\infty}^{\infty} \delta(t - t_0)x(t)\, dt = x(t_0) \tag{1.58}$$

that shows the way how $x(t_0)$ is assigned to $x(t)$ [5]. Similarly, we can solve the integral

$$\int_{-\infty}^{\infty} \delta(f)e^{j2\pi ft}\, df = 1 \tag{1.59}$$

that was solved by substituting zero for the arguments of the sin and cos functions. The earlier equations show that the Fourier transform pair is

$$1 \leftrightarrow \delta(f) \tag{1.60}$$

Similarly,

$$\int_{-\infty}^{\infty} \delta(t - t_0)f(t)dt = f(t_0) \tag{1.61}$$

and

$$e^{j2\pi f_0 t} \leftrightarrow \delta(f - f_0) \tag{1.62}$$

Further applications and the usefulness of the delta function will be seen in many further sections of this book.

1.7.2 Correlation

We will extend the time–frequency analysis and analyze several relationships of the two time functions. This overview section concentrates on the mathematical definitions, while the physical and other applications will be shown later. The reader will find the proofs of many statements in the books on signal processing. Let us consider two periodic functions $f_1(t)$ and $f_2(t)$ that have the same angular frequency ω_1. A function defined as

$$\varphi_{12}(\tau) = \frac{1}{T_1} \int_{-T_1/2}^{T_1/2} f_1(t) f_2(t + \tau)\, dt \tag{1.63}$$

is called correlation function, where τ is a continuous time of displacement of f_2 in the interval $(-\infty, \infty)$. The Fourier transform of this function is

$$\phi_{12}(n) = F_1^*(n) F_2(n) \tag{1.64}$$

where $F_1^*(n)$ is complex conjugate of $F_1(n)$. The earlier relation is called the correlation theorem for periodic functions. Many graphical interpretations are found in books on communication theory [4].

1.7.3 Autocorrelation

If the two functions in the definition of the correlation function are identical, the function is called the autocorrelation function (ACF) $\varphi_{11}(\tau)$. It can be shown that the following relation is valid:

$$\varphi_{11}(\tau) = \frac{1}{T_1} \int_{-T_1/2}^{T_1/2} f_1(t) f_1(t + \tau)\, dt = \sum_{n=-\infty}^{\infty} |F_1(n)|^2 e^{jn\omega_1 \tau} \tag{1.65}$$

where

$$\phi_{11}(n) = |F_1(n)|^2 = \frac{1}{T_1} \int_{-T_1/2}^{T_1/2} \varphi_{11}(\tau)\, e^{-jn\omega_1 \tau}\, d\tau \tag{1.66}$$

is the power spectrum. The relations in the earlier two equations are usually called autocorrelation theorem. These indicate that the autocorrelation function and the power spectrum of periodic functions are a Fourier transform pair. Please note that the power spectrum depends only on the amplitudes of the harmonics and not on their mutual phase. A sine signal has a sinusoid autocorrelation function, whereas the autocorrelation function of a white noise signal is that of a delta function. The autocorrelation function is important in the analysis of noisy signals.

1.7.4 Cross correlation

In the earlier part of this section, we defined the general correlation function by Equation 1.63. Because two different functions of the same frequency are involved, this function is also called cross-correlation function.

Equations 1.63 and 1.64 define these functions and their relation. The subscripts of their definitions are linked to the operations as shown. The change of the function indices results in

$$\varphi_{21}(\tau) = \frac{1}{T_1} \int_{-T_1/2}^{T_1/2} f_2(t) f_1(t + \tau) \, dt$$

$$\phi_{21}(n) = F_2^*(n) F_1(n)$$

(1.67)

The exchange of the indices results in

$$\varphi_{12}(-\tau) = \varphi_{21}(\tau) \quad \text{and} \quad \phi_{12}(n) = \phi_{21}^*(n)$$

(1.68)

Equations 1.64 and 1.67 are the cross-power spectra. The word power is obviously used because we are dealing with a product of two linear functions and quantities that are suitable to calculate power. They can be complex. But after the algebra for the goniometric functions is executed, the result for $\phi_{12}(n)$ is real. What is the physical meaning of the cross and autocorrelations depending on the f functions? They could, for instance, represent sound pressure and particle velocity (which are not in phase). The cross power is acoustic intensity, that is, the sound power per unit of surface. In another example, we could use electrical voltage and current to calculate power.

Another important application is the measurement of sound spaciousness at certain location in an enclosure that is one of the quality factors of sound perception. Hearing uses cross correlation in its binaural analysis of the sounds that enter our ears. Similarly, a signal from two microphones located at the positions of the ears of a manikin can be fed into an analyzer that can calculate the cross-correlation function of the two signals. The lower the cross correlation, the greater the perceived spaciousness of the sound.

1.7.5 Convolution and correlation

Another operation that is very similar to correlation is *convolution*, which is defined as

$$\rho_{12}(\tau) = \frac{1}{T_1} \int_{-T_1/2}^{T_1/2} f_1(t) f_2(\tau - t)\, dt \tag{1.69}$$

$$\rho_{21}(\tau) = \frac{1}{T_1} \int_{-T_1/2}^{T_1/2} f_2(t) f_1(\tau - t)\, dt \tag{1.70}$$

The main difference between these two functions and their correlation is that in their graphical representation, the second function is folded backward. By manipulating the variables, it can be shown that the convolution is a correlation with the first function that has a reversed sign as shown in the following equation:

$$\varphi_{12}(\tau) = \frac{1}{T_1} \int_{-T_1/2}^{T_1/2} f_1(t) f_2(t + \tau)\, dt = \frac{1}{T_1} \int_{-T_1/2}^{T_1/2} f_1(-t) f_2(\tau - t)\, dt \tag{1.71}$$

The Fourier transform of the convolution function for periodic signals is the product

$$F_1(n) F_2(n) \tag{1.72}$$

where $F_1(n)$ and $F_2(n)$ are the Fourier transforms of $f_1(t)$ and $f_2(t)$, respectively. The convolution of the two time signals $f_1(t)$ and $f_2(t)$ is usually written as $f_1(t) * f_2(t)$.

1.7.6 Correlation and convolution of aperiodic functions

As explained earlier, if the period of a function becomes infinitely long, the function becomes aperiodic and most of the sums convert into integrals. The Fourier transform pair of a single function is shown in the previous section. If the Fourier transform of the correlation of aperiodic functions is evaluated, it results in the expression

$$\int_{-\infty}^{\infty} f_1(t) f_2(t + \tau)\, dt = \int_{-\infty}^{\infty} 2\pi F_1^*(\omega) F_2(\omega) e^{j\omega\tau}\, d\omega \tag{1.73}$$

so that

$$\int_{-\infty}^{\infty} f_1(t)f_2(t+\tau)\,dt \quad \text{and} \quad 2\pi F_1^*(\omega)F_2(\omega) \tag{1.74}$$

are a Fourier transform pair. Similarly, the relations for the autocorrelation function using the notations in Equations 1.65 and 1.66 are

$$\varphi_{11}(\tau) = \int_{-\infty}^{\infty} \phi_{11}(\omega)e^{j\omega\tau}d\omega \tag{1.75}$$

$$\phi_{11}(\omega) = \frac{1}{2\pi}\int_{-\infty}^{\infty} \varphi_{11}(\tau)\,e^{-j\omega\tau}\,d\tau = 2\pi|F_1(\omega)|^2 \tag{1.76}$$

As earlier, the second equation is the spectrum of the autocorrelation function. The functions $\varphi_{11}(\tau)$ and $\phi_{11}(\omega)$ are real functions. As a consequence, when we expand the exponential terms in both of these equations, only the terms with cos functions are nonzero. The mean power of the function $f_1(t)$ is obtained for $\tau = (0)$ from

$$\varphi_{11}(0) = \int_{-\infty}^{\infty} f_1^2(t)\,dt \tag{1.77}$$

Another property of the autocorrelation of a random functions is that it is an even function so that

$$\varphi_{11}(-\tau) = \varphi_{11}(\tau) \tag{1.78}$$

As follows from Equation 1.74, the spectrum of the autocorrelation function can be obtained from the multiplication of $F_1^*(\omega)$ and $F_1(\omega)$. A multiplication of a complex spectrum with its complex conjugate removes the exponential terms that carry the information on the phase so that the autocorrelation spectrum depends only on the amplitude of the generating functions. As a consequence, the autocorrelation of functions with differing phase spectra but equal amplitude spectra is the same.

The convolutions of two aperiodic functions and their Fourier transform are

$$\rho_{12}(\tau) = \int_{-\infty}^{\infty} f_1(t)f_2(\tau-t)\,dt \tag{1.79}$$

$$2\pi F_1(\omega)F_2(\omega) \tag{1.80}$$

As before, we can manipulate the variables of the correlation so that we obtain

$$\varphi_{12}(\tau) = \int\limits_{-\infty}^{\infty} f_1(t)f_2(t+\tau)\,dt = \int\limits_{-\infty}^{\infty} f_1(-t)f_2(\tau-t)\,dt \qquad (1.81)$$

This equation shows that the cross correlation of $f_1(t)$ and $f_2(t)$ leads to the same result as the convolution of $f_1(-t)$ and $f_2(t)$.

1.7.7 Convolution integral applied on linear systems

The transfer function $H(\omega)$ between two points in a room contains important information on the transmission of sound in the room. It is affected by the room shape, acoustical properties of the walls, furniture, and other treatments that have some influence on the sound propagation. The transfer function is the Fourier transform of the impulse response. One important way to find the transfer function is to measure the impulse response $h(t)$ function generated at the source point and received at the point of interest. This very short sound impulse should be similar to a Dirac pulse with sound pressure properties represented by

$$p(t) = \begin{cases} 1, & 0 \le t \le \Delta t \\ 0, & \text{otherwise} \end{cases} \qquad (1.82)$$

If the input into the system is $f(t)$, the output $g(t)$ is given by

$$g(t) = \int\limits_{-\infty}^{\infty} f(\tau)h(t-\tau)d\tau = f(t) * h(t) \qquad (1.83)$$

The Fourier transform of the output $G(\omega)$ can be calculated from

$$G(\omega) = H(\omega)\,F(\omega) \qquad (1.84)$$

1.8 SOUND INTENSITY

The sound intensity is a very important acoustic quantity. The intensity is the sound power per unit area that propagates in the sound field from a source to a location where it is absorbed. Relatively recently, in the latter part of the previous century, the progress in signal processing opened the possibility to measure the sound intensity and thus visualize the energy propagation in sound fields. We can measure the details of sound radiation from complex sources and modify their construction to reduce the radiated sound power, which contributes to the progress in noise control. Similarly, the detailed measurement of radiation from audio sources leads to their efficiency improvements and the reduction of sound distortion. The visualization of the sound field intensity in enclosures provides better information on the sound fields than the measurement of sound pressure that provides only static information while intensity carries also a temporal component of the wave progression [9].

This section provides the fundamentals on intensity fields and the background to the common methods of intensity measurements. The instantaneous intensity $\vec{I}_i(r,t)$ at a point r is obtained from the multiplication of sound pressure and particle velocity at some point:

$$\vec{I}_i(r,t) = p(r,t) \cdot \vec{u}(r,t) \ [\text{W/m}^2] \tag{1.85}$$

As can be seen from this definition, the instantaneous direction of the sound intensity vector is the same as the direction of the particle velocity $\vec{u}(r,t)$. The direction of the particle velocity at a point is constant with time only in single plane and spherical waves. Even in two interfering plane waves, the path of the medium molecules is elliptical and, therefore, the direction of the power propagation varies with time. The motion of the medium particles in a complex wave is even more complicated. A more meaningful information on the power propagation in a wave is obtained from a time average of the instantaneous intensity defined by

$$\vec{I}(r) = \frac{1}{T} \int_0^T p(r,t) \cdot \vec{u}(r,t) \, dt \tag{1.86}$$

for tonal sound components with the period $T = 1/f$. For random sound, T is chosen according to the length over which we observe the sound or over other selected time criteria. Because $\vec{I}(r)$ is a quadratic quantity, we use for the calculation a real form for both the sound pressure and the particle velocity so that $\vec{I}(r)$ is

$$I = \frac{1}{T} \int_0^T P \cos(\omega t) \cdot U \cos(\omega t + \theta) \, dt = \frac{1}{2} PU \cos(\theta) \tag{1.87}$$

where θ is the phase shift between p and u. If we are using a complex form, the integrand in the earlier equation becomes

$$p^*(t) \cdot u(t) = Pe^{-j\omega t} \cdot Ue^{(j\omega t + \theta)} \tag{1.88}$$

For measurement of sound intensity, it is necessary to find an estimate of the particle velocity. The particle velocity component $\vec{u}_x$ can be found from 1D Euler's equation

$$\frac{dp(r,t)}{dr} = -\rho \frac{\partial \vec{u}_r(r,t)}{\partial t} \tag{1.89}$$

The left-hand side derivative in the earlier equation can be approximated by small differences Δp and Δr from two microphones (see Figure 14.7 in Chapter 14) so that

$$\frac{dp}{dr} \approx \frac{\Delta p}{\Delta r} = \frac{p_1 - p_2}{\Delta r} \tag{1.90}$$

The sound pressure can be estimated from the average of sound pressure measured by the two small pressure-sensitive microphones:

$$p(t) = \frac{1}{2}\left[p_2(t) + p_1(t)\right] \tag{1.91}$$

The particle velocity is obtained from Euler's equation solution:

$$u(t) = -\frac{1}{\rho_0 \Delta r} \int_{-\infty}^{t} \left[p_2(t) - p_1(t)\right] dt \tag{1.92}$$

By combining the upper equations, the time intensity estimate is

$$I(t) = \frac{1}{2\rho_0 \Delta r} E\left[p_2(t) + p_1(t) \int_{-\infty}^{t} \left[p_2(t) - p_1(t)\right] dt \right] \tag{1.93}$$

where E is the expected value. The analyses of sound fields and power flow are usually expressed in frequency domain. Applying the Fourier transform and its dual channel extensions results in the frequency domain intensity formulations. The sound intensity can be expressed in the complex domain as

$$\hat{I}_T(\omega) = I(\omega) + jQ(\omega) \tag{1.94}$$

where
   $\hat{I}_T(\omega)$ is the complex intensity vector
   $I(\omega)$ is the intensity representing the power flow
   $Q(\omega)$ is called reactive intensity and, as shown further, is linked to sound pressure

The complex intensity is

$$\hat{I}_T(\omega) = \frac{1}{j2\omega\rho_0\Delta r}\left\{ S_{p_2p_2}(\omega) - S_{p_1p_1}(\omega) - j2\,\mathrm{Im}\left[S_{p_2p_1}(\omega) \right] \right\} \tag{1.95}$$

The two quantities $S_{p_1p_1}(\omega)$ and $S_{p_2p_2}(\omega)$ in Equation 1.95 are the auto-spectra of the sound pressures measured by the first and the second microphone. This equation also contains the symbol $S_{p_2p_1}(\omega)$, which is the cross spectrum of the sound pressure obtained by the measurement. Because intensity represents sound power, which is a real quantity, the earlier equation gives us $I(\omega)$ as

$$I(\omega) = \mathrm{Re}\left[\hat{I}_T(\omega) \right] = \frac{1}{\omega\rho_0\Delta r}\,\mathrm{Im}\left[S_{p_2p_1}(\omega) \right] \tag{1.96}$$

This equation shows that to obtain the intensity magnitude, we have to measure the cross spectra of the voltages obtained from the two microphone probes. The reactive intensity $Q(\omega)$ is given by

$$Q(\omega) = \mathrm{Im}\left[\hat{I}_T(\omega) \right] = \frac{1}{2\omega\rho_0\Delta r}\left[S_{p_2p_2}(\omega) - S_{p_1p_1}(\omega) \right] \tag{1.97}$$

The measurement of a sound field by using a two microphone probe and further signal processing permits us to obtain both the potential $V(\omega)$ and the kinetic $T(\omega)$ energy densities as

$$V(\omega) = \frac{P^2(r,\omega)}{4\rho_0 c^2} = \frac{1}{16\rho_0 c^2}\left[S_{p_2p_2}(\omega) + S_{p_1p_1}(\omega) + 2\,\mathrm{Re}\left(S_{p_2p_1}(\omega) \right) \right] \tag{1.98}$$

$$T(\omega) = \frac{\rho_0 U^2(r,\omega)}{4} = \frac{1}{4\omega^2\rho_0\left(\Delta r\right)^2}\left[S_{p_2p_2}(\omega) + S_{p_1p_1}(\omega) - 2\,\mathrm{Re}\left(S_{p_2p_1}(\omega) \right) \right]$$

$$\tag{1.99}$$

Sound intensity data can be used for the visualization of the energy flow in enclosures and, due to the relationship with other acoustical quantities, we can visualize the *structure* of the sound fields. This contributes essentially to the sound phenomena in the enclosures as shown in the following example.

The sound pressure of amplitude $P(r)$ and phase $\phi(r)$, both real quantities, at a point r in a general harmonic sound field can be expressed as

$$p(r) = P(r)e^{-j\phi(r)} \tag{1.100}$$

The particle velocity is

$$\vec{u}(r,k) = j\frac{1}{k\rho_0 c}\nabla p(r) \tag{1.101}$$

The total sound intensity $\vec{I}_T$ amplitude in the complex form (see Equation 1.95) is

$$\vec{I}_T(r) = \vec{I}(r) + j\vec{Q}(r) = \frac{1}{2\omega\rho_0}\left[P^2(r)\nabla\phi(r) - jP(r)\nabla P(r)\right] \tag{1.102}$$

The active intensity amplitude is $\vec{I}(r)$ and the real part of the earlier expression is

$$\vec{I}(r) = \frac{1}{2\omega\rho_0}P^2(r)\nabla\phi(r) \tag{1.103}$$

Because the intensity vector is proportional to the gradient of phase, its direction is normal to the surfaces of the constant phase. We can see how the time-averaged sound power propagates in the sound field.

The reactive intensity $\vec{Q}(r)$ given by

$$\vec{Q}(r) = -\frac{1}{2\omega\rho_0}P(r)\nabla P(r) \tag{1.104}$$

is proportional to the gradient of the surfaces of constant pressure. Because $\vec{Q}(r)$ is proportional to the pressure gradient, the pressure maxima are the source points of reactive intensity vectors and the pressure minima are its sinks. The upper equations show that by using two microphones, we can measure all acoustic quantities of interest as shown schematically in Figure 14.7.

Figure 1.4 illustrates the sound field of two point sources with relative amplitudes +1.0 and −0.5. This figure shows the surfaces of constant phase and active intensity vectors that originate in the stronger source and have sinks in the weaker source. Figure 1.5 shows the distribution of sound intensity in a room with a sound source (left) that radiates the sound power and an electronic absorber that is a sink for the sound power. This illustrates how intensity can be used to reveal the properties of the sound field in an enclosure.

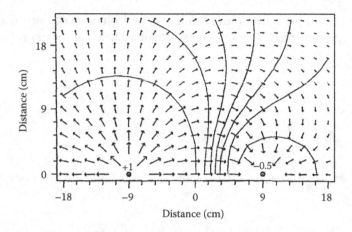

Figure 1.4 Sound intensity flow and wave fronts. (From Elko, G., Frequency Domain Estimation of the Complex Acoustic Intensity and Acoustic Energy Density, Ph.D. thesis 1984, The Pennsylvania State University.)

1.9 SOUND SOURCES

The basic physics and the solutions of the wave equation were discussed in previous chapters. This section concentrates on the modeling of radiation from some real sources such as loudspeakers. These are sources that reproduce music or speech and often can be modified to achieve acoustically optimal performance. Other sources are musical instruments. Their modeling is extremely complicated and is beyond the scope of this book; the reader is directed to References 7,8.

Various types of idealized sources are useful for modeling of real sources. This permits to relate their behavior to the sound perception and determine their optimal placement.

The simplest and most often used idealized source is a small sphere that is at low frequencies much smaller than the wavelength of the radiated sound; such a source is called a point source or monopole. The wave equation for the forced spherical waves is

$$\frac{1}{r^2}\left(\frac{\partial}{\partial r}\left(r^2\frac{\partial p}{\partial r}\right) + \frac{1}{\sin\vartheta}\frac{\partial}{\partial\vartheta}\left(\frac{1}{\sin\vartheta}\frac{\partial p}{\partial\vartheta}\right) + \frac{1}{\sin^2\vartheta}\frac{\partial^2 p}{\partial\varphi^2}\right) - \frac{1}{c^2}\frac{\partial^2 p}{\partial t^2} = -\rho_0\frac{\partial Q}{\partial t}$$

(1.105)

As in the previous chapters, Q represents the volume velocity of a source in m³/s. However, as earlier, our interest particularly for analytical purpose is in harmonic sources. If the source is very small, in the limit a point source, its radiation depends only on one variable that is the distance r from the source. The wave equation reduces into

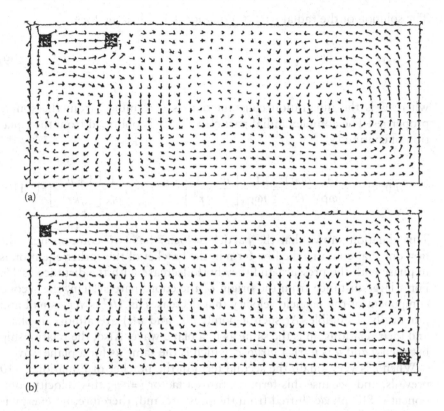

(a)

(b)

Figure 1.5 Sound intensity flow from a source to an absorber for (a and b) two different absorber positions. (From Fahy, F.J., *Sound Intensity*, Elsevier Applied Science, London, 1989.)

$$\frac{\partial^2 p}{\partial r^2} + \frac{2}{r}\frac{\partial p}{\partial r} - \frac{1}{c^2}\frac{\partial^2 p}{\partial t^2} = -\rho_0 \frac{dQ}{dt} \tag{1.106}$$

where the volume velocity of a pulsating sphere is Q that is the product of radial velocity and sphere surface area.

The left side of Equation 1.106 can be reformulated as

$$\frac{\partial^2}{\partial r^2}(rp) - \frac{1}{c^2}\frac{\partial^2}{\partial t^2} = 0 \tag{1.107}$$

Because we are seeking for p, a harmonic function, the solution of the earlier equation can be obtained by substituting for p

$$p(r,t) = p(r)e^{-j\omega t} \tag{1.108}$$

The solution of the radial part of Equation 1.108 has the form

$$p(r) = C\frac{e^{-jkr}}{r}$$ (1.109)

where C is the integration constant that is determined from the source properties. The complex radial particle velocity is obtained from the equation of continuity

$$u_r(r,k) = -\frac{1}{j\omega\rho_0}\frac{\partial p(r)}{\partial r} = \frac{C}{j\omega\rho_0}\left[\frac{jk}{r} + \frac{1}{r^2}\right]e^{-jkr} = p(r)\frac{1}{\rho_0 c}\left[\frac{1+jkr}{jkr}\right]$$ (1.110)

This equation shows that the particle velocity has two components. The first one that contains the first term in the bracket of the earlier equation is in phase with the pressure and they both decrease with the distance as $1/r$. For a large distance from the origin of the coordinate system, the second component of the particle velocity that decreases as $1/r^2$ is very small and practically negligible. At such a point, $p(r)$ and $u(r)$ are almost in phase and their relationship is the same as in a plane wave. Also, the relationship between p and u is $u_r(r) \cong p(r)/\rho_0 c$, which is the same as for a plane wave.

When r is very small, the second term in the bracket of Equation 1.110 prevails, and because this term contains a factor $j = e^{\pi/2}$, this velocity component is 90° phase shifted from the pressure and, therefore, no energy is carried by this component.

1.10 RADIATION OF SOUND POWER FROM SMALL SOURCES

The study of sound radiation from a small source is of utmost importance because it is the fundamental element of large radiators since they can be treated as assemblies of small sources. The size of a source is measured by a ratio of its physical dimension to the wavelength of sound. We have already observed that the wave properties depend on the product $kx = 2\pi x/\lambda$. As stated earlier, the integration constant C can be calculated from the source surface velocity. We will consider a sphere of radius a that pulsates with the surface velocity u_a. Substituting into Equation 1.110, we obtain for $p(a)$

$$p(a,k) = \rho_0 c u_a\left(\frac{jka}{1+jka}\right) = C\frac{e^{-jka}}{a}$$ (1.111)

The constant C from this equation can be substituted into the equation for pressure (1.109), that is,

$$p(r,k) = \rho_0 ca\left[\frac{jka}{1+jka}\right]\frac{u_a e^{-jk(r-a)}}{r} = \rho_0 cQ\left[\frac{jk}{1+jka}\right]\frac{e^{-jk(r-a)}}{4\pi r} \qquad (1.112)$$

where $Q = 4\pi a^2 u_a$ is the volume velocity of the source.

If we decrease the size of the source to a point so that $a \to 0$, the earlier equation will take the form

$$p(r \cdot k) = \frac{j\omega\rho_0 e^{-jkr}}{4\pi r}Q \quad a \to 0 \qquad (1.113)$$

The sound pressure decreases with the distance as $1/r$. The particle velocity in the radial direction is

$$u(r,k) = j\frac{1}{k\rho_0 c}\frac{\partial p(r)}{\partial r} = j\frac{kQ}{4\pi}\left(1-\frac{1}{jkr}\right)\frac{e^{-jkr}}{r} = \frac{p(r,t)}{\rho_0 c}\left(1-\frac{1}{jkr}\right) \qquad (1.114)$$

The decrease of the particle velocity depends on the term $1/kr$ in addition to the pressure decrease. Let us assume that the source is a small pulsating sphere of radius a so that the volume velocity is $Q = 4\pi a^2 u_a$, where u_a is its amplitude.

We will assume that the sound source is very small so that ka is much smaller than one. In the earlier equation for the particle velocity at a large distance where $kr \gg 1$, the second term in the parenthesis is much smaller than one and can be neglected. The relationship between the pressure and the particle velocity at the point r is the same as in the plane wave, which means that the sound pressure is in phase with the particle velocity. Both the sound pressure and the particle velocity decrease with distance as $1/r$. This part of the field is called the far field.

On the other hand, in the region close to the source, called near field, $kr \ll 1$ and the second term in the parenthesis is much greater than one. The particle velocity is equal to

$$u(r,k) = -\frac{kQ}{4\pi kr}\frac{e^{-jkr}}{r} = j\frac{p(r,t)}{\rho_0 ckr} \qquad (1.115)$$

The near-field approximation of the particle velocity shows that there is an additional factor j/kr. The phase of the sound pressure and particle velocity differs by $90°$ and the magnitude decreases as $1/r^2$. Between the near field and the far field is a transition region where the field quantities have values between the discussed limits.

Another important quantity is specific acoustic impedance defined as the ratio of the pressure and particle velocity. After rearranging Equation 1.114 by dividing $p(r)$ by $u(r)$ and separating the real and imaginary parts, we obtain

$$z(r,k) = \frac{p(r,k)}{u(r,k)} = Z_0 \frac{jkr}{1+jkr} = Z_0 \frac{(kr)^2}{1+(kr)^2} + jZ_0 \frac{kr}{1+(kr)^2} = \Re + jX$$

(1.116)

where
$\quad Z_0 = \rho_0 c$ is the characteristic impedance of the medium
$\quad \Re$ is the real part of the impedance
$\quad X$ is the imaginary part of the impedance that a wave is facing at the location r

Figure 1.6 shows how both $\Re$ and X are functions of kr. Please note that $\Re$ increases with kr reaching approaching asymptotically one. The imaginary part peaks at $kr = 1$. The impedance $z(kr)$ is a specific acoustic impedance because the sound pressure is a force per unit area divided by the particle velocity.

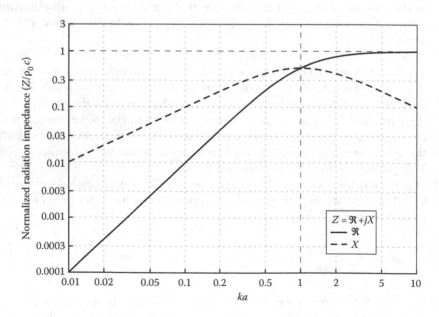

Figure 1.6 Specific acoustic resistance and reactance as a function of wavelength ratio.

The knowledge of the impedance at the surface of a radiator is particularly useful. Both the sound pressure and the particle velocity are uniform if the radiator dimensions are much smaller than the wavelength. Again, let us consider a small sphere with radius a that pulsates with a velocity $u_a(a)$ in infinite space. The sound power is equal to the product of the force needed to move the surrounding medium times its velocity. The power P is given by

$$P = \frac{1}{2} \mathrm{Re} \left\{ p^*(a) u_a 4\pi a^2 \right\} \tag{1.117}$$

Using Equation 1.111, we replace $p(a)$ by u_a and obtain

$$P = \frac{1}{8\pi a^2} \left(4\pi a^2 u_a \right)^2 \mathrm{Re} \left\{ z(a) \right\} \tag{1.118}$$

The product $4\pi a^2 u_a$ is the volume velocity $Q(a)$, and the real part of $z(a)$ is obtained from Equation 1.116 so that the radiated power is

$$P = \frac{1}{8\pi a^2} \rho_0 c |Q|^2 \frac{(ka)^2}{1 + (ka)^2} \tag{1.119}$$

The upper equations and Figure 1.6 reveal the dependence of the radiation efficiency on the source size as measured in a wavelength and frequency as expressed by $ka = 2\pi a/\lambda$. For a small source or a large wavelength, $ka \ll 1$, which shows that the source is an inefficient radiator. For that reason, loudspeakers and other sources for low-frequency radiation must have as large radiation area as possible. A small radiation area with large amplitude may become nonlinear and cause sound distortion. Close to a very small physical source, the sound pressure may be so high that the sound field conditions are nonlinear creating distortion.

1.10.1 Radiation impedance

The radiation impedance is a very useful quantity for analyzing and calculating the sound power that is radiated by sources. Due to their omnidirectional radiation, particularly, the small sources are modeled as point sources or spherical sources of a small radius a. The radiation impedance Z_r is defined as a total force on the surface of a radiator that is needed to move the surrounding medium to the particle velocity of the radiator surface. It is also called mechanical impedance. We obtain for a pulsating sphere of radius a

$$Z_r(ka) \equiv Z_m(ka) = \frac{p(a) \cdot S}{u(a)} = z(a) \cdot S = j\frac{k\rho_0 c 4\pi a^3}{1 + jka}$$

$$= Z_0 \frac{(ka)^2 4\pi a^2}{1 + (ka)^2} + jZ_0 \frac{4\pi ka^3}{1 + (ka)^2} \tag{1.120}$$

The radiated sound power is a real quantity and is obtained from

$$P = \frac{1}{2}\mathrm{Re}\left[p(a)^* Q\right] = \frac{1}{2}\mathrm{Re}\left[Z_r u^2\right] \tag{1.121}$$

1.11 SOUND RADIATION FROM MULTIPLE SMALL SOURCES

The design of optimal sound fields for listening outdoors as well as in rooms usually requires multiple sources. This section will be devoted to the basic physics of mutual effects of sources on their sound radiation into free space. The sound pressure at a point at certain distance from the sources is given by the sum of the pressure from individual sources. However, the sound from a certain source impacts on the radiating surfaces of the other sources and affects their radiation. This mutual coupling has to be analyzed to optimize the complete design.

1.11.1 Coupling of two sources

The basic phenomena will be explained on two sources. The results can be then expanded on multiple sources. Figure 1.7 shows two small sources of radius a and at a distance d. Each source vibrates with different velocity

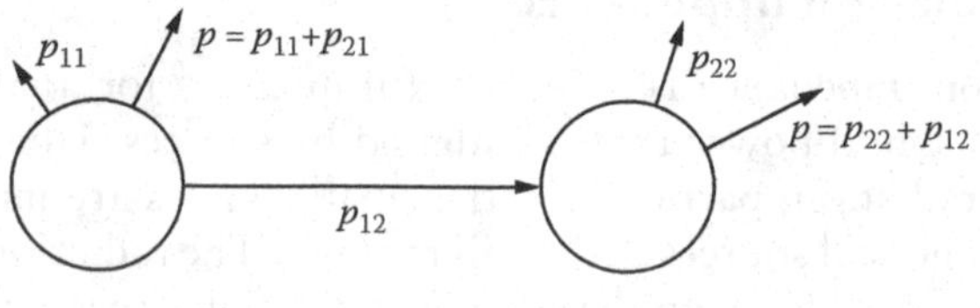

Figure 1.7 Source coupling. p_{11} and p_{22} are pressures when sources radiate alone.

magnitude and phase. Their volume velocities Q are given by the following equations:

$$Q_1 = S_1 u_1 = |Q_1| e^{j(\omega t + \psi)} \quad \text{and} \quad Q_2 = S_2 u_2 = |Q_2| e^{j(\omega t + \varphi)} \tag{1.122}$$

where
  S are the surface areas of the sources
  u are their surface velocities
  ψ and φ are the phase angles of the particle velocities

If we had only one source, the sound pressure on its surface, called self-pressure p_{11}, would be

$$p_{11} = jk\rho_0 c |Q_1| \frac{e^{-jka}}{4\pi a} e^{j(\omega t + \psi)} \tag{1.123}$$

When the second source is present, it creates on the surface of the first source an additional pressure p_{12} called the mutual pressure, that is,

$$p_{12} = jk\rho_0 c |Q_2| \frac{e^{-jkd}}{4\pi d} e^{j(\omega t + \varphi)} \tag{1.124}$$

It is assumed that the sources are small so that the pressure around the sources is the same. The total pressure p_1 on the surface of the first source is given by the sum of the self-pressure and the mutual pressure so that we obtain

$$p_1(k) = p_{11} + p_{12} = jk^2 \rho_0 c \frac{|Q_1|}{4\pi} \left[\frac{e^{-jka}}{4\pi a} e^{j\psi} + \frac{Ae^{-jkd}}{4\pi d} e^{j\varphi} \right] \tag{1.125}$$

where $A = |Q_2/Q_1|$. Due to this mutual coupling, depending on the phases, distance between the sources, and their volume velocities, the pressure can decrease or increase. The important quantity is the power radiated by the sources. The power is obtained from (see the equations in the preceding sections)

$$P_1 = \frac{1}{2} \text{Re} \left[p_1 u_1^* S_1 \right] = \frac{1}{2} \text{Re} \left[p_1 Q_1^* \right] = \frac{k^2 \rho_0 c |Q_1|^2}{8\pi} \left[\frac{\sin(ka)}{ka} + A \frac{\sin(kd - (\varphi - \psi))}{kd} \right] \tag{1.126}$$

For a small source and a low frequency so that $ka \ll 1$ and $ka \ll kd$, the first term in the bracket of the earlier equation is equal to one so that the power is

$$P_1 = \frac{k^2 \rho_0 c |Q_1|^2}{8\pi} \left[1 + A \frac{\sin[kd - (\varphi - \psi)]}{kd} \right] \qquad (1.127)$$

Analogously, the power of the second source is obtained by a very similar formula, after we exchange Q_1 for Q_2. We obtain

$$P_2 = \frac{k^2 \rho_0 c |Q_2|^2}{8\pi} \left[\frac{\sin(ka)}{ka} + \frac{1}{A} \frac{\sin(kd + (\varphi - \psi))}{kd} \right] \qquad (1.128)$$

The sum of the powers radiated by both sources is

$$P = P_1 + P_2 = \frac{k^2 \rho_0 c |Q_1|^2}{8\pi} \left[1 + A^2 + 2A \frac{\sin(kd)}{kd} \cos(\varphi - \psi) \right] \qquad (1.129)$$

The earlier equation shows that if the sources are close in terms of the wavelength, the output power strongly depends on their phases. Should the phases be identical, the cosine term is equal to one and the sinc function will be also close to one. If the sources are, for instance, two identical in-phase loudspeakers, placed next to each other, the radiated sound power is four times the power that would be radiated by a single loudspeaker alone. This effect is due to the mutual coupling that is increasing the radiation efficiency of the individual sources. Loudspeaker staggering is often used by smaller orchestras to increase the loudness of their sound and to achieve the desired directivity.

Another example will show how the total radiated power can be minimized so that the sound of the primary source (unwanted noise) can be suppressed. The total radiated power P is a function of the phase difference between the sources $\psi - \varphi = \gamma$ and the ratio A of the volume velocity of the sources. To establish a minimum of P in the earlier equation, we make the partial derivatives of P with respect to γ and make them equal to zero. From the two equations, we obtain

$$\gamma = \pi \quad \text{and} \quad A = \frac{\sin(kd)}{kd} \qquad (1.130)$$

Substituting these values into Equation 1.129, the minimum of the radiated power is found as

$$P_{min} = \frac{k^2 \rho_0 c}{8\pi} |Q_1|^2 \left[1 - \left(\frac{\sin(kd)}{kd} \right)^2 \right] \tag{1.131}$$

To minimize the radiated power, the sources have to be in opposite phase ($\gamma = \pi$) and A has to be a sinc function of kd. The smaller the product kd, the more the sinc function approaches one and the smaller is the bracket. This configuration of two (or more) sources is used in active control technology to lower the unwanted noise radiated by a source.

REFERENCES

1. Morse, P.M. and Ingard, K.I. (1968) *Theoretical Acoustics*. McGraw-Hill, New York. Reprinted by Princeton University Press (January 1, 1987).
2. Skudrzyk, E. (1971) *The Foundations of Acoustics*. Springer, New York.
3. Cremer, L., Mueller, H.A., and Schultz, T.J. (1982) *Principles and Applications of Room Acoustics*. Applied Science Publishers, New York.
4. Lee, Y.W. (1960) *Statistical Theory of Communication*. John Wiley & Sons, New York. Reprinted by Dover Publications (December 17, 2004).
5. Papoulis, A. (1977) *Signal Analysis*. McGraw-Hill College, New York.
6. Kuttruff, H. (2000) *Room Acoustics*, 4th edn. CRC Press, New York.
7. Fletcher, N.H. and Rossing, T.D. (2010) *The Physics of Musical Instruments*. Springer, New York.
8. Meyer, J. (2004) *Akustik und musikalische Aufführungspraxis (Acoustics and Music Performance*, in the German), Edition Bochinsky, Frankfurt am Main, Germany.
9. Fahy, F.J. (1989) *Sound Intensity*. Elsevier Applied Science, London.
10. Elko, G. (1984) Frequency Domain Estimation of the Complex Acoustic Intensity and Acoustic Energy Density, Ph.D. thesis, The Pennsylvania State University.

Chapter 2

Sound fields in enclosures

2.1 FREE SOUND WAVES IN ENCLOSURES

This chapter is an overview of the basic physical properties of sound waves in enclosures with a focus on low frequencies. The readers that would like to make themselves more familiar with the extended fundamentals will find them in References 1–13.

In Chapter 1, we dealt with the sound fields generated by a shortly acting (impulse) source that created compression and rarefaction in the medium so that sound waves were created. Because there were no medium boundaries, we could concentrate on the phenomena solely within the wave.

Sound that propagates inside enclosures is reflected by the enclosure boundaries. The reflected waves are modified by the reflection properties of the walls and other surfaces. In addition, the size and shape of the surfaces as well as the wave frequency affect the creation and properties of the reflected wave. As before, our interest is in the steady state that will develop after the wave will have reflected so many times, so that further contributions of the reflected wave do not practically affect the sound field. However, due to the hearing properties of the listeners, the analysis of the sound field after a few reflections is very important to the sound perception so that the sound field analysis in small rooms as a function of time must include the time history of the field formation at a listening point. Both the physical analyses and listening experience indicate that the room properties strongly depend on the frequency and that particularly at low frequencies the listening in small rooms is very different from the listening in large rooms (auditoria or concert halls). Therefore, a detailed analyses and acoustical room modifications are essential to achieve satisfactory listening properties.

The earlier short analyses reveal that the construction of satisfactory listening rooms is very demanding and requires an extensive knowledge of acoustical behavior of sound waves in small rooms as well as a good knowledge of their building construction.

Before formulating the steady-state mathematical sound wave descriptors, a basic physical principle will be explained on a solid finite-size body of arbitrary shape. If a mechanical impulse excites this body to vibrations, this body will, after a relatively short transient behavior, keep vibrating only at certain frequencies, usually called its resonance frequencies. The basic law of physics states that after a steady state has been achieved, the vibrational energy exists only at resonance frequencies that depend on the size, shape, and mechanical constants of the body.

The gas in an enclosure behaves according to the same physical principle as the solid body. After some impulse excites the waves in an enclosure, they will settle to exist only at some frequencies and wave fronts that depend on room shape, acoustical properties of the walls, damping, etc. Of course, the solid body as well as the room can be forced to vibration or waves at nonresonant frequencies, but these will shift to resonant frequencies after the excitation will terminate.

These phenomena are particularly pronounced at low frequencies because the resonant frequencies exist at larger frequency intervals. Therefore, the formation of the free waves needs to be examined before the radiation of forced waves is going to be considered.

Figure 2.1 shows schematically an enclosure with a boundary that vibrates by some internal pressure.

The wall impedance is defined as

$$z = \frac{p}{u_n} = R + jX \tag{2.1}$$

As explained earlier, there are two kinds of impedances:

1. Locally reactive impedance that is independent of the angle of wave incidence
2. Impedance with extended reaction where z is a function of the angle of incidence

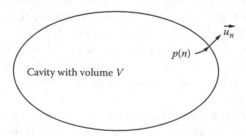

Figure 2.1 A model of a cavity with a vibrating wall.

The kind of impedance depends on the wall material and structure as well as on acoustical phenomena such as refraction and reflection.

The functions that describe the time and spatial distribution of acoustical quantities in the enclosure are called the eigenfunctions. They depend on the volume and shape of the enclosure and its wall impedance. For a locally reactive impedance, the eigenfunctions of the room Ψ_N satisfy the wave equation

$$\Delta \Psi_N + k_N^2 \Psi_N = 0 \tag{2.2}$$

where
   the index N is linked to the integers that characterize the coordinates of
      a particular coordinate system of the eigenfunction Ψ_N
   k_N is the wave number of the particular eigenfunction

A very important property of the eigenfunction is the orthogonality defined as

$$\iiint_V \Psi_M \Psi_N \, dV = 0 \quad \text{for } M \neq N \quad \text{and} \quad V\Lambda_N \quad \text{for } M = N \tag{2.3}$$

where the integral is over the enclosure volume, N and M are two different indices of the eigenfunctions, and

$$V\Lambda_N = \iiint_V \Psi_N^2(r) \, dV \tag{2.4}$$

This orthogonality property is valid only for the walls with locally reactive impedance and other boundary conditions that limit, as shown later, to determine analytically the enclosure response to forced waves.

As mentioned earlier, the free waves in enclosures are standing waves that are created by the wave reflections from the enclosure walls.

Figure 2.2 shows a common rectangular enclosure with the dimensions L_x, L_y, L_z. We use the wave equation for the sound pressure of harmonic waves:

$$\frac{\partial^2 p}{\partial x^2} + \frac{\partial^2 p}{\partial y^2} + \frac{\partial^2 p}{\partial z^2} + k^2 p = 0 \tag{2.5}$$

The sound pressure is a function of three coordinates. In rectangular coordinate system, the eigenfunction can be separated into three functions that

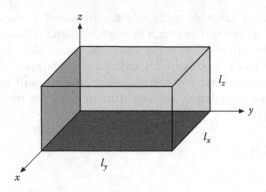

Figure 2.2 Dimensions of a rectangular room.

depend on one coordinate only. We obtain for the spatial dependence of the sound pressure that

$$p(x,y,z) = p_x(x)p_y(y)p_z(z) \tag{2.6}$$

Substituting Equation 2.6 into the wave equation results in the following separation into three spatial functions:

$$\frac{1}{p_x(x)}\frac{d^2 p_x(x)}{dx^2} + \frac{1}{p_y(y)}\frac{d^2 p_y(y)}{dy^2} + \frac{1}{p_z(z)}\frac{d^2 p_z(z)}{dz^2} = -k^2 = -\left(k_x^2 + k_y^2 + k_z^2\right) \tag{2.7}$$

where k_x, k_y, k_z are the wave numbers for wave components that propagate in one direction only. Because Equation 2.7 has terms that depend only on one variable while the equation has to be valid for any values of the variables, this equation can be split into three equations, each depending on one variable only. For the variable x is obtained

$$\frac{d^2 p_x(x)}{dx^2} + k_x^2 p_x(x) = 0 \tag{2.8}$$

The solution of the earlier well-known second-order differential equation is

$$p_x(x) = Ae^{jk_x x} + Be^{-jk_x x} \tag{2.9}$$

The sound pressure $p_x(x)$ represents two waves that propagate in $+x$ and $-x$ mutually opposite directions so that if the amplitudes A and B are equal, this equation represents a pure standing wave in the x direction. Boundary conditions of the walls, perpendicular to the x-axes, can be obtained from both the integration constants A and B. The particle velocity of the waves is generally given by

$$\vec{u}(x,y,z) = \frac{-1}{j\omega\rho_0} \text{grad } p(x,y,z) \qquad (2.10)$$

For the x component is obtained

$$\vec{u}(x,y,z) = \frac{1}{j\omega\rho_0} \text{grad } p(x,y,z) \qquad (2.11)$$

We now apply the boundary condition. For simplicity and physical explanations, it is first assumed that the walls are perfectly reflecting so that the particle velocity normal to the wall is zero. Equation 2.11 can be modified into

$$u_x(x) = -\frac{1}{\rho_0 c}\left[(A-B)\cos(k_x x) + j(A+B)\sin(k_x x)\right] \qquad (2.12)$$

Substituting $x=0$ and $u_x=0$ into the earlier equation results in $A=B$. Again, substituting for $u_x=0$ at $x=L_x$ results in

$$k_x L_x = m\pi \quad m = 0,1,2,3,\dots \qquad (2.13)$$

so that we obtain

$$k_m = \frac{m\pi}{L_x} \quad \text{and} \quad f_m = \frac{c}{2}\frac{m}{L_x} \quad m = 0,1,2,3,\dots \qquad (2.14)$$

The wave number k_m is for a plane standing wave that propagates in the direction of x-axes and is therefore called axial wave. The general name of this wave is a mode and the frequencies $f_{m,n,l}$ are called characteristic frequencies or eigenfrequencies. Whenever the distance between the walls is a multiple of half wavelengths, a standing wave can exist. If the earlier calculation is repeated for the variables y and z, the expressions for the wave number and characteristic frequencies are

$$k_{m,n,l} = \sqrt{\left(\frac{m\pi}{L_x}\right)^2 + \left(\frac{n\pi}{L_y}\right)^2 + \left(\frac{l\pi}{L_z}\right)^2} \quad \text{and}$$

$$(2.15)$$

$$f_{m,n,l} = \frac{c}{2}\sqrt{\left(\frac{m}{L_x}\right)^2 + \left(\frac{n}{L_y}\right)^2 + \left(\frac{l}{L_z}\right)^2}$$

The integers m, n, and l are called quantum numbers. An analysis of the combinations of these numbers results in various wave designations. If two of these quantum numbers are zero, the waves are called axial waves because they propagate in the direction of one of the room axes. A wave with the wave fronts perpendicular to one of the room walls is called a tangential wave (tangential to a wall). These waves have one of the wave numbers equal to zero. If none of the wave numbers is a zero, the waves are called oblique waves. Their wave fronts have some general angle with the enclosure walls.

The sound pressure of an (m, n, l) mode according to Equation 2.6 is

$$p_{m,n,l}(x,y,z) = A\cos\left(\frac{m\pi}{L_x}x\right)\cos\left(\frac{n\pi}{L_y}y\right)\cos\left(\frac{l\pi}{Lz}z\right) \qquad (2.16)$$

where A is the amplitude. The properties of the modes are shown graphically for some axial and tangential modes in Figures 2.3 and 2.4.

Figure 2.3 shows the propagating wave fronts and spatial pressure distribution in an axial mode. Whenever the plane wave frequency is such that the interference of the two waves that travel in opposite direction

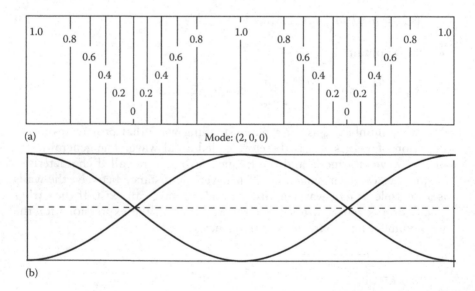

(a)   Mode: (2, 0, 0)

(b)

Figure 2.3 A standing wave in one dimension. A node is a point along a standing wave where the wave has minimum amplitude. (a) Shows wave fronts and pressure magnitude at some time instant and (b) shows maximum and minimum pressure along the standing wave.

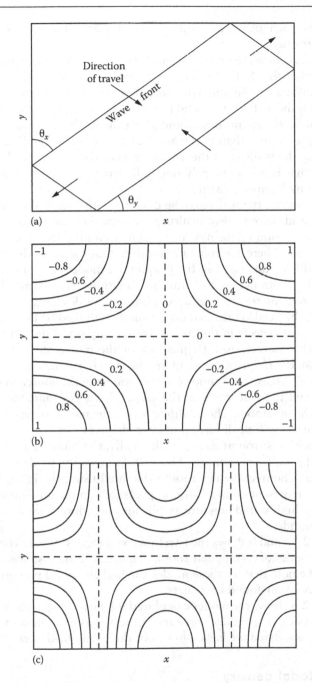

Figure 2.4 (a) The principle of mode generation, (b) the pressure magnitude pattern of
a (1, 1, 0) mode, and (c) the pressure magnitude pattern of a (2, 1, 0) mode.
(After Beranek, L.L. *Acoustics*, McGraw-Hill, New York, 1954.)

reinforces each other, a resonance occurs and a mode of appropriate order is formed.

Figure 2.4 shows the relative spatial pressure distribution of two tangential modes of the (1, 1, 0) and (2, 1, 0) order. The numbers show the maximum amplitude of the sound pressure that can exist at a given location. The minus sign shows that the sound pressure in two areas is 180° out of phase. Note that, in the room corners and along the walls, the sound pressure can reach higher values than in the areas of the room center. The sound pressure along the walls is on the average much higher than in the room middle area that is approximately one-half wavelength away from the walls, as shown by Figures 2.4a and b.

In real rooms, the walls may be covered with sound absorbers or sound diffusers and there is also furniture, loudspeakers and other objects. This makes calculation of the modal frequencies and spatial pressure distribution difficult. There are computational methods such as the finite element method (FEM) discussed in the chapter on modeling that can be used to predict the room modes and transfer functions. As discussed in the next sections, our interest is in other qualities of modes such as damping and, therefore, the needed information is usually obtained by the measurements. Because most objects in the rooms are usually small (in terms of a wavelength), the characteristic frequencies of the modes do not substantially differ from the frequencies calculated for a bare room.

The modes affect the sound transmission from the source to the listening position and the perception of the signal, both the fundamental frequencies and the overtones. The modification of the modes is generally difficult. These matters will be discussed in later chapters.

Although most rooms have parallel walls, the sound field in rooms with nonparallel walls was studied. Figure 2.5 shows the mode pressure distribution for a nonrectangular room calculated using the FEM. The lines in the figure represent constant sound pressure (similar to Figure 2.4). Black areas have high sound pressure amplitude and white areas low sound pressure amplitude.

Figure 2.5a and b shows the two first axial modes. We see their similarity with the modes of rectangular rooms. Figure 2.5c and f shows higher-order modes. We note that for these modes the highest sound pressure amplitude is not necessarily found in corners.

Figure 2.6 shows the modes calculated for a car compartment. As in rectangular rooms, the modes are standing waves as explained in Section 2.1, where we discussed the basic physics of the sound field in enclosures.

2.1.1 Modal density

To discuss the effects of the modes on listening, an analysis of their number and frequency intervals will be made.

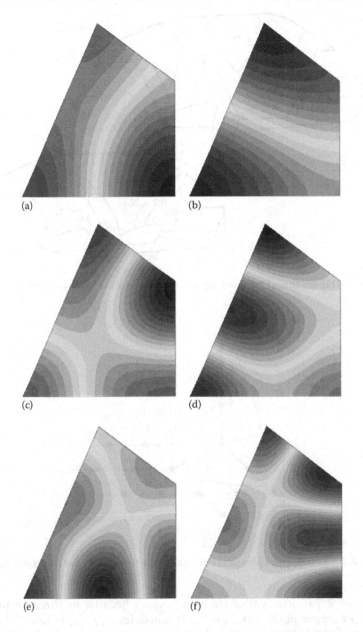

Figure 2.5 Sound pressure magnitude distributions for waves tangential to the floor in a prism-shaped room with rigid walls, as found by the finite element method. Light regions have low sound pressure and dark regions have high pressure. Room wall lengths shown clockwise from top are 2.5 m, 3 m, 4 m, and 5.85 m. The resonance frequencies of the modes shown are (a) 52 Hz, (b) 43 Hz, (c) 77 Hz, (d) 82 Hz, (e) 99 Hz, and (f) 113 Hz.

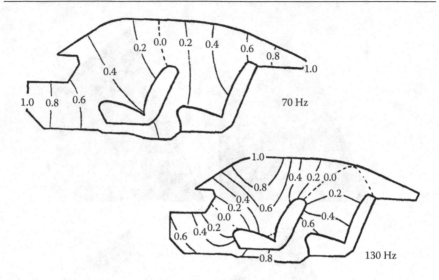

Figure 2.6 Modes shapes for low-frequency modes in a car compartment.

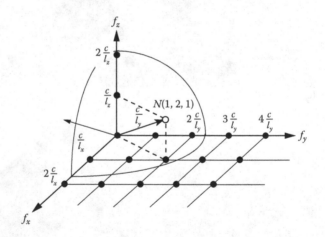

Figure 2.7 The *frequency space* of mode resonance frequencies for a rectangular room.

Figure 2.7 is often called frequency space because its three perpendicular axes represent the axial modal frequencies f_n, f_m, f_l as calculated from Equation 2.14. Each point in the space represents one mode. The length of the vector starting at the origin and terminating at the point representing a particular mode is the frequency as can be understood from the sum of squares in Equation 2.15. If we draw a sphere of the radius f (see Figure 2.7) and determine how many modal points are within this

1/8 of the total sphere volume, we obtain the number of modes N within the frequency interval $0-f$. We obtain

$$N = \frac{4}{3}\frac{\pi V}{c^3}f^3 + \frac{\pi S}{4c^2}f^2 + \frac{L}{8c}f \qquad (2.17)$$

where
   $V = L_x L_y L_z$ is the room volume
   $S = 2(L_x L_y + L_x L_z + L_y L_z)$ is the room surface
   $L = 4(L_x + L_y + L_z)$ is the length of the edges

The first term in Equation 2.17 is larger than the second and third term. At low frequencies, particularly at low volume, the number of modes is small but grows rapidly with f^3. For example, in a small room of volume $7 \times 5 \times 3 = 105$ m³, the number of modes, between 0 and 100 Hz approximately, is $N = 22$ modes. The first mode (axial) has the frequency 24 Hz. In a hall that has a volume of 36,000 m³, the number of modes under 100 Hz is approximately $N = 3,800$ and the frequency of the first mode is 4.8 Hz. An important conclusion from this simple example is that the modes in small rooms are well separated in frequency and that the signal transmission is very irregular because at each modal frequency, the resonance effect elevates the sound pressure. As discussed later, the effects of mode separation depend on damping.

The mode separation can be followed on modal density n, that is, the number of modes in a unit frequency interval. It can be calculated as

$$n = \frac{dN}{df} = \frac{4\pi V}{c^3}f^2 + \frac{\pi S}{2c^2}f + \frac{L}{8c} \qquad (2.18)$$

For larger f, the first term dominates the other terms. To continue with the earlier example, we find out that at 100 Hz, the calculated modal density is $n = 0.545$ modes per 1 Hz and at 50 Hz $n = 0.202$ modes/Hz. The frequency interval between the modes at 100 Hz is 1.83 Hz and at 50 Hz $n = 4.95$ Hz. When measuring the frequency response of the room, we will find the frequency intervals between the modes irregular. However, the calculated values provide good average numbers and a good help to estimate the *smoothness* of the frequency response.

2.1.2 Energy in modes

As explained in the previous sections, the free waves in an enclosure exist only at certain frequencies. We identified three distinguished types of modes that depend on their way of propagation: axial, tangential, and oblique modes. The modes are standing waves that propagate in different directions. As the calculation of the energy density reveals, the axial modes carry more energy than the tangential modes and the oblique modes are weaker than the tangential modes. This is true for some random excitation by a short pressure pulse. The information about different amplitudes of the different type of modes is important for the formation of forced waves as it is discussed in the next section.

The calculation of the energy density ratios of different types of modes is performed for a rectangular enclosure. As shown in the previous section, the spatial average for harmonic waves can be obtained by averaging the pressure squared. We obtain for an axial mode

$$w_{\text{axial}} = \frac{1}{L_x} \int_0^{L_x} \cos^2\left(\frac{m\pi}{L_x}x\right) dx = \frac{1}{2} \tag{2.19}$$

The energy density of tangential modes is

$$w_{\text{tan}} = \frac{1}{L_xL_y} \int_0^{L_x}\int_0^{L_y} \cos^2\left(\frac{m\pi}{L_x}x\right)\cos^2\left(\frac{n\pi}{L_y}y\right) dxdy = \frac{1}{4} \tag{2.20}$$

and for the oblique modes

$$w_{\text{obl}} = \frac{1}{L_xL_yL_z} \int_0^{L_x}\int_0^{L_y}\int_0^{L_z} \cos^2\left(\frac{m\pi}{L_x}x\right)\cos^2\left(\frac{n\pi}{L_y}y\right)\cos^2\left(\frac{l\pi}{L_z}z\right) dxdydz = \frac{1}{8} \tag{2.21}$$

The axial modes carry twice the energy of tangential modes and four times the energy of oblique modes. Therefore, the peaks in the frequency response of the room are likely to appear at the frequencies of axial modes. The room frequency response is in detail analyzed in later chapters.

2.2 MODES IN ENCLOSURES WITH DAMPING

The previous section concentrated on the fundamental physics of the wave propagation in enclosures. However, when designing and constructing particularly small rooms with acoustical mission, the damping of modes at low frequencies can be essential to achieve optimum acoustical condition.

This section and the next ones show the basic effects of damping and absorption on the modes and the sound transmission in small rooms.

The sound absorption is caused by natural absorbers such as the structural properties of the building, attenuation of air, objects in the room, and people. As always, the sound energy that escapes through the walls can be expressed by using the wall impedance. The sound attenuation by air in small rooms is usually negligible compared with other losses. The important acoustical quantities will be explained by analyzing a 1D wave that propagates in a rectangular enclosure. The variables can be separated (see Section 2.1) so that the wave equation can be expressed as

$$\frac{1}{X(x)}\frac{d^2 X(x)}{dx^2} = \frac{1}{c^2}\frac{d^2 T(t)}{dt^2} = -\bar{k}_x^2 = -\left(\frac{\omega + j\delta}{c}\right)^2 \tag{2.22}$$

where
   $X(x)$ is the spatial sound pressure component in the x direction
   $T(t)$ is the time component
   $\bar{k}_x^2$ is the complex wave number that permits to express the losses by the damping factor δ (which will later be expressed by the wall impedance)

The complex wave number is

$$\bar{k}_x = \frac{\omega + j\delta}{c} = k + j\frac{\delta}{c} = \beta_x + j\alpha_x \tag{2.23}$$

The damping shifts the mode frequency, but for damping that is usually small, this shift is not significant.

For a harmonic wave, the time equation is

$$T(t) = Ae^{j(\omega+j\delta)t} + Be^{-j(\omega+j\delta)t} = Ae^{j\omega t}e^{-\delta t} \tag{2.24}$$

In this equation, the negative frequency has been suppressed ($B = 0$). The term $e^{-\delta t}$ represents the decreasing amplitude with time.

2.2.1 Effects of boundary conditions as expressed by wall impedance

Most absorbers that are used to control the sound fields are placed on walls. Their effects are shown on a rectangular room with the dimensions L_x, L_y, L_z. Figure 2.8 shows the 1D situation where the sound wave incident perpendicularly on the walls.

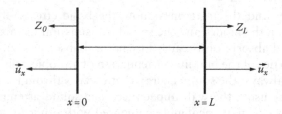

Figure 2.8 Sound propagation between two walls.

The related relationships between p, u, and z and complex $\bar{k}_x$ are

$$p(x) = z_x u_x \quad u_x = -\frac{1}{j\omega\rho}\frac{\partial p}{\partial x} \quad \frac{z_x}{\rho c} = \zeta_x = \xi_x + j\eta_x \quad \bar{k}_x = \beta_x + j\alpha_x \qquad (2.25)$$

The conditions at the boundaries for $x = 0$ and $x = L_x$ are

$$jkp = \zeta_x \frac{\partial p}{\partial x} \quad \text{at } x = 0$$

$$jkp = -\zeta_x \frac{\partial p}{\partial x} \quad \text{at } x = L_x \qquad (2.26)$$

Similar equations can be written for the other walls.

First, we concentrate on the relationship between the wall impedance and the wave number. The boundary conditions will be applied on the general wave equation solution (for harmonic waves):

$$p(x,y,z) = X(x)Y(y)Z(z) \qquad (2.27)$$

So that we obtain for one dimension

$$p(x) = X(x) = Ce^{-jk_x x} + De^{jk_x x} \qquad (2.28)$$

After we substitute the earlier equation into Equation 2.26 for $x = 0$ and $x = L_x$, we obtain after a simple manipulation

$$C(k + k_x\zeta_x) + D(k - k_x\zeta_x) = 0 \quad \text{for } x = 0$$

$$C(k - k_x\zeta_x)e^{-jk_x L_x} + D(k + k_x\zeta_x)e^{jk_x L_x} = 0 \quad \text{for } x = L_x \qquad (2.29)$$

For a nonvanishing solution, the following determinant formulated from the unknowns C and D must be zero:

$$\begin{vmatrix} k + k_x \zeta_x & k - k_x \zeta_x \\ (k - k_x \zeta_x) e^{-jk_x L_x} & (k + k_x \zeta_x) e^{jk_x L_x} \end{vmatrix} = 0 \tag{2.30}$$

The unknown in this determinant is $k_x = \beta_x + j\alpha_x$, a complex wave number of the mode between the parallel walls shown in Figure 2.8. The determinant solution is a transcendental equation that has the form

$$e^{jk_x L_x} = \pm \frac{k - k_x \zeta_x}{k + k_x \zeta_x} \tag{2.31}$$

This equation represents two equations that are usually expressed in a tangent form as

$$\tan(u) = j \frac{2u \zeta_x}{kL_x}, \quad \tan(u) = j \frac{kL_x}{2u \zeta_x} \quad \text{with} \quad u = \frac{1}{2} k_x L_x \tag{2.32}$$

The solution has to be found numerically or graphically [4].

2.3 FORCED WAVES IN AN ENCLOSURE

2.3.1 General formulation

As explained in the previous sections, the free sound field in an enclosure consists of modes. These are standing waves that exist only at certain frequencies. If a source is placed into an enclosure, the sound waves are forced by the radiation from the source. The time and frequency function of the waves depends on the operation of the source and the properties of the enclosure such as volume, shape, and sound absorption. The wave equation for the free waves is satisfied by eigenfunctions Ψ_N as shown in Equation 2.2. A source will be characterized by a density function (distribution function) $q(r_0)$ that describes its volume velocity at all source points that are located at a coordinate r_0. As explained in Appendix 2.A of this chapter, the continuity equation is expanded by the sound source term so that the general wave equation for the sound pressure is

$$\Delta p = \frac{1}{c^2} \frac{\partial^2 p}{\partial t^2} - \rho_0 \frac{\partial q}{\partial t} \tag{2.33}$$

For a harmonic source $q(r_0,t) = q(r_0)e^{j\omega t}$, the wave equation changes to

$$\Delta p(r) + k^2 p(r) = -j\omega\rho_0 q(r_0) \tag{2.34}$$

where
  r is the coordinates of the field point for which we calculate the pressure
  r_0 is the coordinates of the source

As derived in Appendix 2.A, the sound pressure $p(r)$ in the enclose generated by a point source of the volume velocity Q located at the point r_0 is

$$p(r,k) = -jk\rho_0 c \frac{Q}{V} \sum_{N=0}^{\infty} \frac{\Psi_N(r_0)\Psi_N(r)}{\Lambda_N(k^2 - k_N^2)} \tag{2.35}$$

where
  k_N is the wave number of the mode N
  k is the wave number of the source
  $\Psi_N(r_0)$ is the eigenfunction evaluated at the source point r_0
  $\Psi_N(r)$ is the eigenfunction evaluated at the field point r
  V is the volume of the room

The orthogonality is

$$V\Lambda_N = \iiint_V \Psi_N^2(r)dV \tag{2.36}$$

2.3.2 Effects of damping

Because most rooms are rectangular with the dimensions L_x, L_y, L_z, the principal effects will be shown and analyzed for this type of room. Also, it is assumed that most of the sound energy is absorbed by the room walls but that their impedance is rather large. The relationship between the wall impedance and the wave number of the modes is rather complicated as shown in Equations 2.32. The wave number $\bar{k}_N$ is complex for a nonzero wall impedance. The effect of damping can be determined by analyzing the term $\left[k^2 - k_N^2\right]$ in Equation 2.35. If the numerical value of k is approaching k_N, the sound pressure is growing and the sound transmission has usually a peak, particularly at low frequencies where the modes are well separated. The complex $\bar{k}_N$ can be expressed as

$$\bar{k}_N = \frac{\omega_N}{c} + j\frac{\delta_N}{c} = \beta_x + j\alpha_x \tag{2.37}$$

The term in the denominator of Equation 2.35 is

$$k^2 - \bar{k}_N^2 = k^2 - \left(\frac{\omega_N}{c}\right)^2 - j2\frac{\omega_N}{c}\frac{\delta_N}{c} - \left(\frac{\delta_N}{c}\right)^2 \tag{2.38}$$

If a damping expressed by δ_N/c is small, the last term in the earlier equation can be neglected. Equation 2.38 becomes

$$p(r,k) \approx -jk\rho_0 c \frac{Q}{V} \sum_{N=0}^{\infty} \frac{\Psi_N(r_0)\Psi_N(r)}{\Lambda_N \left(k^2 - k_N^2 - j2k_N\frac{\delta_N}{c}\right)} \tag{2.39}$$

Because the eigenfunctions Ψ_N is due to $\bar{k}_N$ complex but Equation 2.36 has to be real, the orthogonality is calculated from

$$V\Lambda_N = \iiint_V \Psi_N(r)\Psi_N^*(r)dV \tag{2.40}$$

2.3.3 Bandwidth of modes

When designing the room, the bandwidth of modes (among other things) has to be sufficiently large. Figure 2.9 shows the energy E of a mode as a function of the frequency. The bandwidth of a mode is measured by the magnitude of its response in terms of $\Delta\omega$ or Δf at one-half of the energy in the mode. Because the energy of a mode is a function of pressure squared,

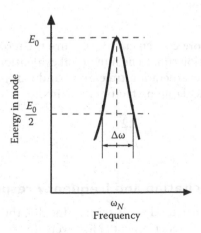

Figure 2.9 Bandwidth of a mode.

we can use the denominator of Equation 2.39 to calculate the bandwidth using the ratio of the one-half to the full energy to obtain the bandwidth of the mode as

$$\Delta\omega = 2\delta_N \tag{2.41}$$

If the modes are well separated, this equation allows measurement of the modal damping from the frequency response or transfer function of the room. However, when the modes are close, the modal curves are not symmetrical and the measurements suffer from errors.

2.3.4 Reverberation and modal bandwidth

As shown earlier, the energy in the mode is proportional to the pressure squared. If the room is fully excited by the sound and the sound source is turned off, the sound decay depends on the damping factor. The time dependence of the potential energy is given by

$$p^2(t) = p_0^2 e^{-2\delta t} \tag{2.42}$$

where p_0^2 is the initial amplitude of the potential energy. The decay curve in the room can be measured and the reverberation time T_{60} can be determined from the slope of the decay curve. By definition, the reverberation time is the time needed for the sound energy to decrease to 10^{-6} (60 dB) of its original value. Substituting into the previous equation results in

$$\frac{p^2(T_{60})}{p^2(0)} = 10^{-6} = e^{-2\delta T_{60}} \tag{2.43}$$

Sometimes, it is more convenient to measure the mode bandwidth Δf [Hz] than the reverberation time. The numerical evaluation leads to the relationship between the reverberation time and bandwidth that can be measured and from which the damping factor calculated:

$$\delta = \frac{6.91}{T_{60}} \quad \text{(i.e.,)} \quad T_{60} = \frac{2.2}{\Delta f} \tag{2.44}$$

2.3.5 Room excitation and frequency response

Most small rooms are used for listening. Ideally, the response of the room should be frequency independent. However, due to the sound reflections

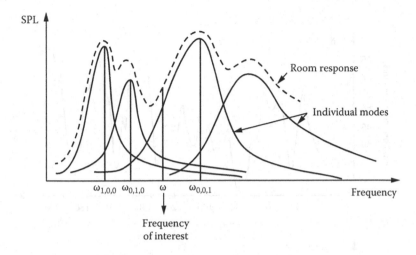

Figure 2.10 Schematic modal contribution to room response.

from the walls and objects in the room and the formation of the modes, the transmission of sound from the source to the listening position differs from point to point. Although many aspects of listening in rooms are discussed in later sections, the steady-state room response is discussed in this section.

The properties of small rooms are particularly affected by modes. Equation 2.39 provides the information on the steady-state conditions and permits the analysis of sound transmission from a source point to a listening point. Figure 2.10 shows schematically some modes at low frequencies and the total frequency response. Due to a small damping and particularly the large frequency intervals between the modes, the room response is very irregular. Because the response of an individual mode varies from point to point in a room, the total response will differ in a similar way.

Figure 2.11 shows the effects of modal damping. The damping of sound in the room doubles from Figure 2.11a to b and from Figure 2.11b to c. When the damping is low, most of the individual modes can be identified. In Figure 2.11c, many modes overlap because of the higher damping and only some modes are securely identifiable. Figure 2.12 shows a wider frequency range of the modal response in the same room. We can notice the increase of modal density with increasing frequency.

Finally, Figure 2.13 compares the calculated and measured response in the same room and location. Although there is a high degree of similarity, the details of the curves differ. Usually, there is little interest in the exact

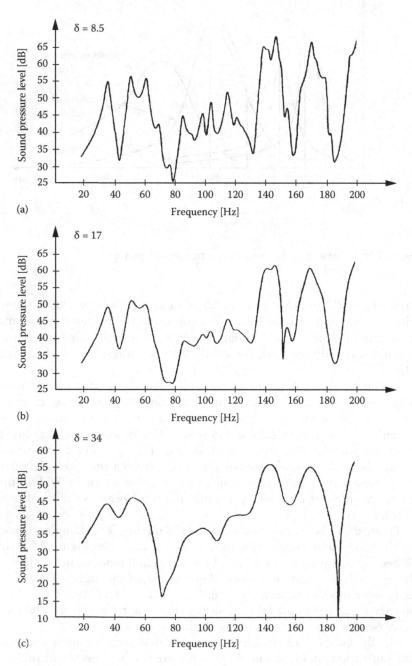

Figure 2.11 Effect of damping on room response. (a) Little damping, (b) medium damping, and (c) much damping.

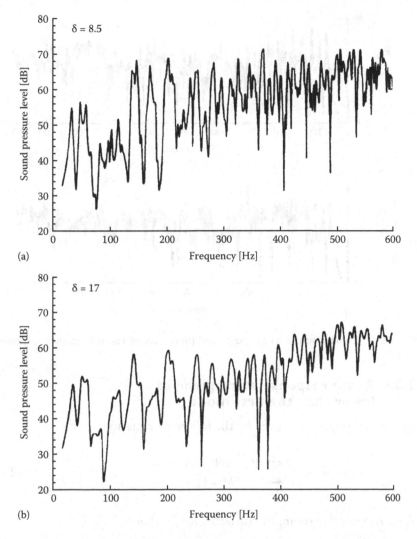

Figure 2.12 Effect of damping on wider frequency range curve. (a) Little damping, and (b) medium damping.

values of the sound pressure at a certain frequency or a point. Above certain frequency, the modal composition is better described by statistical laws. We would like to know the probability of the sound pressure fluctuation of the room response, the frequency intervals between the frequency curve peaks, the sound pressure average levels at frequency bands, and similar others. This will be discussed now.

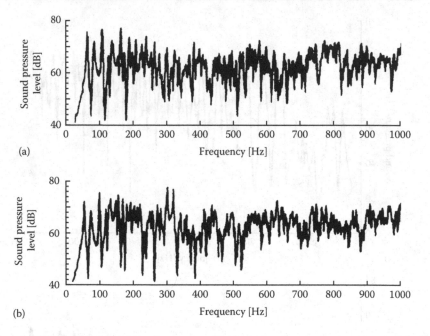

Figure 2.13 Comparison of (a) measured and (b) calculated room frequency responses.

2.3.6 Room response at frequencies lower than the first mode

The modal response is given by the following equation:

$$p(r,k) = -jk\rho_0 c \frac{Q}{V} \sum_{N=0}^{\infty} \frac{\Psi_N(r_0)\Psi_N(r)}{\Lambda_N\left(k^2 - k_N^2\right)} \qquad (2.45)$$

For a rectangular room, the variables for $f \to 0$ are

$$k_N = 0 \quad \Psi_N = 1 \quad \Lambda_N = 1 \qquad (2.46)$$

The average wall impedance is

$$\frac{z}{\rho_0 c} = \xi + j\eta = \frac{r}{\rho_0 c} + j\frac{x}{\rho_0 c} \quad R_a = \frac{r}{S} \qquad (2.47)$$

where
R_a is acoustic wall resistance, considered to be uniform
S is the total surface of the walls

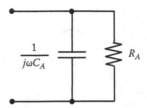

Figure 2.14 Equivalent circuit for modal response under first modal frequency.

The complex wave number is

$$\bar{k}_N^2 \simeq k_N^2 + j2k\xi\left[\frac{1}{\xi_x L_x} + \frac{1}{\xi_y L_y} + \frac{1}{\xi_z L_z}\right]$$

$$\simeq j\frac{k}{L_x L_y L_z}\left[\frac{2S_x}{\xi_x} + \frac{2S_y}{\xi_y} + \frac{2S_z}{\xi_z}\right] \simeq j\frac{k}{V}\frac{\rho_0 c}{R_a} \tag{2.48}$$

where $V = L_x L_y L_z$ and $S_x = L_x L_y$.

When substituting the earlier equation into Equation 2.35, it results in

$$\frac{p}{Q} = \frac{k\rho_0 c}{V}\frac{1}{jk^2 + \dfrac{k}{V}\dfrac{c\rho_0}{R_a}} = \frac{R_a}{1 + j\omega\dfrac{V}{\rho_0 c^2}R_a} \tag{2.49}$$

where

$V/(\rho_0 c^2) = C_a$ is the acoustic compliance of the room volume

R_a is the acoustic wall resistance

These combine in parallel in the equivalent circuit for the Q volume velocity point source input as shown in Figure 2.14. The frequency response is shown schematically in Figure 2.15.

2.4 STATISTICAL PROPERTIES OF THE SOUND FIELDS

We have seen in previous sections that the sound field in enclosures is created by interfering modes. Because the modes exist only at certain frequencies, the frequency response at a point in the room is not smooth and reflects the resonance-like behavior of the modes even for a source that has a flat spectrum in a reflection-free environment. Particularly at low frequencies,

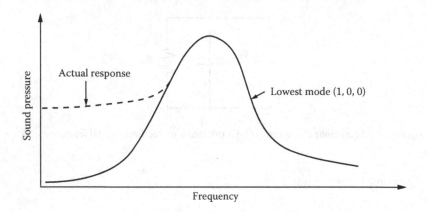

Figure 2.15 Room frequency response under first modal frequency.

the character of the sound fluctuations depends on the damping and density of modes. The modal response can be obtained from Equation 2.39 and the modal density from Equation 2.18.

Let us examine what happens with the frequency response when the frequency is increasing. At very low frequencies, only a few modes are excited and their frequencies may be identified from the frequency response curve. Because the modal density depends on the square of the frequency, the modal overlap is growing. However, the frequency response does not become smooth and is characterized by alternating maxima and minima due to wave interference. The individual modes can no longer be identified. The frequency response at different points varies. The room response to the sound radiation can be expressed by statistical laws that were formulated by Schroeder [14] and further extended by Kuttruff [6] and others.

Although there is no sharply defined frequency that separates the low- and high-frequency response parts of the room, we often use the *Schroeder* frequency as a defined division between low- and high-frequency responses of the enclosure. At this frequency, three modes overlap in one mode frequency bandwidth. Obviously, the Schroeder frequency f_S depends on modal density linked to the room volume and modal damping related to the reverberation time. It is defined in metric units as

$$f_S = 2000\sqrt{\frac{T_{60}}{V}} \qquad (2.50)$$

where
   T_{60} is the reverberation time [s]
   V is the enclosure volume [m³]

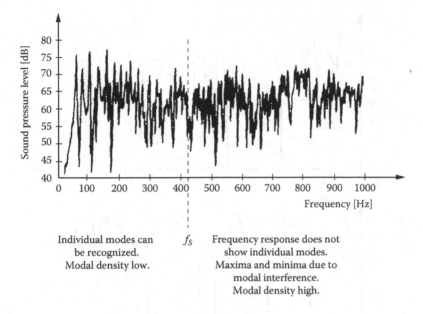

Individual modes can f_S Frequency response does not
be recognized.   show individual modes.
Modal density low.   Maxima and minima due to
                                 modal interference.
                                 Modal density high.

Figure 2.16 Frequency response curve and the Schroeder frequency.

The constant 2000 applies to metric units. Figure 2.16 shows the separation of the two frequency regions. This book is dealing with small rooms, many of which are highly damped. In these rooms, the Schroeder frequency can be low and affect the frequency properties of the enclosure. For instance, a room 200 m³ with $T_{60} = 1$ s has a Schroeder frequency of about 141 Hz. For a small living room of 50 m³ and $T_{60} = 0.5$ s, the Schroeder frequency f_S is 200 Hz.

This section will summarize the essential statistical properties of the sound fields in enclosures above the Schroeder frequency. These results are derived for a steady state that means that after the source is turned on and kept steady, the time needed to reach steady state is at least as long as the reverberation time.

In doing sound field measurements for some purpose, we have to distinguish between two different curves: the *frequency curve* that is measured at certain point with the source driven by a very slowly varying frequency and the *spatial curve* taken at a single frequency by a slowly moving microphone along a straight line. Both curves have a similar pattern because they are created by interfering modes with random amplitude and phase. Figure 2.17a shows the sound pressure level curve obtained along a line in a room and Figure 2.17b is the curve obtained for the spectrum at some point [6].

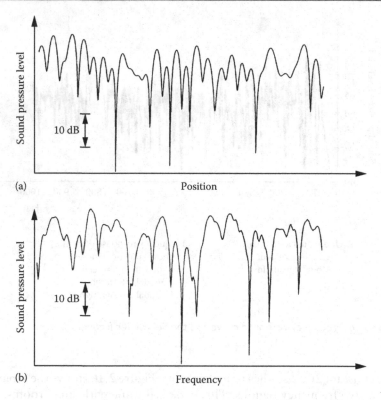

Figure 2.17 (a) Spatial and (b) frequency response curves for a room.

Both curves originate from Equation 2.39 with the variables either k or r. Their statistical distributions are the same [14]. The variable $\xi = \dfrac{p^2}{\langle p^2 \rangle}$ is an exponential distribution with the distribution function w and the cumulative distribution F. The variable p is the sound pressure and the value in the denominator is the average value:

$$w(\xi) = e^{-(\xi)} \quad \text{and} \quad F = 1 - e^{-(\xi)} \tag{2.51}$$

The graphs of both distributions are shown in Figure 2.18. The logarithmic representation of twice the variance that contains 70% of the sound amplitudes is (nonsymmetrically) within 11 dB as shown in Figure 2.19.

It is important to recognize that the damping does not make the space curve flat. The peaks are created by the waves that meet at a given point

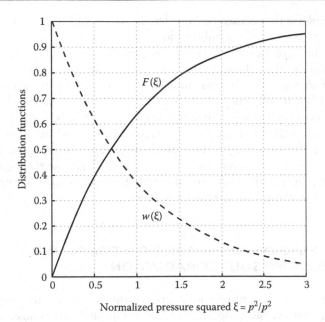

Figure 2.18 Statistical distribution of pressure square in a room. (After Schroeder, M.R. and Kuttruff, K.H. On frequency response curves in rooms. Comparison of experimental, theoretical, and Monte Carlo results for the average frequency spacing between maxima, *J. Acoust. Soc. Am.*, 34(1), 76, 1962.)

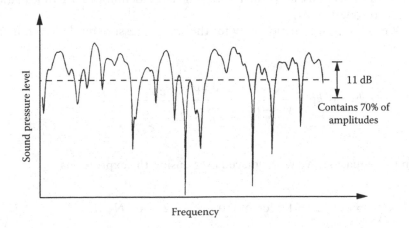

Figure 2.19 Frequency response curve in a room and its mean and ±σ limits. (After Schroeder, M.R. and Kuttruff, K.H. On frequency response curves in rooms. Comparison of experimental, theoretical, and Monte Carlo results for the average frequency spacing between maxima, *J. Acoust. Soc. Am.*, 34(1), 76, 1962.)

with the phase that causes the interference to be additive, while in the minima, the phase is causing subtraction. In a plane standing wave, the adjacent maxima are $\Delta x = 0.5\lambda$ apart. However, the average distance of adjacent maxima above the Schroeder frequency in a room space curve is $\Delta x = 0.79\lambda$ [6,14]. The adjacent maxima on the frequency curve are affected by the damping. Their frequency distance is $\Delta f \simeq (4/T_{60})$. The higher the damping, the greater is the frequency interval between the peaks.

There are other statistical properties of the sound fields in enclosures that are useful for steady-state applications such as sound power measurements, noise transmission, and room response. Because the signals we are concerned with in the book are of a rather transient nature, only the most important statistical properties are covered.

2.5 EFFECTS OF BOUNDARIES ON THE SOUND FIELD AND SOURCE RADIATION

The sound radiation into a room depends on the source, its properties and location, and on the properties of the room such as its size, shape, and sound absorption. The fundamentals of the radiation from a small source into a room were covered in Section 2.3.6. This section provides deeper analysis of the effects of the room on the sound radiation and the sound field. Also, the sound power radiated by a small source into a free space is covered in Sections 1.10 and 1.10.1. The basic definitions are in Equations 1.119 through 1.121.

We will repeat Equation 2.39 for the sound pressure inside a rectangular room:

$$p(r,k) \approx -jk\rho_0 c \frac{Q}{V} \sum_{N=0}^{\infty} \frac{\Psi_N(r_0)\Psi_N(r)\varepsilon_N}{\left(k^2 - k_N^2 - j2k_N \frac{\delta_N}{c}\right)} \tag{2.52}$$

In this equation, Λ_N was replaced by ε_N using the expressions

$$\frac{1}{\Lambda_N} = \varepsilon_N \quad \varepsilon_N = 1 \quad \text{for } N = 0, \quad \varepsilon_N = 2 \quad \text{for } N \geq 1 \tag{2.53}$$

The radiated power P from a small source is

$$P = \frac{1}{2}\text{Re}\{Z_{rad}\}Q^2 \tag{2.54}$$

where

Z_{rad} is the radiation impedance

Q is the volume velocity of the source

The radiation impedance depends on the incidence of reflected waves on the source and, for a rectangular room, is given by [16]:

$$Z_{rad} = -\frac{j\omega\rho_0}{V} \sum_N \frac{\Psi_N^2(r_0)\varepsilon_N}{k^2 - k_N^2 - j\dfrac{2k_N\delta_N}{c}} \qquad (2.55)$$

where $\Psi_N(r_0)$ is the eigenfunction at the source location r_0.

Figure 2.20 shows the radiation impedance for the source location in the room corner at $(0, 0, 0)$, where the source excites a maximum of modes [17]. The room size in this case was $7.49 \times 6.56 \times 3.56$ m³. The

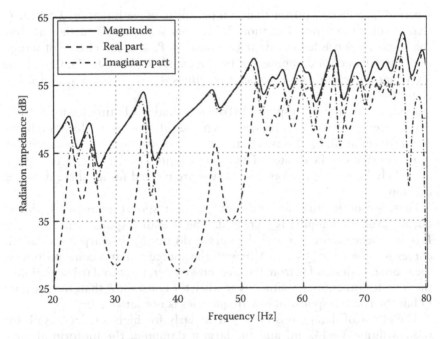

Figure 2.20 Example of a real and imaginary part of sound pressure in a room. (After Loubeau, A. Reduction of low frequency sound pressure variability in a small enclosure by use of a source with frequency-independent radiation power. MSc thesis, The Pennsylvania State University, Graduate Department of Acoustics, College of Engineering, University Park, PA, 2003.)

figure shows both the real and imaginary part of the radiation impedance as well as the total magnitude. The peaks of the curves appear at the
modal frequencies as long as the modes are well separated. The numera-
tor of the earlier equation consists of the product of cosine function that
interferes and results in maxima for the modal frequencies. The magnitude of the denominator is the smallest for $k^2 - k_N^2 = 0$ and determines the
size of the peaks.

The effect of wall damping, source, and receiver position will be further demonstrated for a constant, frequency-independent radiated sound
power. This can be achieved by placing an electric filter before the input
to the loudspeaker. Computationally, the sound pressure can be calculated
from [16,17]

$$p(r,k) = -\frac{jk\rho_0 c}{V}\sqrt{\frac{2P_0}{\mathrm{Re}\{Z_{rad}\}}}\sum_{N=0}^{\infty}\frac{\Psi_N(r_0)\Psi_N(r)\varepsilon_N}{\left(k^2 - k_N^2 - j2k_N\dfrac{\delta_N}{c}\right)} \tag{2.56}$$

where P_0 is constant power that will permit to solve Equation 2.54 for Q
that is substituted into Equation 2.52. This section will further analyze
the room response for constant Q, constant P, and the effects of damping. Three values of damping will be presented: $(\delta_N/c) = 0.0075$, 0.015, and
0.03 that correspond to reverberation times $T_{60} = 2.7$ s, 1.3 s, and 0.7 s,
respectively.

First, the frequency response between a single radiating location and a
single receiving location is shown in Figure 2.21. The source location in the
room corner at $(0, 0, 0)$ was selected to show a response for the maximum
of modes that can be excited. The receiver point at $(3, 2, 1.2)$ was selected
as a likely listening position. The curves are plotted for all three damping
constants.

First, we notice that for constant Q, the response increases with frequency even as damping is increased. The overall response with constant
P is in essence more flat and the curves do not have sharp peaks as the
curves for constant Q have. Although the source radiates constant power,
the room response is far from flat. As shown later, the standard deviation of
the transfer function amplitude is smaller for constant P than for constant
Q but the room response at low frequencies is not satisfactory.

The effect of damping is large particularly for higher values. With the
room volume $V = 147$ m^3 and the largest damping, the uniform absorption coefficient $\alpha = 0.18$. The curves of individual modes are wider and,
as we see from the curves for constant Q, some distinct modal response
disappears.

So far, we have shown the transfer function for the source position in
the room corner. This might be a good location for a subwoofer from the

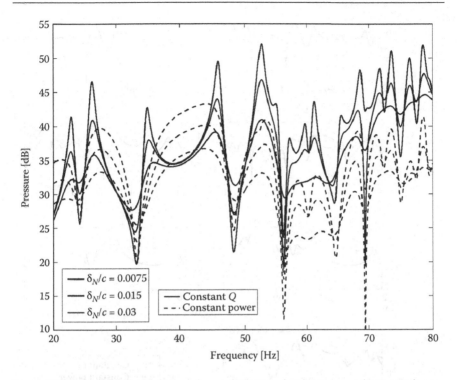

Figure 2.21 Frequency dependence of sound pressure levels source radiation with constant volume velocity and constant power on room damping. Comparison of effect of damping terms on frequency response at point (3, 2, 1.2) due to a point source at (0, 0, 0). Damping values for δ_N/c are denoted by different width lines. Solid lines indicate pressure with constant Q and dashed lines indicate pressure with constant radiated power. (After Loubeau, A. Reduction of low frequency sound pressure variability in a small enclosure by use of a source with frequency-independent radiation power. MSc thesis, The Pennsylvania State University, Graduate Department of Acoustics, College of Engineering, University Park, PA, 2003.)

viewpoint of maximum output. However, there are other locations of interest for full-range speakers that also are effective for radiating very low frequencies. Figure 2.22 shows the frequency responses for both constant Q and P from 7 source locations in a room [17]. The importance of this figure is that it presents the sound radiation from source locations that do not excite certain modes and, as a result, the frequency responses are uneven, particularly with constant Q. However, the response for constant P is much more even.

So far, the transfer of signals was shown from point to point. Ideally, we would like to achieve good (flat) frequency responses over larger floor area of the room. Therefore, several grids of 49 points each were selected and an

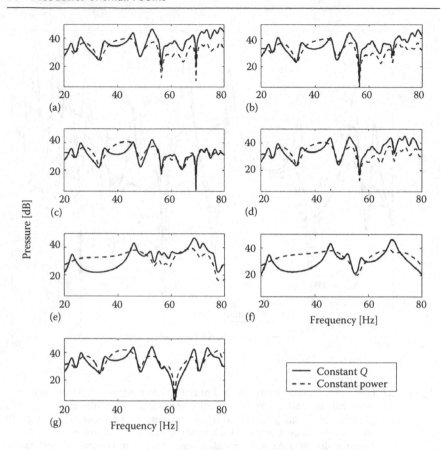

Figure 2.22 Frequency response for constant Q and constant power at receiver location (3, 2, 1.2) for seven source locations with $\delta_N/c = 0.015$. (a) Source location (0, 0, 0); (b) source location (1, 0, 0); (c) source location (1, 1, 1); (d) source location (0, 1.2, 0); (e) source location (0, $L_y/2$, 0); (f) source location (0, $L_y/2$, 1.5); and (g) source location (0, 1.2, 1.5). (After Loubeau, A. Reduction of low frequency sound pressure variability in a small enclosure by use of a source with frequency-independent radiation power. MSc thesis, The Pennsylvania State University, Graduate Department of Acoustics, College of Engineering, University Park, PA, 2003.)

average of transfer functions calculated and the results for one such grid, located in the room center, is shown in Figure 2.23. Comparing Figures 2.22 and 2.23 reveals some degree of similarity given by the source location that radiates certain composition and damping of modes that affects the whole sound field in the room. Although this figure is presented for one damping coefficient, the lack of excited modes has its effects on the whole room sound field.

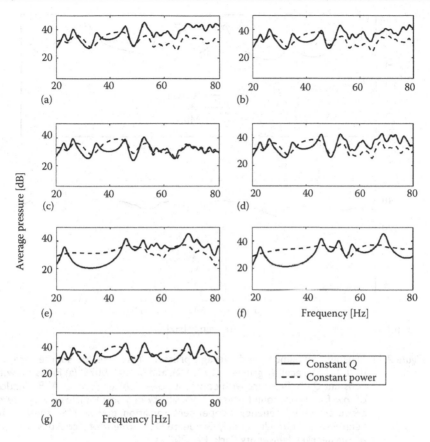

Figure 2.23 Average frequency response for constant Q and constant power for receiver grid (3, 2, 1.2) for seven source locations with $\delta_N/c = 0.015$. (a) Source location (0, 0, 0); (b) source location (1, 0, 0); (c) source location (1, 1, 1); (d) source location (0, 1.2, 0); (e) source location (0, $L_y/2$, 0); (f) source location (0, $L_y/2$, 1.5); and (g) source location (0, 1.2, 1.5). (After Loubeau, A. Reduction of low frequency sound pressure variability in a small enclosure by use of a source with frequency-independent radiation power. MSc thesis, The Pennsylvania State University, Graduate Department of Acoustics, College of Engineering, University Park, PA, 2003.)

The average curves shown so far do not show the range of differences among the 49 points that were averaged. Figure 2.24 shows the upper and lower limits of the transfer functions to individual grid points. The differences can be quite large.

A good measure of any fluctuating function is its standard deviation, which was calculated for many source and receiver grid locations and modal damping of the room. Figure 2.25 shows an example of these calculations

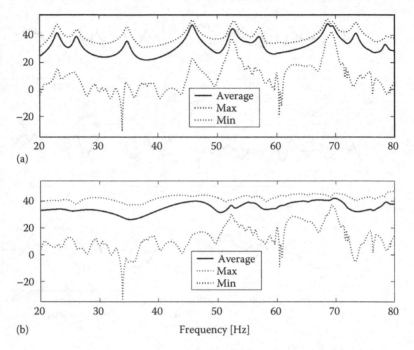

Figure 2.24 Minimum, maximum, and average pressure response for source location (0, 1.2, 1.5), receiver grid (4.5, $L_y/2$, 1.2), and $\delta_N/c = 0.0075$. (a) Pressure with constant Q. (b) Pressure with constant power. (After Lubeau, A. Reduction of low frequency sound pressure variability in a small enclosure by use of a source with frequency-independent radiation power. MSc thesis, The Pennsylvania State University, Graduate Department of Acoustics, College of Engineering, University Park, PA, 2003.)

for the source position at (0, 1.2, 1.5) that is one of the probable loud-speaker placements. A grid at the room center was selected for listening. The figure shows the average frequency responses for three values of damping. The horizontal lines mark the standard deviations (2σ) for constant Q and P. Comparing the values for σ_Q and σ_P reveals that they all decrease with increasing damping.

The important conclusion from this data is that the spatial pressure distribution of the modes substantially affects the smoothness of the transfer function and that the room damping has to be large enough to secure adequate modal overlap. Although one would intuitively hope that the frequency response would be flatter by radiating constant power with frequency, the improvement is relatively small and other means such as damping materials have to be used to control the transfer function. This will be discussed in Chapter 10.

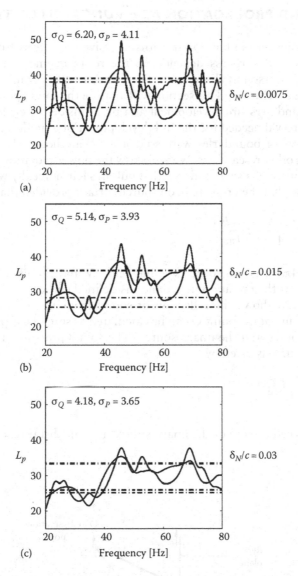

Figure 2.25 Frequency response with standard deviation for source location (0, 1.2, 1.5) and $\delta_N/c = 0.0075$. Dash-dotted lines represent one standard deviation above and below the mean pressure level. (a) Receiver grid location (3, 2, 1.2). (b) Receiver grid location ($L_x/2$, $L_y/2$, 1.2). (c) Receiver grid location (4.5, $L_y/2$, 1.2). (After Lubeau, A. Reduction of low frequency sound pressure variability in a small enclosure by use of a source with frequency-independent radiation power. MSc thesis, The Pennsylvania State University, Graduate Department of Acoustics, College of Engineering, University Park, PA, 2003.)

2.6 SOUND PROPAGATION AS A FUNCTION OF TIME

The modes in a room develop due to successive reflections from the room boundaries. This process depends on the room geometry and the wall absorption and sound dispersion. Therefore, it takes some time until a steady state is achieved. This section will analyze the sound power radiation and the sound pressure development at a point inside the enclosure during the time interval needed for the full development of a mode.

The effects of boundaries were studied by Waterhouse [18] and Adams [19] among other researchers. We will show the power radiation from a small spherical source of a small radius a that pulsates harmonically with a surface velocity u so that the pressure p_s on source surface provided that $k^2 a^2 \ll 1$ is

$$p_s = U\left[\frac{\rho c k}{4\pi} + j\frac{\rho c k}{4\pi a}\right] \tag{2.57}$$

where $U = 4\pi a^2 u$ is the source volume velocity and the expression in the parentheses is the radiation impedance as defined earlier.

Figure 2.26 shows the reflection of sound from an infinite plane. The sound pressure at the point O can be calculated as sum of the pressure from the main source and the image source. The sound pressure at a distance r from the source is given by

$$p(r) = j\frac{U f \rho e^{-jk(r-a)}}{2r} \tag{2.58}$$

For $a \ll r$, the pressure of the image source at some distance x is

$$p_i = j\frac{U f \rho e^{-jk2x}}{4x} \tag{2.59}$$

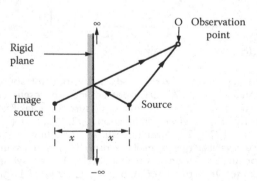

Figure 2.26 Reflections from a wall.

The sum of the pressures at the point O is $p_s + p_i$ and the sound power radiated by the main source after some simple algebra [19] is

$$\frac{W}{W_s} = 1 + \sum_{i=1}^{\infty} \frac{\sin(kd_i)}{kd_i} \qquad (2.60)$$

where $W_s = U^2((\rho c k^2)/(4\pi))$ is the sound power radiated by the small source into a free space. The summation term in the earlier equation is the contribution of each image source and d_i are the distances from each image source to the main source. Because for distant image sources d_i becomes large, the contributions of the high-order images to the total power are small.

In a real room, the intensity of a wave at each reflection is somewhat decreased due to the wall absorption. Let us assume that the walls have a uniform power reflection coefficient β so that at each reflection the strength of the wave is decreased by this factor. After the first reflection, the ratio W/W_s becomes

$$\frac{W}{W_s} = 1 + \beta \frac{\sin(kx)}{kx} \qquad (2.61)$$

The summation in Equation 2.60 will be modified so that we obtain

$$\frac{W}{W_s} = 1 + \sum_{i=1}^{m} \beta^m \frac{\sin(kd_i)}{kd_i} \qquad (2.62)$$

The sum is calculated for the sound reflected m times and the power of the reflection coefficient β will increase by one for each successive reflection order.

2.6.1 Sound power output dependence on time

If a sound source is placed in a room and tuned on, it takes some time until a steady state is achieved at some listening point. Even the source may need some time to reach a steady state from the instant of its turn on. There are many musical instruments, such as pipe organs, in which the sound power output grows relatively slowly, comparable to the travel time that the sound wave needs to reach the first reflector. These transient phenomena are important for the perception of the sound that can be obviously different from room to room.

To calculate the time dependence of the sound power radiation from a source that has a much shorter time to reach a steady state of its radiation,

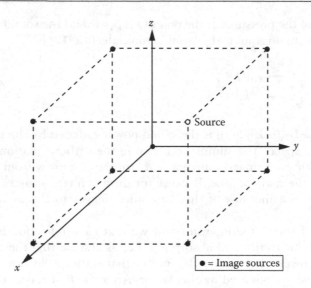

Figure 2.27 A source and its images at a right angle corner between three planes.

we will modify Equation 2.62. Figure 2.27 shows schematically a rectangular room with a single source and its images in a plane. The wave front of the sound wave radiated by the source is a sphere of radius that depends on the sound speed and selected propagation time. At some instant, the sound field in the room will be created by the contributions of the image sources that are located inside the arrival time *sphere* that contains the sources in the figure.

Equation 2.62 must be modified to express the time dependence:

$$\frac{W(t)}{W_s} = 1 + \sum_{i=1}^{N(t)} \beta^{N(t)} \frac{\sin(kd_i)}{kd_i} \tag{2.63}$$

The summation and the exponent of the reflection factor increase by one and count only the image sources that are in the sphere of radius where the sound wave front could reach within the selected time.

The following examples were adapted from a publication of Adams [19]. The room is a parallelepiped with the dimensions $6 \times 5 \times 3.5\,m$ and the power reflection coefficient $\beta = 0.9$. The reverberation time is then approximately 1 s. The source was located at the distance 1 m from the walls of a room corner. Figure 2.28 shows the results of calculations for different time delays within the frequency interval 0–200 Hz.

The first Figure 2.28a was calculated for the sum of the direct sound and one (first) reflection from the walls. The dip around 100 Hz shows the sound

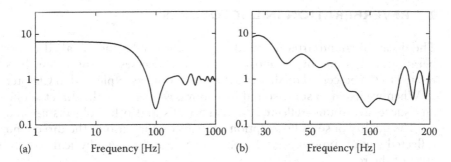

Figure 2.28 Power output of main source with (a) only the seven first mirror images taken into account and (b) all mirror image contributions within 50 ms. (From Adams, G. The dependence of loudspeaker power output in small rooms, JAES, 37(4), 203, 1989.)

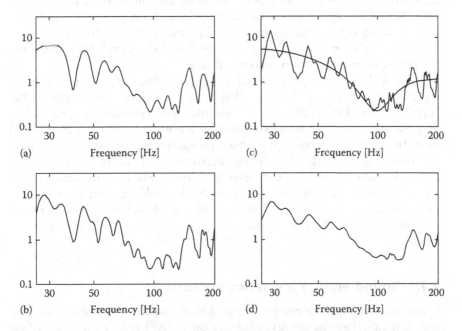

Figure 2.29 Power output of main source with all mirror image contributions within time t taken into account: (a) t = 100 ms, (b) t = 150 ms, (c) t = 500 ms (β = 0.9), and (d) t = 500 ms (β = 0.8). The solid line in (c) is the same as in Figure 2.28a for comparison. (After Adams, G. The dependence of loudspeaker power output in small rooms, JAES, 37(4), 203, 1989.)

power decrease due to the approaching first axial resonance. Figure 2.28b was obtained for a time delay of 50 ms. Figure 2.29a–d shows results calculated for delays of 100, 150, 500 ms (β=0.8), and 500 ms (β=0.9). It is interesting to see the strong influence of the first reflections and that the steady state was nearly obtained except for sum fluctuation of the frequency curve.

2.7 REVERBERATION IN ENCLOSURES

The decay of sound after a sound source stops radiating is called reverberation and is one of the most important processes in any room. It is related to the perceived quality of sound in a room as explained in Chapter 8. When we listen to some sound in a room, we perceive the direct sound and successive sound reflected from the walls and other objects around us. The quality of sound perception depends on the ratio of the direct and reflected sound, type of signal, sound decay, and other physical parameters of the room.

The reverberation time is measured by the time that elapses for the sound pressure level to decrease by 60 dB from its initial steady-state level. The reverberation process is complicated and depends on the room shape, distribution of the absorbing surfaces in the room, and internal sound propagation. The derivation of reverberation time depends on various assumptions on the sound field in the room and, therefore, several different formulas for the calculation of reverberation time have been developed. The most important formulas are summarized in here.

Historically, the importance of the sound decay for listening quality was recognized by Wallace Clement Sabine. He even measured the reverberation time although in a simple and consequently inaccurate way. Sabine's thinking even established what we call today diffuse sound field. In essence, this concept is based on spatial uniformity of basic field descriptors such as the directional distribution of the plane waves that create the sound field, zero time-averaged sound intensity, uniform spatial energy distribution, and similar others. Such a field in strict sense cannot exist (think of room corners). However, the equation for the reverberation time derived by Sabine is in use due to its simplicity in many practical situations.

2.7.1 Sound energy incidence on walls

Before we calculate the energy removed by the sound-absorbing walls, we will first calculate the energy incidence on walls. Let us assume a diffuse sound field that has a uniform energy density w so that the amount of energy in a volume dV is wdV.

Figure 2.30 shows a ring of volume dV that contains the energy $wdV = w2\pi r\sin\theta rd\theta dr$. This energy spreads over a sphere of radius r so that the energy incident on a wall element dS is

$$wdV \frac{dS\cos(\theta)}{4\pi r^2} \qquad (2.64)$$

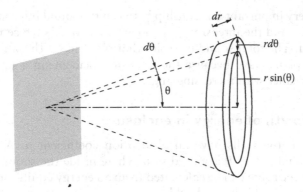

Figure 2.30 The observed ring of volume V.

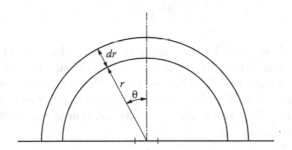

Figure 2.31 The variables used in the analysis of the sound incident from a hemisphere surrounding the observation point.

The total energy dE that incidents on dS from a hemisphere layer of thickness dr shown in Figure 2.31 is equal to

$$dE = w\frac{dS2\pi r^2 dr}{4\pi r^2}\int_0^{\pi/2}\sin(\theta)\cos(\theta)d\theta = \frac{wdS}{4}dr \qquad (2.65)$$

The energy arrives during the time interval $dt = (dr/c)$, where c is the speed of sound, so that the earlier equation becomes

$$\frac{dE}{dt} = \frac{wc}{4}dS \qquad (2.66)$$

The sound intensity I_{inc} incident on the wall is

$$I_{inc} = \frac{dE}{dSdt} = \frac{wc}{4} \qquad (2.67)$$

This is a very important relationship between the sound intensity in a plane wave $I_0 = wc$ and the intensity I_{inc} incident on a room surface for a diffuse sound field. The total sound intensity that incident on the walls is smaller than in a plane wave because the intensity incident on a unit wall area becomes smaller with increasing angle θ.

2.7.2 Growth of energy in enclosure

Assume that the walls have an absorption coefficient α. A source that radiates sound power P is placed somewhere inside the room. The energy increase in the room can be calculated from an energy equilibrium equation that assumes a diffuse sound field

$$P = V \frac{dw}{dt} + I_{inc}\alpha S = V \frac{dw}{dt} + \frac{wc}{4}\alpha S \qquad (2.68)$$

The power P supplied by the source is consumed by the energy increase inside the room, expressed by the first right-hand side time-dependent term of the earlier equation. At the same time, some energy is absorbed by the room walls of area S that have absorption coefficient α as expressed by the second term of the earlier equation. The solution of this differential equation for $w = 0$ for $t = 0$ is

$$w(t) = \frac{4P}{c\alpha S}\left[1 - e^{-(c\alpha S / 4V)t}\right] \qquad (2.69)$$

Figure 2.32 shows the exponential time dependence of the energy density $w(t)$ increase that approaches asymptotically the value $4P/c\alpha S$.

2.7.3 Decay of energy in enclosure

The dependence of the energy density on time after the source has been turned off can be found by solving Equation 2.69 for $P = 0$. We then obtain

$$w(t) = w_0 e^{-(c\alpha S / 4V)t} \qquad (2.70)$$

where w_0 is the initial energy density. Figure 2.33 shows the energy density (and sound intensity) decay with time in a logarithmic scale that is more practical because of the large range of time decay values that are usually pursued. This idealized decay is represented by a smooth line. As we show later, the actually measured time decay is affected, particularly in small rooms, by modes, various other resonance effects, and phenomena that make the decay curve irregular.

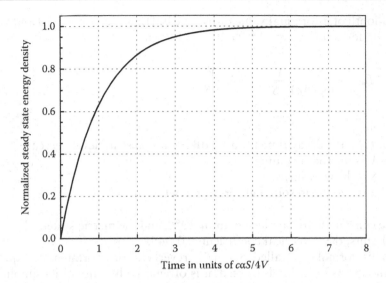

Figure 2.32 The energy density increases asymptotically to the value where the energy supplied by the source equals the energy lost by sound absorption.

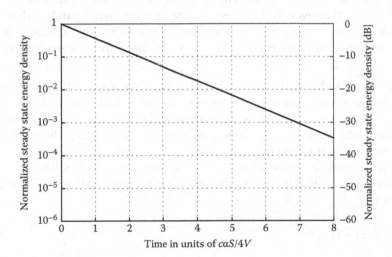

Figure 2.33 The decay of energy using Sabine's model shown on a logarithmic scale and in decibels.

2.7.4 Reverberation time

As mentioned earlier, the reverberation process is very important for the quality of listening in rooms. The quantitative definition of reverberation time is the time T in which the energy density of sound will decrease to one millionth of its original value or by 60 dB. Using Equation 2.70 and substituting for $w(T)/w_0 = e^{-(c\alpha S/4V)T} = 10^{-6}$ results in

$$T_{60} = \frac{24}{c\log_{10}(e)}\frac{V}{\alpha S} = 0.161\frac{V}{\alpha S} \qquad (2.71)$$

where
    T_{60} is the decay time for a 60 dB sound pressure level decrease
    V is the room volume
    S is the wall surface
    α is a uniform sound absorption coefficient

The constant 0.161 applies to air at 20°C and the metric system. For imperial units, the numerical coefficient is 0.049.

This formula is called Sabine's (reverberation) formula or equation. Although its precision is limited, it is often used because of its simplicity.

The product αS is often called total absorption and is given in units of metric sabin [m²S] or if imperial units are used in sabin [ft²S]. In Equation 2.71, the absorption coefficient is expected to be uniform over the wall surface S. However, in real rooms, the surfaces have patches that have different absorption coefficients. The total absorption is then achieved by summing the products of areas with the same absorption so that $\alpha S = \sum_{i=0}^{n} \alpha_i S_i$.

If the sound absorption by the walls is small, approaching zero in the limit, the reverberation time increases correctly to infinity. However, for large absorption where $\alpha \rightarrow 1$ (anechoic chamber), the reverberation time is finite although it should approach zero. As shown later, there are other formulas that improve the reverberation time prediction.

As mentioned earlier, there is some sound attenuation when the sound propagates in air. The attenuation expressed in terms of sound intensity is given by

$$I(x) = I_0 e^{-mx} \qquad (2.72)$$

where m is an attenuation coefficient that depends primarily on sound frequency, temperature, and humidity. The characteristics of m are shown in Figure 2.34 [13].

Taking this sound absorption into account Sabine's formula has to be modified so that T_{60} is

$$T_{60} = 0.161\frac{V}{S\alpha + 4mV} \qquad (2.73)$$

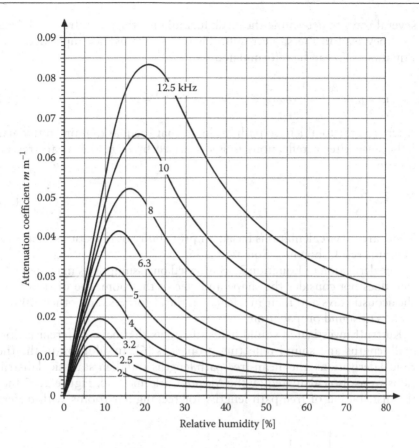

Figure 2.34 Values of the attenuation coefficient *m* for air at a temperature of 20°C, for various values of relative humidity and frequency. The attenuation coefficient *m* depends primarily on sound frequency, temperature, and humidity. (After Harris, C.M. Absorption of sound in air vs. humidity and temperature, *J. Acoust. Soc. Am.*, 40, 148, 1966.)

2.7.5 Mean free path

For frequencies where the wavelength is much shorter than the dimensions of the room and its objects, the sound in the room can be thought of as propagating in rays. The length of each ray between two successive reflections is different. However, there are very many reflections during the sound decay process. Therefore, summing the length of all possible path lengths and dividing them by their number will give the average length of a ray, which is called mean-free-path length $\bar{l}$. This length is also defined as the mean value of the probability density of the distribution of the path length l that a ray travels between subsequent wall reflections. There are several ways to determine the basic formula for the mean-free-path length but they lead to the same result. Therefore, we use a simple derivation. Equation 2.70 can be reformulated into

$$w = w_0 e^{\frac{-\alpha(ct/4V)}{S}} = w_0 e^{-\alpha(l_t/\bar{l})} = w_0 e^{-\alpha(n\bar{l}/\bar{l})} = w_0 e^{-\alpha n} \tag{2.74}$$

where l_t is the total length path in time t that is equal to n mean free paths $\bar{l}$. Because after n reflections w is decreased by $e^{-\alpha}n$, the mean free path must be

$$\bar{l} = \frac{4V}{S} \tag{2.75}$$

There are other calculations based on probability of reflections that lead to the same result shown in Equation 2.75.

The derivation of Equation 2.73 is based on assumptions made by Sabine that does not consider the shape and size of the room, the distribution of the acoustic rays and their incidence on the absorbers, and other unknown factors of sound propagation.

Kuttruff investigated sound fields of rooms that have different ratios of wall lengths and sound reflecting properties [6]. In addition to the theoretical derivations, the mean-free-path lengths and their statistical distributions were also calculated by the Monte Carlo method. Figure 2.35 shows the distribution of free path lengths for rectangular rooms with different

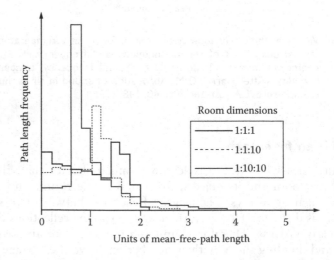

Figure 2.35 The mean-free-path distributions for three different room geometries. (From Kuttruff, H., *Room Acoustics*, Applied Science Publishers, London, U.K., 1973.)

Table 2.1 Relative room

Dimensions	$\bar{l}_{MC}$	σ^2
1:1:1	1.0090	0.342
1:1:2	1.0050	0.356
1:1:5	1.0066	0.412
1:1:10	1.0042	0.415
1:2:2	1.0020	0.363
1:2:5	1.0085	0.403
1:2:10	0.9928	0.465
1:5:5	1.0035	0.464
1:5:10	0.9993	0.510
1:10:10	1.0024	0.613

ratios of the wall lengths as obtained by the Monte Carlo method. It is very interesting to see that for some dimension ratios a certain length l prevails but not for a cubical room.

Table 2.1 assembled by Kuttruff [6] shows the ratios of the mean free paths $\bar{l}_{MC}$ obtained from the Monte Carlo method and $\bar{l}$ calculated from Equation 2.75. The numbers for different room sizes deviate from one very little that shows a very good validity of Equation 2.75. The table also shows the variance σ^2 of the lengths l calculated from

$$\sigma^2 = \frac{\overline{l^2} - \bar{l}^2}{\bar{l}^2} \tag{2.76}$$

Sabine's equation was derived under the assumption that the sound field is diffuse and that the sound power is absorbed by the walls continuously. In Eyring's approach, the sound is absorbed stepwise at each reflection and not continuously [6]. As before, we will calculate first the buildup of the energy in room. If a source is turned on at $t = 0$ to radiate sound power P, it takes the wave time τ to travel the one mean-free-path length $\bar{l}$. The radiated energy E_0 is

$$E_0 = P \cdot \tau \quad \text{and} \quad \tau = \frac{1}{n} \tag{2.77}$$

where n is the number of reflection in a unit of time (see Equation 2.74). At each reflection, a fraction α of energy is absorbed. If the enclosure remains between $t = \tau$ and $\tau = 2\tau$ of the original energy,

$$\tau = \frac{1}{n} \tag{2.78}$$

During the considered time interval τ to 2τ, the source will radiate the energy $E = P \cdot \tau$ so that the energy in the room will be

$$E_1 = E_0 + E'_{rem} = P\tau[1 + (1-\alpha)] \tag{2.79}$$

During the next time interval 2τ and 3τ, the energy will be

$$E_2 = P\tau[1 + (1-\alpha) + (1+\alpha)^2] \tag{2.80}$$

After the time $\tau = kt$, in which there are $k - 1$ reflections, the energy will be

$$E_{k-1} = P\tau[1 + (1-\alpha) + \cdots (1+\alpha)^{k-1}] \tag{2.81}$$

After summing the geometrical progression in the earlier equation, the energy density w is

$$w = \frac{E_{k-1}}{V} \frac{P\tau}{V} \frac{1 - (1-\alpha)^k}{\alpha} \tag{2.82}$$

Using the relationships derived earlier, this equation can be reformulated as

$$w = \frac{4P}{c\alpha S}\left[1 - (1-\alpha)^{(cS/4V)t}\right] = w_0\left[1 - e^{\ln(1-\alpha)(cS/4V)t}\right] \tag{2.83}$$

where $w_0 = (4P/c\alpha S)$ was shown in Equations 2.69 and 2.70 as the energy density reached after $t \to \infty$. Examining the earlier equation for $t = 0$ and $t \to \infty$, we obtain for w the correct values 0 and w_0.

Next, we will calculate the reverberation time. The energy density after the first reflection time τ is $w_\tau = w_0(1 - \alpha)$ and after k reflections is

$$w_{kt} = w_0(1-\alpha)^k = w_0(1-\alpha)^{(cS/4V)t} = w_0 e^{\ln(1-\alpha)(cS/4V)t} \tag{2.84}$$

The reverberation time T_{60} can be obtained by first using the upper equation substituting T_{60}

$$\frac{w}{w_0} = 10^{-6} = e^{\ln(1-\alpha)(cS/4V)T_{60}} \tag{2.85}$$

so that the reverberation time is

$$T_{60} = -\frac{24\ln(10)}{c}\frac{V}{S\ln(1-\alpha) + 4mV} = -0.161\frac{V}{S\ln(1-\alpha) + 4mV} \tag{2.86}$$

This equation extended by the air attenuation term is called Eyring's formula. It differs from Sabine's formula (Equation 2.71) by the denominator. Numerically, the results are practically equal for $\alpha \leq 0.1$ and, therefore, Sabine's equation is often used for rooms with hard walls. For $\alpha \geq 0.1$, the magnitude estimations of the two reverberation time equations differ.

The reverberation time prediction using these two formulas may be reasonably usable for enclosures with absorption distributed over all walls and a sound field, which is reasonably diffuse.

2.7.6 Fitzroy's formula

The calculation of the reverberation time using Sabine's or Eyring's formula is reasonably exact when the wall absorption across the room is sufficiently diffuse. That means that the product of the area and average absorption coefficient of each wall are not too much different. But there are rooms with the sound absorbers that are concentrated on one or two walls and where the reverberant sound field is less diffuse.

As an example, the ceiling could be suspended and consist of sound-absorbing tiles so that the sound absorption is concentrated on one wall. Usually, in such cases, the measured reverberation time is much longer than the one calculated by Eyring's formula. Fitzroy speculated that the vertically propagating sound waves are well absorbed while the horizontal wave survives much longer [20]. This nonuniformity of sound absorption causes the reverberation time predicted by his formula to differ very little from its measured value. Fitzroy's formula for the reverberation time in a rectangular room with air at 20°C is in metric units:

$$T_{60} = \left(\frac{S_x}{S}\right)\left[\frac{0.161V}{-S\ln(1-\alpha_x)}\right] + \left(\frac{S_y}{S}\right)\left[\frac{0.161V}{-S\ln(1-\alpha_y)}\right] + \left(\frac{S_z}{S}\right)\left[\frac{0.161V}{-S\ln(1-\alpha_z)}\right]$$

(2.87)

A form similar to Sabine's equation can be obtained by replacing $-\ln(1-\alpha)$ in the respective denominators by α for small values of α.

The units are metric (meters). In the first term, the ratio S_x/S is obtained by dividing the total area of two walls perpendicular to x-axes by the total wall area of the room $S = S_x + S_y + S_z$. The absorption coefficient α_x is the spatial average over the x walls. The other terms are similar for the other directions. The reverberation time consists of three ensembles that travel in three perpendicular directions. Each term is a modification of Eyring's formula that contributes proportionately to the wall areas that are parallel.

2.8 DIRECT AND REVERBERANT SOUND FIELD IN ENCLOSURE

This section will analyze the steady-state sound field generated by a source radiating a wide frequency band signal. The energy density of the sound field in a reverberant room can as a first approach be considered as consisting of two components, the direct w_d and reverberant w_r sound fields.

The total energy density w at the observation point is then

$$w = w_d + w_r \tag{2.88}$$

Let us assume that the sound field is diffuse. Figure 2.36 shows schematically the sound pressure level as a function of distance from the source. This curve is often referred to as the *draw-away* curve and shows the idealized behavior of the averaged sound pressure level for a wide-spectrum sound in a room, as a function of distance to the sound source. Assuming a small omnidirectional source, the sound pressure will decrease with the distance. However, the reverberant field will be about constant beyond some distance where the reverberant field energy density dominates.

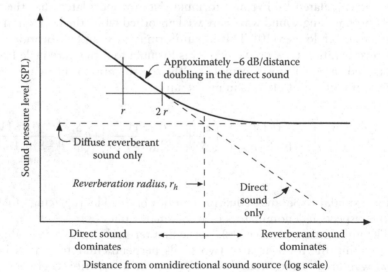

Figure 2.36 The draw-away curve (solid line) shows the theoretical behavior of the steady-state sound pressure level as a function of distance from an omnidirectional sound source in a room with an ideally diffuse sound field. The *reverberation radius* r_h is where the direct and the reverberant fields are equally strong.

The sound pressure squared of the direct field p_d^2 of a small source that has a directivity factor Q, defined as a ratio of the power radiated in certain direction to average power radiated in all direction, is at a distance r:

$$p_d^2 = \rho c \frac{PQ}{4\pi r^2} \tag{2.89}$$

where P is the radiated power. An omnidirectional source has $Q = 1$.

The sound power that creates the reverberant field is the sound power that exists in the room after the first reflection. Therefore, the original power P has to be multiplied by $(1 - \bar{\alpha})$. The energy density is of the reverberant field is

$$w_r = \frac{4P(1-\bar{\alpha})}{c\bar{\alpha}S} = \frac{4P}{cR} \tag{2.90}$$

where
  $\bar{\alpha}$ is the average absorption coefficient
  S is the surface of the walls
  $RC = (\bar{\alpha}S)/(1-\bar{\alpha})$ is called the room constant because its quantities characterize the room properties

The total sound pressure squared is

$$p^2 = \rho_0 cP\left[\frac{Q}{4\pi r^2} + \frac{4}{RC}\right] \tag{2.91}$$

This equation can be divided by the reference levels for sound pressure $p_0 = 2 \cdot 10^{-5}$ Pa and sound power $P_0 = 10^{-12}$ W so that we obtain

$$L_p = L_P + 10\log_{10}\left[\frac{Q}{4\pi r^2} + \frac{4}{RC}\right] \tag{2.92}$$

The distance at which the energy density of the direct field and the reverberant field are equally large is called the reverberation radius r_h or sometimes the critical distance. We obtain

$$r_h = \sqrt{\frac{Q \cdot RC}{16\pi}} = \sqrt{\frac{Q}{16\pi}\frac{1-\bar{\alpha}}{\bar{\alpha}S}} \tag{2.93}$$

When listening binaurally to a source emitting transient noise, the effective reverberation radius will be much larger because of the signal analysis capabilities of hearing.

In addition to the directivity factor Q, the directivity index (DI) is also used to characterize the level of sound in the main lobe direction I_{max} compared to that of an equally strong omnidirectional sound source I_0 at the same distance

$$DI = 10\log(Q) = 10\log\left(\frac{I_{max}}{I_0}\right) \tag{2.94}$$

Taking the directivity of the sound source into account using the DI as parameter, the draw-away curve will behave as shown in Figure 2.37.

The curves were drawn for a room that is assumed to have a reverberation time of about 0.5 s and a volume of 50 m³ typical of many living rooms [22–24]. The figure shows that the reverberation radius is at about 0.5 m in such a room for a $DI = 0$ dB, which is typical for small closed-box low- and midfrequency loudspeakers. A high-frequency loudspeaker driver can reach $DI = 12$ at high frequencies. At such frequencies, the room absorption is usually higher than that assumed for the graph so the listener may well be in inside the near field radius of the driver/room combination.

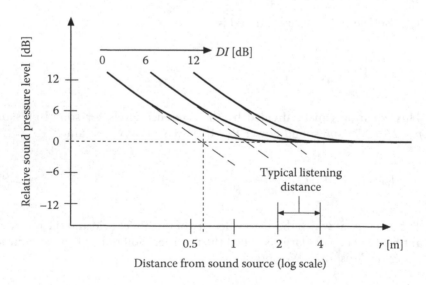

Figure 2.37 Draw-away curves for some values of DI in a room with dimensions 2.5 × 4 × 5 m³ and $A_S = 16$ m²S, typical for many living rooms $RT_{60} = 0.5$ s. Ideally diffuse reverberant field assumed.

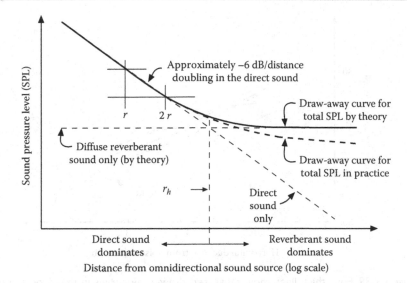

Figure 2.38 The draw-away curve in a room with scatterers and nonnegligible sound absorption by walls and scatterers.

In real rooms, the reverberant field is not ideally diffuse, which results in a draw-away curve that drops in level also outside the reverberation radius limit as shown in Figure 2.38.

It has been shown that the draw-away curve will depend on the room shape, dimensions, room surfaces, presence of sound scattering objects, their size, and characteristics [15]. The reference gives equations for the draw-away curves using such variables. The smaller the room, the more will the draw-away curves resemble the one shown in Figure 2.37. Ceiling height turns out to be an important variable assuming that scatterers are large and mainly close to the floor. The effect of sound scattering objects was found for large flat rooms to slightly raise the sound pressure level close to the source, up to typically 10 m distance, and to decrease the levels at greater distances. Close to the source, the scatterers tend to obstruct sound propagation and raise the sound pressure level compared with the empty room.

The drop rate of the draw-away curve is about −3 dB per distance doubling in the region from the reverberation radius to a distance of about three times the ceiling height [16,21]. In small listening rooms, the room width and length are typically less than three times the height.

Measurements in four American living rooms ca. 1980 showed that an average decay rate was about −3 dB per distance doubling in the frequency range around 1 kHz at a distance up to about three times the reverberation radius [16]. Figure 2.39 shows results from these measurements for the ⅓ octave band centered on 1 kHz.

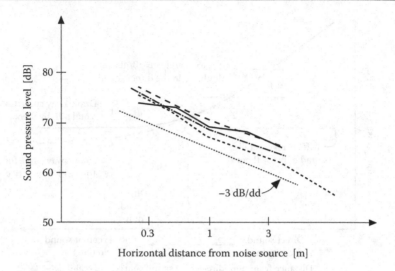

Figure 2.39 Four thick lines show measured draw-away curves in the ⅓ octave band centered on 1 kHz, using an omnidirectional noise source, in four American living rooms ca. 1980. (After Schultz, T.J. Improved relationship between Sound Power Level and Sound Pressure Level in Domestic and Office Spaces, Report 5290, *Am. Soc. of Heating, Refrigerating, and Air-Conditioning Engineers (ASHRAE)*, Prepared by Bolt, Beranek, and Newman, Inc., Cambridge, MA, 1983.)

2.A APPENDIX: FORCED WAVES IN ENCLOSURES

There are two mechanism principles that generate sound. A vibrating object, solid state or liquid, sets the gas into motion that propagates as sound. Examples are vibrating walls, some musical instruments, and loudspeakers. The other mechanism consists of setting a part of the gas directly into motion such as in wind instruments and human voice irregular air stream. As we have seen in previous sections, the sound behavior is described by the wave equation. However, the equation for free waves has to be modified to include the action of the source.

First, we amend the equation of continuity (see Equation 1.14) by a term that will express fluid motion. This additional term expresses generated mass per unit of volume flow q at a rate q volume unit per unit of volume per unit of time. The simplified equation of continuity is

$$\rho_0 \frac{\partial u_x}{\partial x} = -\frac{d\rho_t}{dt} + \rho_0 q \qquad (2.A.1)$$

where $\rho_t = \rho_0 + \delta$ with ρ_t is the total density, ρ_0 is the constant equilibrium density, and δ is the density increment due to the sound. As stated in the section on thermodynamic law,

$$\frac{d\rho_t}{dt} = \frac{1}{c^2}\frac{\partial p}{\partial t} \tag{2.A.2}$$

Following the procedure in Section 1.1 on the wave equation, derivation is obtained for 1D wave equation

$$\frac{\partial^2 p}{\partial x^2} = \frac{1}{c^2}\frac{\partial^2 p}{\partial t^2} - \rho_0\frac{\partial q}{\partial t} \tag{2.A.3}$$

and for 3D equation

$$\Delta p = \frac{1}{c^2}\frac{\partial^2 p}{\partial t^2} - \rho_0\frac{\partial q}{\partial t} \tag{2.A.4}$$

The quantity $q(r_0)$ is the source distribution function that depends on the position r_0. Its dimension is 1/s. For a harmonic point source at a location r_0, the wave equation becomes

$$\Delta p + k^2 p = -j\omega\rho q(r_0) \tag{2.A.5}$$

The modes in an enclosure for the sound pressure are described by the eigenfunctions $\Psi(r)$. Their numerical value depends on the coordinate r. The source with a location r_0 can be expanded in a series of these eigenfunctions that can be expressed as

$$q(r_0)e^{j\omega t} = \sum_{N=0}^{\infty} Q_N\Psi_N(r)e^{j\omega t} \tag{2.A.6}$$

where N is the quantum number of the eigenfunction $\Psi_N(r)$ that depends on the coordinate system of the enclosure. For a rectangular enclosure, N is a function of three integers m, n, and l. Q_N is the amplitude of the individual terms of the series of the source expansion. The upper equation can be multiplied by another eigenfunction $\Psi_M(r)$ and integrated over the enclosure volume V. We obtain

$$\iiint_V q(r_0)\Psi_M(r)dV = \sum_{N=0}^{\infty} Q_N \iiint_V \Psi_N(r)\Psi_M(r)dV \tag{2.A.7}$$

We assume that the functions Ψ are orthogonal and satisfy the condition

$$\iiint_V \Psi_N(r)\Psi_M(r)dV = V\Lambda_N \text{ for } N = M \text{ and } 0 \text{ for } N \neq M \tag{2.A.8}$$

Applying this condition on the integral in Equation 2.A.6 reveals that this integral is zero except for $N = M$ so that the right-hand side of the equation will consist of terms from $N = 0$ to $N \to \infty$. From each term, we can obtain the series coefficient Q_N given by

$$Q_N = \frac{1}{V\Lambda_N} \iiint_V q(r_0)\Psi_N(r)dV \qquad (2.A.9)$$

with Λ_N calculated from

$$V\Lambda_N = \iiint_V \Psi_N^2(r)dV \qquad (2.A.10)$$

The analytical evaluation for a realistic source characterized by the function $q(r_0)$ probably cannot be found; however, a numerical solution by a computer is feasible.

Quite often, there is interest in the low-frequency sound pressure spatial distribution when the source is small approximated by a point source so that $q(r_0)$ is expressed by

$$q(r_0) = Q\delta(r - r_0) \qquad (2.A.11)$$

where

Q is the (frequency-dependent) source amplitude
δ is the Dirac function
r_0 is the source location

The coefficient Q_N becomes

$$Q_N = \frac{Q}{V\Lambda_N} \iiint_V \delta(r - r_0)\Psi_N(r)dV = \frac{Q}{V\Lambda_N}\Psi_N(r_0) \qquad (2.A.12)$$

Free waves in an enclosure. As explained in Section 2.1, the free waves in an enclosure are characterized by the eigenfunctions $\Psi_N(r)$. The differential equation for these eigenfunctions is

$$\Delta\Psi_N(r) + k_N^2\Psi_N(r) = 0 \qquad (2.A.13)$$

The differential equation for forced pressure waves (see Equation 2.A.5) is

$$\Delta p(r) + k^2 p(r) = -j\omega\rho_0 q(r_0) \qquad (2.A.14)$$

The sound pressure $p(r)$ can be expanded into a series of eigenfunctions $\Psi_N(r)$

$$p(r) = \sum_{N=0}^{\infty} A_N \Psi_N(r) \qquad (2.A.15)$$

where A_N is the series coefficient that has to be determined. Substituting this series into Equation 2.A.13 for the forced pressure results in

$$\sum_{N=0}^{\infty} A_N \Delta \Psi_N(r) + \sum_{N=0}^{\infty} A_N k^2 \Psi_N(r) = -j\omega\rho_0 \sum_{N=0}^{\infty} Q_N \Psi_N(r) \qquad (2.A.16)$$

We replace in the upper equation $\Delta\Psi_N(r)$ by $-k_N^2\Psi_N(r)$ from the Equation 2.A.13 for the free waves and obtain

$$\sum_{N=0}^{\infty} A_N \Psi_N(r) \left[k^2 - k_N^2 \right] = -j\omega\rho_0 \sum_{N=0}^{\infty} Q_N \Psi_N(r) \qquad (2.A.17)$$

If we compare the Nth terms on both sides of the upper equation, we obtain for A_N

$$A_N = -j\omega\rho_0 Q_N \frac{1}{k^2 - k_N^2} \qquad (2.A.18)$$

For a point source where Q_N is given by Equation 2.A.12, the coefficient A_N is given by

$$A_N = -j\omega\rho_0 Q \frac{\Psi_N(r_0)}{k^2 - k_N^2} \frac{1}{V\Lambda_N} \qquad (2.A.19)$$

The sound pressure at any location r in the enclosure is

$$p(r,k) = \sum_{N=0}^{\infty} A_N \Delta\Psi_N(r) = -j\omega\rho_0 \frac{Q}{V} \sum_{N=0}^{\infty} \frac{\Psi_N(r_0)\Psi_N(r)}{\Lambda_N \left(k^2 - k_N^2 \right)} \qquad (2.A.20)$$

where
  V is the volume of the enclosure
  r_0 is the coordinate of the point source
  r is the coordinate of the point for which we calculate the sound pressure
  k is the wave number of the frequency of interest
  k_N is the wave number of the Nth mode

The analysis and application of the earlier equation is in Section 2.2.

2.B APPENDIX: APPROXIMATE SOLUTION OF $\bar{k}_x$

It is useful to calculate the approximation of the room response $p(x)$ for the wall impedance real and large. Using Equation 2.31 and substituting for the complex $\bar{k}_x$ from Equation 2.23, we can simplify the following expression:

$$e^{j(\beta_x + j\alpha_x)L_x} \simeq e^{-\alpha_x L_x} \qquad (2.B.1)$$

because $\cos(\beta_x L_x) \simeq 1$ and $\sin(\beta_x L_x) \simeq 0$. Using the earlier equation, we can now approximate Equation 2.31. After simplifications by neglecting small terms or approximating arithmetic operations,

$$e^{-\alpha_x L_x} \simeq 1 - \alpha_x L_x \simeq \left(1 - \frac{k}{k_x \xi}\right)^2 \simeq 1 - \frac{2k}{k_x \xi_x} \simeq 1 - \frac{2k}{\beta_x \xi_x} \qquad (2.B.2)$$

This equation permits us to calculate the product

$$\alpha_x \beta_x \simeq \frac{2k}{\xi_x L_x} \quad \beta_x \neq 0 \qquad (2.B.3)$$

The wave number approximation is Equation 2.15:

$$k_{m,n,l} = \sqrt{\left(\frac{m\pi}{L_x}\right)^2 + \left(\frac{n\pi}{L_y}\right)^2 + \left(\frac{l\pi}{L_z}\right)^2} \quad m,n,l = 0,1,2,\ldots \qquad (2.B.4)$$

In case the mode is axial or tangential and some of these quantum numbers m, n, or l are zero, Equation 2.B.2 becomes

$$e^{-\alpha_i L_x} \simeq 1 - \alpha_i L_i \simeq \left(1 - \frac{k}{k_i \xi_i}\right)^2 \simeq 1 - \frac{2k}{j\alpha_i \xi_i} \qquad (2.B.5)$$

and

$$j\alpha_i^2 = \frac{2k}{\xi_i L_i} \qquad (2.B.6)$$

The wave number approximation is further calculated from

$$\bar{k}_N = \left[(\beta_x + j\alpha_x)^2 + (\beta_y + j\alpha_y)^2 + (\beta_z + j\alpha_z)^2 \right]^{1/2} \qquad (2.B.7)$$

Squaring the terms, neglecting the small damping terms squared, and approximating the square root results in

$$\bar{k}_N \simeq k_N + j \frac{1}{k_N}\left(\alpha_x\beta_x + \alpha_y\beta_y + \alpha_z\beta_z\right) \quad \beta_i \neq 0 \qquad (2.B.8)$$

For $\beta_i = 0$, the term $\alpha_i\beta_i$ is replaced by Equation 2.B.6 so that the final approximation for $\bar{k}_N$ is

$$\bar{k}_N \simeq k_N + j \frac{k}{k_N}\left[\frac{\varepsilon}{\xi_x L_x} + \frac{\varepsilon}{\xi_y L_y} + \frac{\varepsilon}{\xi_z L_z}\right] \varepsilon = 1 \;\; \beta_i = 0 \;\; \text{or} \;\; \varepsilon = 2 \;\; \beta_i \neq 0$$

$$(2.B.9)$$

We can further show that the reverberation time as derived by Sabine is linked to the wave description of the sound field. After substituting into Equation 2.40 the sine terms, we obtain for

$$\Lambda_N = \frac{1}{\varepsilon_m \varepsilon \varepsilon_l} \quad \varepsilon_i = 1 \;\; \text{for } i = 0, \quad \varepsilon_i = 2 \;\; \text{for } i \neq 0 \qquad (2.B.10)$$

It has been found that for a high value of the wall impedance (real), the wall absorption coefficient α is

$$\alpha \simeq \frac{8}{\xi} \qquad (2.B.11)$$

The expression for the sound pressure (Equation 2.35) after squaring the wave number, neglecting the squared damping term, and substituting for Λ_N is

$$p(r,k) \simeq -jk\rho_0 c \frac{Q}{V} \sum_{N=0}^{\infty} \frac{\Psi_N(r_0)\Psi_N(r)\varepsilon_m\varepsilon\varepsilon_l}{k^2 - k_N^2 - j2k\left[\varepsilon_m \dfrac{\alpha_x}{8L_x} + \varepsilon_n \dfrac{\alpha_y}{8L_y} + \varepsilon_n \dfrac{\alpha_z}{8L_z}\right]}$$

$$(2.B.12)$$

The damping term in the denominator in the upper equation can be further developed for oblique waves ($\varepsilon = 2$):

$$j = \frac{2k}{L_x L_y L_z}\left[\varepsilon_m \frac{\alpha_x}{8} L_y L_z + \varepsilon_n \frac{\alpha_y}{8} L_x L_z + \varepsilon_n \frac{\alpha_z}{8} L_y L_y\right]$$

$$= j \frac{k}{4V}\left[\alpha_x S_x + \alpha_y S_y + \alpha_z S_z\right] = j \frac{k}{4V} A_N \qquad (2.B.13)$$

where $A_N = \sum \alpha_i S_i$ is the total absorption of the walls. Comparing Equation 2.B.13 with the damping term in Equation 2.39 and replacing k by k_N for the Nth mode results in

$$2k_N \frac{\delta_N}{c} = \frac{k_N}{4V} A_N \qquad (2.B.14)$$

From this equation, based on wave acoustics, the same equation is obtained for the reverberation time as from Sabine's equation based on the energy flow in the enclosure:

$$T_{60} = \frac{6.91}{\delta_N} = 0.161 \frac{V}{A_N} \qquad (2.B.15)$$

REFERENCES

1. Morse, P.M. and Ingard, K.U. (1953) *Theoretical Acoustics*. McGraw-Hill, New York.
2. Cremer, L., Mueller, H.A., and Schultz, T.J. (1882) *Principles and Applications of Room Acoustics*. Applied Science, New York.
3. Beranek, L.L. (1954) *Acoustics*. McGraw-Hill, New York.
4. Morse, P.M. (1976) *Vibration and Sound*. Acoustical Society of America, Woodbury, New York.
5. Kinsler, L.E. and Frey, A.R. (1962) *Fundamental of Acoustics*. John Wiley & Sons, New York.
6. Kuttruff, H. (1973) *Room Acoustics*. Applied Science Publishers, London, U.K.
7. Skudrzyk, E. (1971) *The Foundations of Acoustics*. Springer-Verlag, New York.
8. Toole, F.E. (2009) *Sound Reproduction*. Elsevier, New York.
9. Morse, P.M. and Bolt, R.H. (1944) Sound waves in rooms. *Rev. Mod. Phys.*, 16(2), 69–150.
10. Vorländer, M. (2008) *Auralization*. Springer-Verlag, Berlin, Germany.
11. Blauert, J. and Xiang, N. (2008) *Acoustics for Engineers*. Springer-Verlag, Berlin, Germany.
12. Kleiner, M., Klepper, D.L., and Torres, R.R. (2010) *Worship Space Acoustics*. J. Ross Publishing, Ft. Lauderdale, FL.
13. Harris, C.M. (1966) Absorption of sound in air vs. humidity and temperature. *J. Acoust. Soc. Am.*, 40, 148–159.
14. Schroeder, M.R. and Kuttruff, K.H. (1962) On frequency response curves in rooms. Comparison of experimental, theoretical, and Monte Carlo results for the average frequency spacing between maxima. *J. Acoust. Soc. Am.*, 34(1), 76–80.
15. Stangham, E. (1983) Noise attenuation in factories. *Appl. Acoust.*, 16(3), 83–214.

16. Schultz, T.J. (1983) Improved relationship between Sound Power Level and Sound Pressure Level in Domestic and Office Spaces, Report 5290, Am. Soc. of Heating, Refrigerating, and Air-Conditioning Engineers (ASHRAE), Prepared by Bolt, Beranek, and Newman, Inc., Cambridge, MA.
17. Lubeau, A. (2003) Reduction of low frequency sound pressure variability in a small enclosure by use of a source with frequency-independent radiation power. MSc thesis, The Pennsylvania State University, Graduate Department of Acoustics, College of Engineering, University Park, PA.
18. Waterhouse, R.V. (January 1958) Output of a sound source in a reverberation chamber and other reflecting environments. *JASA*, 30, 4–13.
19. Adams, G. (1989) The dependence of loudspeaker power output in small rooms. *JAES*, 37(4), 203–209.
20. Fitzroy, D. (1959) Reverberation formulae which seems to be more accurate with non-uniform distribution of absorption. *J. Acoust. Soc. Am.*, 31, 893–897.
21. Peutz, V.M.A. (1968) The sound energy distribution in a room. Sixth International Congress on Acoustics, Tokyo, August 21–28, paper E-5-2, pp. E165–E168.
22. Burgess, M.A. and Uteley, W.A. (1985) Reverberation time in British living rooms. *Appl. Acoust.*, 18, 369–380.
23. Bradley J.S. (1986) Acoustical measurements in some Canadian homes, Institute for Research in Construction, National Research Council, Ottawa, K1A 0R6.
24. Diaz, C. and Pedrero, A. (2005) The reverberation time of furnished rooms in dwellings. *Appl. Acoust.*, 66, 945–956.

Chapter 3

Geometrical acoustics

3.1 HEARING AND IMPULSE RESPONSE

One of the reasons for the study of room acoustics is to find tools to predict the sound transmission properties sufficiently well from the viewpoint of hearing, so that rooms can be designed that have optimum properties for the transmission and enjoyment of speech and music. With proper tools, problems and faults in already existing rooms can then also be pinpointed and possibly corrected.

The wave equation for sound and its solutions for a rectangular room are introduced in Chapter 2. It is clear from the discussion in that chapter that it is impossible to keep track of all the room modes at high frequencies since the modal density increases in proportion to frequency squared. It is important to remember that the wave approach formulations in that chapter primarily apply to resonant systems that have relatively little damping, whereas real rooms may have very high damping. Additionally, to provide sufficient information from the viewpoint of room acoustics, we need to account for even small details of source, receiver, and room geometry. We also need to use the phase relationships of the modes if we want to assemble the resulting sound pressure field correctly.

The voice and acoustical music instruments generate primarily transient sounds. Our hearing analyzes these sounds with particular interest in the onset transients. This makes it necessary to analyze both signal frequency response and time history. Since our ears are separated by some distance, we can use the phase difference between ear signals at low frequencies and the time delay of signal amplitude envelopes at high frequencies to determine the direction of arriving sounds as explained in Chapter 6. Such differences between the signals at the ears are also a result of reflection and diffraction of sound by the torso and cause frequency-dependent sound pressure level differences that depend on the arrival angle. At medium and high frequencies, the onset time differences of direct, reflected, and reverberant sound at the ears are essential for our appreciation of room acoustic quality. For hearing, earlier sounds mask and make inaudible later sounds so the reverberation or rather decay time of sounds is important.

We expect rooms to have reverberation time suitable for their use. Speech and music require different reverberation times for the best speech intelligibility or musical effect. The reverberation time varies with frequency since the sound absorption of room surfaces and objects varies with frequency.

The properties of human hearing point toward the importance of analyzing the transient properties of sound transmission in rooms at medium and high frequencies, while the frequency response is more important for low frequencies. While the transient performance of a room can be found by adding all mode contributions and then using the inverse Fourier transform to obtain the time response of the room, this approach is impractical in the design office.

For practical work in room acoustics, it is instead necessary to use *geometrical acoustics*, an analysis method that is sufficiently accurate and precise so that room acoustic analysis can be done intuitively and quickly and still yield useful results for the architect and acoustician. The architect's initial design considerations usually include elementary manual tracing of *sound rays* and reverberation time calculation using Sabine's or Fitzroy's formulas. Sabine's formula essentially requires the sound field in the room to be diffuse, which is seldom the case. Fitzroy's formula is particularly useful in predicting the reverberation time for those rectangular rooms where sound absorption is unevenly distributed between room surfaces.

Geometrical acoustics draws on our extensive experience in using mirrors for visual purposes to become an intuitive first choice in room acoustics. In much practical work in room acoustics, geometrical acoustics is the preferred method and can be used whether analysis is made by hand or computer. The possibility to easily measure psychoacoustically motivated room acoustics quality metrics and to be able to compare measurement to prediction is also an important aspect in the choice of room acoustic analysis methods, as described in Chapter 14.

We use the impulse response of a room to study both its early reflections and its reverberation. The impulse response can be thought of as an acoustical signature of the room and is usually defined as the sound pressure as a function of time at the point of interest in the room for an impulsive source at some other point in the room. The impulse response will be a result of the time for the sound to reach the room surfaces, be reflected, and reach the listener. While omnidirectional impulse response analysis is often done if the response at one point is sufficient, using computers, directional and binaural impulse response analysis can be readily implemented [1–4]. The ISO 3382(1) standard on measurement of the acoustical properties of auditoria requires a minimum of two sources and five receiver positions [7]. Convolving the room impulse response with source sound signals by using computers, auralization, we can listen to the influence of the room *filter* on the sound emitted by the source [5].

With geometrical acoustics, we find an approximation to the impulse response useful for the analysis of speech and music transmission in the room. Two restrictions are important however: (1) geometrical acoustics is only applicable for mid- and high frequencies of sound where the walls and objects are large compared to wavelength, and (2) geometrical acoustics is primarily useful for the study of the behavior of the initial reflections of sound in the room, typically the sound that arrives reflected at the listener within 30 ms of the direct sound.

To use geometrical acoustics, we start with the ray and mirror image approaches of geometrical optics.

3.2 GEOMETRICAL ACOUSTICS

3.2.1 Ray tracing and mirroring

Two methods are commonly used in geometrical acoustics for prediction of impulse response: the ray tracing method (RTM) and the mirror image method (MIM). Both are based on the use of Fermat's principle that states that a wave travels by the quickest route and, in homogeneous air sound follows straight lines, thought of as rays.

When RTM is used, the sound power can be thought to move away from the source as rays or particles that carry a part of the total power radiated by the source. Usually, each ray or particle carries a fraction of the power that is inversely proportional to their number. The ray path is a straight line unless sound is reflected or scattered by planes and objects. To find the ray paths, we send out an imaginary ray or sound particle and follow it as it moves in space as time grows. Figure 3.1 shows an application of simple optical ray tracing.

An alternative to ray tracing is mirror image analysis. Mirror images are a result of sound reflection by plane rigid surfaces. The sound that arrives at the listener may be thought to come from the mirrored copies of the source visible to the listener. When the ray hits a mirror, whether plane or curved, it will be reflected. For a mirror, the incident and reflecting rays as well as the normal to the surface are in the same plane, and the angles of incidence and reflection are equal. Once the mirrored copies of the source are found, we can draw rays from the mirror images of the source to the listener as shown by Figure 3.2. The number of times a reflection has been mirrored, or a ray path reflected, is called the mirror image or ray reflection order. Mirror image analysis will give the same result as analysis by ray tracing if a large number of rays are studied.

Impulse response measurements in real rooms will typically show large deviations from the calculated impulse responses. Nevertheless, the RTM and MIM are important practical tools in analyzing the reflection paths of a room

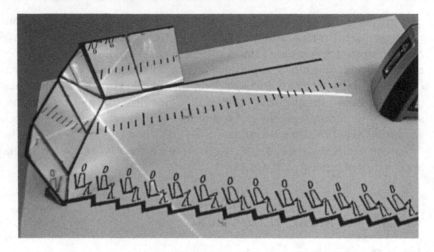

Figure 3.1 Simple ray tracing can be done with mirrors on a room drawing using a laser tracker. (Photo by Mendel Kleiner.)

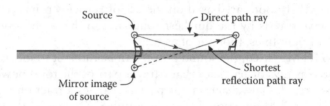

Figure 3.2 RTM and MIMs give the same results providing an infinite number of rays or image sources are used.

since both methods often give sufficient results for practical work in room acoustics in contrast to the wave theory for the same frequency range.

3.2.2 Sources and receivers

The following approximations and assumptions are typical for geometrical acoustics:

- Sources and receivers are small.
- Surfaces (of room boundaries and objects in the room) are plane and rigid and have dimensions much larger than the wavelengths of sound considered.
- Sources are unaffected by their surroundings.
- Source signal is assumed to be a *short pulse* or *wavelet*.

A sound source that is small compared to wavelength can be regarded as a point source but—by suitable approximation—also directional sources such as voice, musical instruments, and loudspeakers can be studied using geometrical acoustics. In the use of the MIM, the receiver is assumed to be a point, whereas in ray tracing, small spheres, polyhedra, or simply cubes are used as targets to collect the rays. In computer simulation of ray tracing and mirror imaging, both sources and receivers can be assigned complicated directivity such as that of voice, musical instruments, microphones, and hearing [4].

3.2.3 Planes

Elementary ray tracing and mirror imaging is assumes surfaces that are infinite, rigid, plane, and smooth. However, few physical surfaces can be considered as perfectly reflecting planes in the sense of geometrical acoustics. Physical surfaces are neither infinitely large, perfectly plane, nor rigid.

At low frequencies, the room dimensions are too small for the room surfaces to be large compared to wavelength, whereas at high frequencies, the irregularities caused by poor workmanship and unavoidable lack of precision are so large as to make the surfaces more scattering than mirroring. In practice, it is customary to consider as ideal mirrors such nominally plane surfaces that have dimensions larger than about three wavelengths and that have random surface irregularities smaller than 1/16 wavelength. A further requirement is rigidity, which is assumed to be present when the impedance of the surface is much larger than the sound field impedance. When the angle of incidence is larger than about 80° (*grazing incidence*), most surfaces will appear mirroring.

RTM and MIM are only useful to study the major paths of sound propagation. When surfaces are curved, ray tracing is intuitively simpler to use than mirror imaging. Additionally, RTM does not require the user or software to keep track of the mirror images and find their visibility. MIM is however more exact than RTM, but since building practice is not exact, this advantage is of little practical use.

A curved surface may be subdivided into smaller surface patches and analyzed as previously or one can use the mirroring laws of optics. Subdividing large surfaces into small surfaces to take into account small irregularities will result in difficulties both in the case of ray tracing and in image analysis. Patches must be several wavelengths large to act as mirrors in either case. Keeping track of the visibility of many mirror images is cumbersome as is tracing the myriad of rays necessary to hit all small surface patches. Because of the multitude of rays or images, it is tiresome to analyze the sound propagation in small rooms by manual ray tracing for higher orders than the first or second order of reflection.

Higher orders can be calculated using computers, but both the resulting accuracy and precision will be low because of the previously mentioned inexactness of building practice that along with limited wall size and wall unevenness leads to scattering rather than specular reflection of sound. The higher the reflection order, the more destructive will their influence be on the accuracy of geometrical acoustics prediction.

3.2.4 Multiple reflecting planes

The impulse response at the listener from a sound source in a room is determined fundamentally by the geometry of the room in the order by which the sound is reflected by the room surfaces.

The case of two opposing surfaces shown in Figure 3.3 is fundamental. Note that the surfaces are mirrored by one another an infinite number of times although this is not indicated in the figure. The sound will bounce back and forth, reflected an infinite number of times. If the planes are not parallel, the mirror images will be located along curves instead of along a straight line.

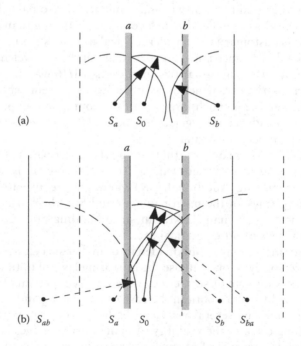

Figure 3.3 Mirroring by two parallel surfaces a and b. (a) shows the sound source and the first-order mirror image sources and (b) shows conditions after some additional time and the additional second-order mirror images necessary to study the wave fronts. For each time instant, the radius of curvature is determined by the speed of sound and the time from the start of the sound.

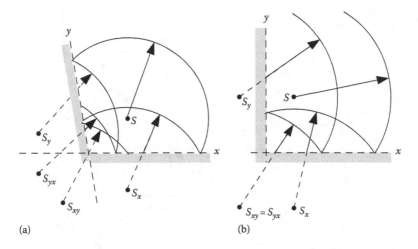

Figure 3.4 Mirroring of sound at the corners of intersecting surfaces, showing first- and second-order image sources for two cases: (a) surfaces not at right angle and (b) surfaces at right angle.

In the case of nonparallel surfaces, the mirroring will be more complicated, as exemplified in Figure 3.4. In this case, there are four different mirror image sources since rays will be reflected twice (note that two of the image sources have identical locations in the case of the 90° corner).

Figure 3.5 shows how the mirror images are distributed in a section through a rectangular room, parallel to the floor. Each reflected room is a mirror image of the original room. The source and the receiver in this example are assumed to be at the same height over the floor. There will be an infinite number of mirror images. One can show mathematically that the infinite sum of the impulse response pressure contributions from all the image sources will result in the room modes discussed in Chapter 3.

Within each frame $c\Delta t\Delta\phi$ shown in Figure 3.5, determined by times t, Δt, and angle $\Delta\phi$, there will be an increasing number of images. The number of mirror images of the source increases rapidly with time from the start of the source sounding. At some time t after the sound has left the source, the reflection density dN/dt (number of mirror images per time unit) will be

$$\frac{dN}{dt} = \frac{4\pi c^3}{V}t^2 \tag{3.1}$$

The intensity of each ray or image source is determined by the increase in surface area that is associated with the expanding surface of the spherical wave front. The intensity of the contribution by each source can be approximated by investigating the number of times the rays traverse a wall.

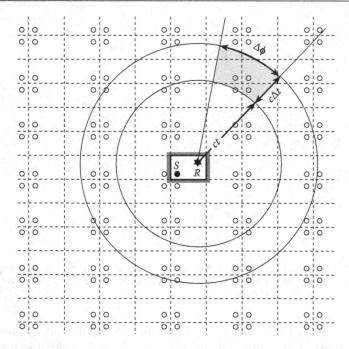

Figure 3.5 Mirroring by the walls of a rectangular room. The sound source is marked by the filled circle at S and the receiver is positioned at the star at R. The figure shows the location of mirror sources of various orders (empty circles). Only the sources in the gray zone contribute to the impulse response from the angle segment Δφ during the time segment Δt.

Assume the plane room walls to have an incidence angle-independent sound-absorption coefficient α. Each time the wave is reflected by a wall surface, it will lose a part α of its energy, as heat or by transmission to some other acoustic system.

The distances in 3D space between the mirror sources and the listener are easily calculated since here each path is the space diagonal between the image source and the receiver. For a nonrectangular room, however, the MIM is cumbersome to apply. Additionally, in a real room, the images will become diffused (and the associated rays scattered).

3.2.5 Visibility

A more complicated situation arises when a plane has patches of different reflectance. Theoretically, a plane may, for example, be infinite and still be assigned sound-absorptive or sound-diffusive patches, or the plane may be finite. In this case, the mirror image will not be visible.

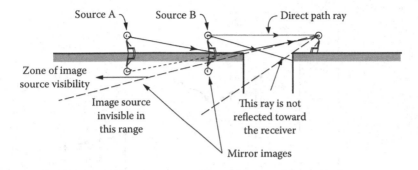

Figure 3.6 Principle of visibility.

Figure 3.6 shows how a geometric discontinuity in a plane can reduce image source visibility. The mirror image of source A is visible from the receiver but not the mirror image of source B.

If an incoming ray with intensity I_0 is specularly reflected by a surface that has a sound-absorption coefficient α, its reflected intensity I_R will be reduced:

$$I_R = I_0(1 - \alpha) \tag{3.2}$$

Note that many absorbers will scatter some of the incoming sound because of material inhomogeneity and/or limited size. Resonant absorbers such as the Helmholtz and membrane absorbers always scatter their reflected energy. For ray tracing, there are many ways to simulate scattering of sound, such as giving the reflected rays random reflection angles or to reflect a multitude of weaker rays.

3.2.6 Mean-free-path length

The mean-free-path length, $\bar{l}$, between two successive ray wall collisions depends on the room geometry, the diffusivity of its walls, and the presence of sound-diffusing objects in the room. The mean-free-path length can be derived in various ways as discussed in Section 2.7.5. It is intuitive, however, to use the approach of thermodynamics theory for the statistical treatment of randomly bouncing gas particles. The mean-free-path length $\bar{l}$ of diffuse reflections in a volume V, having a total surface area S, is given by

$$\bar{l} \approx \frac{4V}{S} \tag{3.3}$$

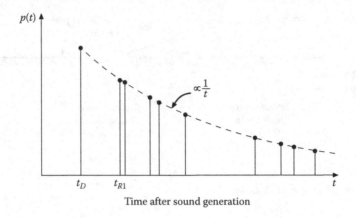

Figure 3.7 An example of the GIR of a room.

The distribution of path lengths in shoebox-shaped rooms with hard walls will depend on the room geometry as shown in Figure 2.35. An even distribution of path lengths is typical of rooms that do not have an *extreme* shape or distribution of sound-absorptive materials over their surfaces. Wall diffusivity will tend to reduce the differences between rooms.

3.2.7 Geometrical impulse response

The geometrical impulse response (GIR) of a room is the impulse response determined by the mirror image (or ray) distribution of reflected sound and thus is determined by room geometry and size. An example of such a *geometrical* impulse response is shown in Figure 3.7. The region between t_D and t_{R1} can be called the geometrical acoustics anechoic or free time as discussed in the previous section.

Assume that source radiates a power W_S. The intensity I of sound at distance d will then be $W_S/4\pi d^2$ since the sound power is spherically distributed around a small nondirectional source. If the source has been reflected N times with a mean-free-path length l_m and the reflecting surfaces have a mean sound-absorption coefficient α_d, the intensity of the arriving sound will be

$$I_N \approx \frac{W_s}{4\pi(N \cdot l_m)^2}(1-\alpha_d)^N \qquad (3.4)$$

3.3 DIFFRACTION AND SCATTERING

Assume a speaker and a listener sitting on a plane rectangular patch of smooth concrete that is set in a sound-absorbing lawn as shown in

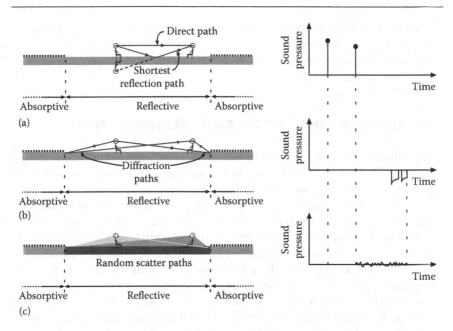

Figure 3.8 (a) Source and listener on a sound-reflecting plane set in a sound-absorbing lawn. The sound pressure shows the three impulse response contributions that are added to give the total impulse response. (b) Diffraction around the source and the listener is neglected here but shows up in measurement. (c) The diffracted sound components are reduced in level—not shown here—by the scatter. The total impulse response is to the first approximation the linear sum of the three contributions.

Figure 3.8. The sound from the source will move toward the concrete patch to be reflected there by its very high impedance. The spherical wave front will eventually reach the boundaries of the patch where the impedance of the surface drops to low value and the wave will be diffracted and absorbed.

As the transient spherical wave front expands over the sound-reflecting surface, it sets up new source points according to Huygens' principle, and these create the reflected wave front that to the observer seems to come from a source on the other side of the reflecting slab as shown in Figure 3.8a. As the wave front reaches the edges of the concrete patch, it will no longer be reflected but absorbed by the outside lawn. In a more exact treatment, there will be sound diffracted from the lines where the sound-reflecting and sound-absorbing surfaces meet. Some of this diffracted sound will arrive at the listener as shown in Figure 3.8b. Some diffracted sound will also propagate to the other borders and set up higher orders of diffraction.

In a more exact treatment, also the random scattered sound from minute surface irregularities needs to be taken into account. The scattered sound will result in a primarily high-frequency random impulse response as shown in Figure 3.8c.

3.4 SURFACE DIMENSIONS AND FRESNEL ZONES

An infinite plane wave front creates a new infinite plane wave front as implied by Huygens' principle, but what happens when the size of the reflecting surface is finite and the incoming wave is spherical?

To answer this question, study the case of a first-order reflection by a limited size surface. Figure 3.9 shows a case of a listener on the normal to a wall and an omnidirectional mirror image source on the other side of a wall on the same normal. We note that waves along some of the paths away from the direct path from the mirror image must arrive out of phase with the direct wave from the mirror image. According to Huygens' principle, the resulting sound pressure at the microphone is due to constructive and destructive interference between the sound pressure contributions by the various circular zones, called Fresnel zones.

Distance differences ¼ λ shorter than d_direct do not cause destructive interference. This defines the innermost Fresnel zone. As the zones grow larger, the negative and positive interference between them becomes weaker. The conditions for out-of-phase and in-phase reflection, that is, when the

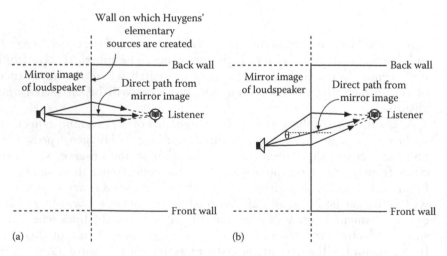

Figure 3.9 Paths having different distance in the creation of Fresnel zones using Huygens' principle: (a) perpendicular incidence and (b) incidence at an angle. For clarity, the loudspeaker causing the mirror images is not shown.

distance differences between the loudspeaker driver and the listener are multiples n of one-half of the wavelength λ is,

$$\Delta l = \sqrt{l_1^2 + r_n^2} + \sqrt{l_2^2 + r_n^2} - l_1 - l_2 = n\frac{\lambda}{2}$$

$$(n = 1, 2, 3, \ldots)$$

(3.5)

Here, l_1 and l_2 are the distances over various paths between source and receiver via the reflecting surface as found by Huygens' principle. For the case of Figure 3.9a, the Fresnel zones will be circles. If we assume $l_1, l_2 \gg r_n$, the radii r_n are

$$r_n^2 \approx \frac{n\lambda}{\dfrac{1}{l_1} + \dfrac{1}{l_2}}$$

(3.6)

Figure 3.9b shows a more common case of the mirror image displaced relative the listener and now the circles become ellipses. The larger the circles or ellipses, the less do they influence the pressure pattern. It is obvious that the maximum response is obtained for a patch that only contains the first Fresnel zone and a useful approach is to only consider this innermost zone. The contributions by the outer rim sources can be considered approximately as the diffraction by the edges of the reflecting surface.

Figure 3.10 shows the Fresnel zones over the sound reflecting wall in Figure 3.9a. Light patches indicate areas that result in positive interference while dark areas show patches of negative interference with the direct path sound. The higher the frequency, the smaller the center zone has to be to reflect sound efficiently. (For the situation shown in Figure 3.9b, the Fresnel zones become ellipses as mentioned and thus a longer path on the wall is required for cancelation to occur.)

This means that, by this principle, a surface will reflect, that is, mirror, irrespective of its size once the size is over some limit and the *reflection point* is sufficiently within the surface area. We can formulate a condition for how small the surface can be to function as an acceptable reflector of a plane wave of sound for a certain wavelength and source/receiver geometry.

We can formulate a condition for how small the surface can be to function as an acceptable reflector of a plane wave of sound for a certain wavelength λ, under condition of source and receiver distances being very large compared to wavelength. Assuming, for example, that the source is so far away that its sound field is almost plane at the mirror, the only variable will then be the distance d between receiver and mirror. It is reasonable to

Figure 3.10 Fresnel zones on a rectangular mirror where the source and the receiver are on the normal to the surface but offset from the center of the surface.

assume that the surface must contain at least the inner half Fresnel zone. This approximation results in the equation

$$l_{limit} = \frac{d^2}{2\lambda} \qquad (3.7)$$

where
    d is the smallest surface dimension
    l_{limit} is the distance to the observation point for including only the inner
    Fresnel zone in the summation of sound reflection by the mirror

This condition is sometimes used when one needs to design sound reflectors for auditoria. Note that a reflecting sheet must have sufficient surface mass (mass per unit area) so that it remains in place and is not moved by the sound field. An object acts as a diffuser if it is small compared to the wavelength.

Using the geometry shown in Figure 3.9a and assuming a frequency of 1 kHz and that the loudspeaker is placed 1 m beyond the wall and the listener at 2 m from the wall, we find that the diameter for ¼ λ difference is about 0.7 m large. For an angle θ = 45°, the largest ellipse axis will be about 1.2 m long. This means that a diffuser treatment of the wall needs to be of the order of 1.2 m by 0.7 m to function well. A small plane object parallel to the wall that can change the path length by ½ λ can help reduce

Figure 3.11 Is this to be considered plane wall with mirroring boxes or a diffusing wall with sparse diffusers? (Photo by Mendel Kleiner.)

a wall reflection if it is small compared to the wavelength; the Fresnel zone criterion implies that it needs to be about 0.8 m by 0.5 m for the geometry shown in Figure 3.9b. To change the path length by this amount, the object needs to stand out about ⅜ λ from the wall, in this case about 0.13 m. This example applies to sound of a frequency of 1 kHz but it indicates the size of the relative wall unevenness that is required for diffusion. Figure 3.11 shows a wall treated by reflective cubes that stand out.

A surface that has a random depth distribution function with an rms unevenness of about 0.15 m will have good diffusion properties from about 1 kHz and upward in frequency. A second necessary criterion is that the peak of the autocorrelation function of the height irregularity sideways across the object is smaller than about ¼ λ for the highest frequencies involved.

3.5 IMPULSE RESPONSE OF REAL ROOMS

Because of the wave-related phenomena described in the previous sections, real room impulse responses are not as simple as shown in Figure 3.7. Figure 3.12 shows a measured pressure impulse response of the room used for Figure 3.7. The extra impulse response components are due to

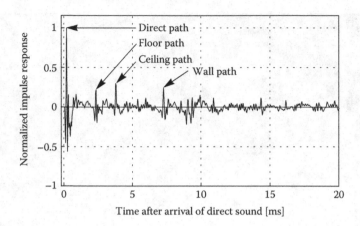

Figure 3.12 A measured room impulse response corresponding to the GIR shown in Figure 3.7.

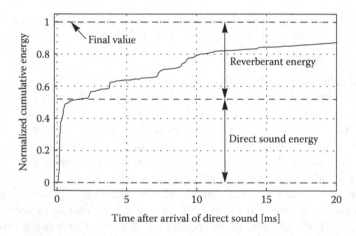

Figure 3.13 The time integral of the effective value of the room impulse response shown in Figure 3.12. The curve is normalized to its final value. We note that the ratio between direct *D* and reverberant sound *R* is *D/R* ≈ 1.

the presence of scattering, diffraction, complex surface impedance, loudspeaker and microphone impulse responses, etc. The energy behavior, however, shown in Figure 3.13 can often be quite well simulated by calculating the time integral of effective value of the sound pressure in octave or—better—⅓ octave bands. Measurement and analysis of impulse response is further discussed in Chapter 14.

3.6 ROOM FREQUENCY RESPONSE

The measured impulse response contains all the information needed to characterize the room as a filter between sound source and receiver as long as there is no noise or nonlinear distortion. Typically, the response is measured using a dodecahedron loudspeaker and a small microphone, thus approximating a point source and point receiver.

As shown by Equation 1.53 the transfer function of a room "filter" is the Fourier transform of the room's impulse response. The frequency response of a filter or transducer is the level of its transfer function magnitude as a function of frequency. The frequency response of this room filter will depend not only on the properties of the room but also on the sound source and receiver. The sound source directivity of a loudspeaker or microphone is virtually impossible to characterize over the full audio range, but for low frequencies, loudspeakers can often be assumed omnidirectional. For laboratory purposes, a semi-omnidirectional sound source such as a dodecahedron loudspeaker is often used. Such a loudspeaker will have considerable directivity over a limiting frequency that depends on the size and, geometry of the enclosure, the loudspeaker driver characteristics, the way drivers are mounted on the enclosure, and the distance between drivers. Typically, omnidirectionality only extends to about 1 kHz for a dodecahedron loudspeaker, such as the one shown in Figure 14.8. For frequencies over a few hundred hertz, most sound sources—including large loudspeakers and musical instruments—have considerable directivity, which affects the measured response. Some loudspeakers and many musical instruments have a dipole-like directivity at low frequencies.

The properties of the steady-state sound field are also a function of the sound signal generating the field. The properties of the measured impulse response allow determination of the room response to a sine tone since the impulse response is related to the room transfer function over the Fourier transform. By convolving the room's impulse response with the sound signal as in Equation 1.83, one obtains the room's response to the sound signal. Figure 3.14 shows the frequency response of the impulse response shown in Figure 3.12. (A 0.5 s length of the measured impulse response was used for the Fourier transform used to find the transfer function for this graph.)

We see from the behavior of the frequency response curve in Figure 3.14 that it has very different characteristics at low and high frequencies. The resonance peaks of some low-frequency room modes below the Schroeder frequency, defined in Equation 2.50, are clearly seen. For frequencies above about 350 Hz in this room, the frequency response curve has a more random-like character as a result of the interference between many modes that are active at the same frequency, which causes the peaks and dips in the curve due to positive and negative interference. Because of the quite

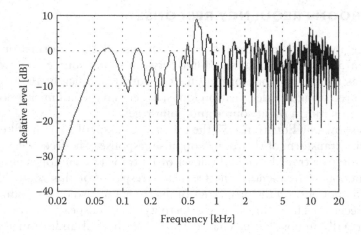

Figure 3.14 The frequency response of the impulse response shown in Figure 3.12.

broadband signal characteristics of most sounds such as voice and music, the bandwidth of the critical filters of our hearing, and our binaural signal analysis system that will all be discussed in Chapter 7, we do not hear the frequency response irregularities the way that they appear visually in the narrowband Fourier analysis shown in Figure 3.14. Experiments have shown that the effective bandwidth at medium and high frequencies is such that the frequency response is better studied using ⅙ octave-band analysis.

The sound field in the frequency range above the Schroeder frequency is often considered *diffuse* but this is not necessarily so. Consider, for example, a shoebox-shaped room with rigid floor and ceiling but that has very sound-absorptive walls. The sound field in such a room will be dominated by sound bouncing between floor and ceiling. In the high-frequency approximation of geometrical acoustics, we can consider the sound field as the result of rays (representing plane waves) periodically reflected by these surfaces. The sound field is not diffuse and the frequency response very periodic, but one can still calculate a Schroeder frequency. Hearing can usually easily hear the rattle characteristic of the *flutter echo* sound that is a result of the short period reflected sound as will be discussed in Chapter 8.

Care is advised when using the Schroeder frequency to discuss the subjectively perceived diffuseness of room sound fields.

3.7 DIFFUSE SOUND FIELD

Geometrical acoustics shows that all the contributions of the various reflections, after some time, set up very complex sound field in the room. The sound waves will come from many directions and will have bounced off many walls and objects. In a room that has diffusive walls with similar sound-absorption properties all over, the arriving sound will on the average be equally strong from all directions. The resulting sound field is said to be diffused, sometimes called a Sabine space, and is characterized by the following physical properties:

- Energy density is equal at all places.
- All directions of incidence are equally probable.
- Incident sound has random phase.

Sabine's equation (2.71) is often used for calculating the approximate reverberation time of a room if the following approximate conditions for diffuse sound fields are met:

- The room averaged sound-absorption coefficient $<\alpha_m>$ should be less than 0.3.
- The sound-absorbing surfaces are approximately evenly distributed over the room.
- The shape of the room is not *extreme* (long, flat, etc.). The criterion $l_{max} < 1.9V^{1/3}$ is used in the ISO 354 standard [6]. Here, l_{max} is the longest straight line that fits within the boundary of the room (in a rectangular room, this will be the space diagonal) and V is the volume of the room.

The earlier conditions are formulated from the high-frequency time domain point of view. In the corresponding low-frequency domain description of the sound field in Chapter 2, we regard the steady-state sound field at each room as created by modal resonance. If one has a large number of modes excited by a wide bandwidth noise, the room does not have an extreme shape, and the modes have approximately the same damping, one can regard the resulting sound field as diffuse.

This corresponds to the intuitive definition used in geometrical acoustics since if there are many modes, there will be many directions of incidence. If, in addition, the modes are equally damped, the intensity of the sound from each direction will be approximately the same. The requirement for random phase will be satisfied if we excite the room by a wide bandwidth noise signal.

When the walls of the room under study have low average sound absorption, the sound absorption is well distributed over the different walls, and the room does not have an *extreme* shape, the sound field in the room will have properties similar to those of an ideally diffuse sound field.

In reality, there are no ideally diffuse sound fields even in very reverberant rooms. When studying the sound fields of rooms, one can always find places where the sound is not diffuse, for example, close to plane and rigid walls where the sound field is very coherent because of the nearly ideal reflection of sound. If there are major wall areas, which have markedly higher sound absorption and size than others, the conditions are also not met. Additionally, the field is not diffuse where the sound intensity of the source is high compared to the reflected sound. In practical engineering, however, one can often assume the sound field diffuse at positions being more than one-half wavelength away from wall surfaces (see Chapter 2) and beyond some distance from the sound source as shown in Figure 2.36.

3.7.1 Rooms having extreme shape and/ or absorption distribution

Equation 2.92 shows that it is possible to reduce the sound pressure level in the diffuse field of a room by increasing the room absorption. This level decrease will of course be large only if the initial room absorption is small. However, one often finds that the subjective improvement of a room's acoustic, due to added room absorption, is large even if the sound pressure level decrease is small. This subjective impression is due both to the reduced reverberation time and the selective removal of unwanted wall or ceiling reflections. Speech communication is much easier if the reverberation time is short and early reflections are emphasized over late reverberant sound.

A room having little room absorption and a correspondingly long reverberation time is often called a *live* or *hard* room. This is the case for reverberation chambers for sound absorption and sound power measurement.

A room that is characterized by short reverberation time is said to be *dead* or *dry*. An extreme case is the anechoic test chamber used for electroacoustic measurements, that needs its walls to have a reflection coefficient $|r| \leq 0.1$, that is, a mean absorption coefficient $\alpha \geq 0.99$. A sound wave that is reflected by such a wall has only 1% of its energy left, which corresponds to a sound level drop of -20 dB because of the wall's sound absorption.

REFERENCES

1. Cremer, L., Müller, H.A., and Schultz, T.J. (1982) *Principles and Applications of Room Acoustics*, Vol. 1. Applied Science Publishers, New York.
2. Krokstad, A., Strøm, S., and Sørsdal, S. (1968) Calculating the acoustical room response by the use of a ray tracing technique. *J. Sound Vib.*, 8(1), 118–125.
3. Baxa, D.E. (1976) A strategy for the optimum design of acoustic space. Dissertation, University of Wisconsin, Madison, WI.
4. Dalenbäck, B.-I. (1996) Room acoustic prediction based on a unified treatment of diffuse and specular reflection. *J. Acoust. Soc. Am.*, 100(2), 899–909.

5. Kleiner, M., Dalenbäck, B.-I., and Svensson, P. (1993) Auralization—An overview. *J. Audio Eng. Soc.*, 41(11), 861–875.
6. ISO 354:2003, Acoustics – Measurement of sound absorption in a reverberation room. ISO, International Organisation for Standardisation, Geneva, Switzerland.
7. ISO 3382 Acoustics – Measurement of room acoustic parameters. Part 1: Performance rooms; Part 2: Ordinary rooms. ISO, International Organisation for Standardisation, Geneva, Switzerland.

Chapter 4

Sound absorption in rooms

4.1 ABSORPTION MECHANISMS

The goal of the acoustical design of small rooms is in many cases to create optimum sound fields for listening. Therefore, the propagation and reflection of sound has to be appropriately adjusted. As discussed in later chapters, one part of this optimization process consists of making some reflecting surfaces sound absorbing or diffusing to control the sound transmission from the source to the listening point. This section presents an overview of the most frequently used sound-absorbing materials and structures. Chapter 5 discusses the properties of diffusers.

Sound is the organized motion of the fluid molecules that carry the sound. The energy decrease of the wave is most often caused by a partial conversion of the organized molecular motion into a random motion so that heat is created. This mechanism exists in the boundary layer where the wave touches a solid surface. This is particularly the case for open-pore materials since they consist of a skeleton of tiny hard fibers with the mutually connected pores in which the surface area for the interaction with the wave is large. The losses are proportional to the particle velocity of the incoming wave. If, for instance, a porous material is placed on the surface of a solid wall, the partially reflected wave interferes with the incoming wave and the major losses occur at locations distant a quarter wavelength from the backing wall.

Although the physics of sound conversion of energy into heat is in most absorbers the same, their mechanical structure differs. Open-pore absorbers that consist of fine fibers are (if possible) placed into regions of high particle velocity. On the other hand, absorbers based on acoustical or mechanical resonance achieve high losses by constructing mechanism that elevate particle velocity naturally such as plates, membranes, and tubes or are using the Helmholtz acoustical mass–spring system. In addition to these absorbing systems, sound is absorbed through various functional objects placed in rooms. People absorb sound by virtue of the textiles used for clothing that typically function as an open-pore absorber.

The noise control literature provides information on other ways of energy losses, particularly useful are References 1–3.

4.2 FUNDAMENTAL QUANTITIES AND THEIR MEASUREMENT

The sound absorption of a flat sound absorbing surface is characterized by its sound-absorption coefficient α defined as the ratio of the sound power absorbed W_{abs} to the sound power incident W_{inc} per unit area equal to

$$\alpha = \frac{W_{abs}}{W_{inc}} \tag{4.1}$$

The absorbing surface also reflects sound that is characterized by a sound pressure reflection factor R defined as the complex ratio of the reflected and incident sound pressures. The relationship between α and R is

$$\alpha = 1 - \left|\frac{p_{refl}}{p_{inc}}\right|^2 = 1 - |R|^2 \tag{4.2}$$

Occasionally, a sound power reflection coefficient β is used where $\beta = 1 - \alpha$ that should not be confused with sound pressure reflection factor R. Both the absorption and reflection coefficients are the function of incidence of a plane wave on a unit of area and the frequency. If the absorber is large compared to wavelength and has an area S, the total absorbed energy A, usually called absorption, is equal to

$$A = S \cdot \alpha \tag{4.3}$$

Some sound absorbers cannot be characterized by the absorption per unit area but rather by their total absorption A in m². (The absorption area is sometimes given in square meter Sabine.) Such absorbers are, for example, people, chairs, the Helmholtz resonators, and other irregular objects.

The sound absorption can be also expressed in terms of the impedance Z of the absorbing surface. The impedance can be considered having a real and an imaginary part:

$$z = \frac{1}{\rho_0 c} Z = \frac{1}{\rho_0 c} \frac{p}{u} = \frac{\mathrm{Re}(Z) + j\,\mathrm{Im}(Z)}{\rho_0 c} \tag{4.4}$$

where
    Z is the specific acoustic impedance
    z is acoustic impedance normalized to the wave impedance of the air $\rho_0 c$

The reflection factor R is

$$R = \frac{Z - Z_0}{Z + Z_0} \tag{4.5}$$

The absorption coefficient α can be shown to be

$$\alpha = \frac{4\,\mathrm{Re}(Z)Z_0}{\left[\mathrm{Re}(Z) + Z_0\right]^2 + \mathrm{Im}(Z)^2} \tag{4.6}$$

where $Z_0 = \rho_0 c$. These generalized notations are used throughout this chapter for perpendicular sound incidence on the absorber. However, the absorption coefficient is angle dependent. If the plane sound wave incidents under the angle θ with the normal direction to the absorber, the normal component of the particle velocity is multiplied (Figure 4.1) by $\cos(\theta)$ so that the upper equations are modified and result in

$$R(\theta) = \frac{Z\cos(\theta) - Z_0}{Z\cos(\theta) + Z_0} \tag{4.7}$$

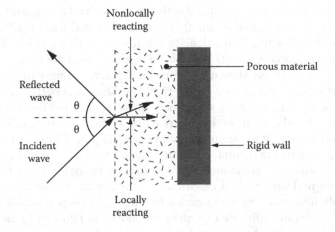

Figure 4.1 Locally and nonlocally propagating wave in a sound absorber. The difference between the locally and nonlocally reacting materials is explained under Equation 4.10.

so that

$$\alpha(\theta) = \frac{4\operatorname{Re}\big(Z\cos(\theta)\big)Z_0}{\big[\operatorname{Re}\big(Z\cos(\theta)\big)+Z_0\big]^2+\big[\operatorname{Im}\big(Z\cos(\theta)\big)\big]^2} \tag{4.8}$$

We also can define the absorption coefficient for random incidence. If we place the absorber into a random sound field where the sound intensity incidents on the material with equal amplitude from all directions, then the random absorption coefficient α_R is and

$$\alpha_R = \frac{I_a}{I_i} = 2\int_0^{\pi/2}\alpha(\theta)\cos(\theta)\sin(\theta)\,d\theta \tag{4.9}$$

where I_a and I_i are the absorbed and incident intensity, respectively, and $\alpha(\theta)$ is defined by Equation 4.8. The impedance of the material can be measured, for example, by using an intensity measurement probe and suitable software. If the material is locally reacting, Kuttruff [4] presents the results as

$$\alpha_R = \frac{8}{|z|}\cos(\beta)\left\{1+\frac{\cos(2\beta)}{\sin(\beta)}\arctan\left(\frac{|z|\sin(\beta)}{1+|z|\cos(\beta)}\right)\right.$$

$$\left.-\frac{\cos(\beta)}{|z|}\ln\big(1+2|z|\cos(\beta)+|z|^2\big)\right\} \tag{4.10}$$

where $z = Z/\rho_0 c$ is the normalized wall impedance and $\beta = \arctan[\operatorname{Im}(z)/\operatorname{Re}(z)]$. The maximum value of $\alpha_R(\text{max}) = 0.955$ is obtained for a wall impedance $|z| = 1.6$ and $\beta \approx \pm 30°$. This calculation leads to a realistic value of the maximal absorption coefficient. Note that the absorption coefficient cannot be larger than unity. As shown later in this chapter, the measured values in a reverberation room can be larger than one due to incorrect assumptions about the diffusivity of the sound field in the room used for measurement.

Sound absorbers can be locally reacting or nonlocally reacting. Locally reacting means that the sound cannot propagate parallel to the absorber surface. Because this tangential component cannot exist, a wave that is incident on the absorber at some angle θ continues its propagation in the material in the normal direction of the surface. This can be, for instance, achieved by subdividing the absorber interior by thin walls perpendicular to the surface. In the nonlocally reacting absorber shown in Figure 4.1, the wave that is incident on the absorber under some angle θ continues to travel in the absorber but under a different angle that depends on the speed of sound inside the absorber.

4.2.1 Measurement of absorption coefficient for normal sound incidence

When developing or comparing the performance of absorbers, the absorption coefficient is usually measured for perpendicular incidence of a plane wave. This can be done for low frequencies and homogeneous materials using a waveguide with a rectangular or circular cross section [5]. To prevent the generation of cross modes, the size d of the rectangular waveguide side must be smaller than one-half of the wavelength of the lowest cross mode, and for a circular waveguide, the diameter must be less than 0.63 of the wavelength. This obviously limits the use of these measurements to low frequencies although special microphone assemblies can be used to increase the frequency range of measurement. The material sample is placed at one end in the wave guide and the sound is supplied from a loudspeaker located at the other end of the tube.

As the standard *ISO 9614-3:2002* specifies in detail, the complex ratio of the incident and the reflected sound power can be determined by measuring by means of two sound pressure microphones [19]. The major advantage of measuring the sound intensity is that the absorption coefficient can be measured at short frequency increments using digital signal processing. The older analog technique required the measurement of sound pressure at the points of pressure maxima and minima, which are at different locations for each frequency. The pressure maxima and minima and their location with respect to the absorber sample had to be found for each frequency, and therefore, high-frequency resolution measurements were impractical.

4.2.2 Measurement for random incidence

When the sound-absorbing material is installed in a room, the sound field that incidents on this material can be very complicated. Major normal sound field incidence is very rare since most sound fields in rooms are diffuse or at least the sound wave is incident at some angle. The sound incidence depends on the room shape, objects in the room, acoustical properties of the room, and location, size, and shape of the absorber, to name a few variables. These are also different for each frequency.

In order to compare the absorption coefficient of different absorbers, the measurements according to the standard *ISO 354* are made in a large standardized room of at least 200 m³ volume that has bare, highly sound-reflective walls [5]. Due to the high wall reflectivity and the density of reflections, the sound field can be considered similar to an idealized diffuse sound field. A diffuse field consists of plane waves that travel in all directions with a uniform probability distribution. Also, all waves must have the same intensity. To achieve a diffuse sound field at both low and high frequencies, a variety of reflectors are usually placed in the room interior. The sound field diffusivity can be investigated by measuring the pressure correlation function ρ, which, for a perfectly diffuse field, is given by [6]

$$\rho(kr) = \frac{\sin(kr)}{kr} \tag{4.11}$$

where
    k is the wave number
    r is the distance in some selected direction

The deviation from perfect field diffusivity is given by the deviation from the sinc function. However, there is not a scale for the consequences of the pressure deviation from the earlier equation.

    A 10 m² large sample of the absorber is placed in a central position of the floor. The reverberation time is measured for a bare room and for the room with the sample of the sound absorber. The absorption coefficient α_R is calculated using Sabine's equation so that

$$\alpha_R = \frac{0.161V}{S}\left[\frac{1}{T_{60,S}} - \frac{1}{T_{60,0}}\right] \tag{4.12}$$

where
    V is the volume of the chamber in metric units
    S is the area of the measured sample in metric units
    $T_{60,S}$ is the reverberation time with the sample
    $T_{60,0}$ is the reverberation time of the empty room

This equation was derived under the assumption made by Sabine that the sound field is perfectly diffuse and the energy density and the sound incidence on the sample are uniform. However, this is not the case in a real room. Particularly, the sound incidence on the absorbing sample is affected by diffraction on the sample boundary, sample position, frequency, and other field distortions. Different room geometries give different results. Therefore, the calculated value of the absorption coefficient for highly absorbing materials can incorrectly exceed unity. Such α_R values are usually corrected to 1 when published. When the measured values of α_R are used for the calculation of reverberation time (using Sabine's equation), we have to expect that the calculation is subject to an error that primarily depends on the position of the absorbing material. This error is particularly large for highly absorbing materials placed in the room corners as well as on the equation used to calculate the reverberation time, as discussed in Chapter 14.

4.3 POROUS ABSORBERS

The porous materials most commonly used for commercial sound absorbers consist of very thin fibers that are typically compressed and glued together to form a plate of required thickness [3]. The internal structure of materials that are suitable for sound absorption permits the air and thus the sound to travel through the open space present in the solid skeleton. The air friction in the boundary layers on the surface of the fibers converts the acoustic energy into heat. The materials used are synthetic such as mineral, glass, and polyester wools, ceramic, metal, graphite, polypropylene, Kevlar, or natural such as cotton, compressed hay, flax, and jute. The choice of material depends on the desired absorption properties as a function of frequency, required thickness, mechanical properties, application, and similar criteria. As will be shown later, porous absorbers are practical to use in a variety of layer configurations, inside the cavities of resonant absorbers and as flow resistance elements. Porous absorbers are effective in absorbing medium- and high-frequency sound.

4.3.1 Principal characteristics of porous materials: Flow resistance

The flow resistance is the force per unit area required to move a steady flow of air through a porous material at velocity v. The flow resistance R_f is defined as [3]

$$R_f = \frac{\Delta p}{v} \tag{4.13}$$

where Δp is the static pressure difference across the porous material. The unit of flow resistance is called rayl [N s/m]. We often use the flow resistance per unit of the material thickness Δx that is called the flow resistivity R_1 and defined as

$$R_1 = \frac{R_f}{\Delta x} \tag{4.14}$$

There is a variety of equipment to measure the flow resistivity. In essence, measurement follows the definition given in Equation 4.13. A micromanometer is used for measurement of the air pressure difference at both sides of the material sample that is clamped in a mechanical device. The air flow is measured by a mechanically driven piston or another type of a flow meter that permits to measure the air stream. Usually, several samples have to be measured because porous materials are usually not homogeneous.

4.3.2 Principal characteristics of porous materials: Porosity

Because the porous material is in the path of the wave propagation, the flow inside the material is constricted. Due to the continuity of the flow, the air flow speed inside the material is increased. The quantity that permits the relevant calculations is called porosity and is defined as the ratio of the volume voids V_a to the total volume of the material V_m. The porosity σ is defined as

$$\sigma = \frac{V_a}{V_m} \tag{4.15}$$

There are several methods available to measure the porosity. The weight method measures the empty volume by filing the material sample with water or some other liquid so that the empty volume can be determined from the weight difference of the filled and empty sample. In another technique, the sample is placed in a cavity where the volume can be changed by a piston [7]. The actual air volume is changed because of the pressure increase when compressing the volume.

The selection of the measurement technique depends on the type and internal structure of the solid part of the material. There are many theories to predict the acoustic properties of the porous materials [1,7,8]. We will present some empirical data obtained for the most commonly used materials to provide guidance for their application in small rooms.

4.3.3 Effects of flow resistance and porosity on porous materials

There are many models for the analysis of the physical properties of porous materials. Before we show some empirical results of these models, some effects of the presented basic material constants on one of the simple model will be presented.

Due to the porosity, the particle velocity u inside the porous material is higher by a factor $1/\sigma$ so that the 1D equation of continuity (see Chapter 1) is

$$\frac{\partial}{\partial x}\left(\frac{u}{\sigma}\right) = -j\omega\rho_0\kappa_p \tag{4.16}$$

where $\kappa_p = \left(1/\rho_0 c^2\right)$. The equation of force extended by a term representing the effect of flow resistance results in

$$-\frac{\partial p}{\partial x} = \rho_0\frac{\partial}{\partial t}\left(\frac{u}{\sigma}\right) + R_1 u = \left[j\omega\rho_0\frac{1}{\sigma} + R_1\right]u = j\omega\rho_e\frac{u}{\sigma} \tag{4.17}$$

The quantity ρ_e is the effective density of air inside the material. It is equal to

$$\rho_e = \rho_0 \left(1 - j\, \frac{\sigma R_1}{\omega \rho_0} \right) \tag{4.18}$$

We see that the air density ρ_e is complex and has been amended by a term that depends on flow resistivity and frequency. Similarly, the effective speed of sound is

$$c_e = \frac{1}{\sqrt{\rho_e \kappa_p}} = \frac{c}{\sqrt{1 - j\, \dfrac{\sigma R_1}{\omega \rho_0}}} \tag{4.19}$$

The knowledge of the sound speed is essential to optimize the thickness of the material particularly if the design goal is the maximum absorption. The characteristic impedance Z_m of the material is

$$Z_m = \rho_e c_e = \rho_0 c \sqrt{1 - \frac{\sigma R_1}{\omega \rho_0}} \tag{4.20}$$

The last two quantities also show the dependence on flow resistivity and frequency.

Figure 4.2 shows the effect of the flow resistivity R_1 on the absorption coefficient. The upper part shows the absorption coefficient of mineral wool as a function of frequency. The parameter in the curves is the bulk density. We do not see any useful consistency with this representation. In the lower curve, the frequency axes were transformed into $\rho_0 f / R_1$ and the resulting curve is consistent with many cases of glass or mineral wool applications. The important conclusion is that the absorption is increasing with frequency and that this type of absorber is not practical for low-frequency absorption unless backed by a substantial airspace as described in a later section.

4.3.4 Acoustic impedance and propagation constant

The propagation of a plane wave in a porous material can be expressed by using a complex propagation constant k_m to express the energy losses

$$k_m = \beta - j\alpha \tag{4.21}$$

where
   β is the phase constant that would be measured for a free plane wave
   α is the damping constant

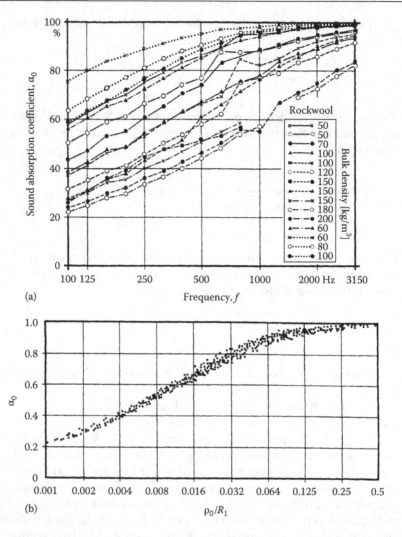

Figure 4.2 Normal-incidence sound absorption coefficients of different rock wool materials of practically infinite thickness (0.5–1 m) measured in an impedance tube plotted as a function of (a) frequency with bulk density ρ_A as parameter and (b) nondimensional frequency parameter. (After Ver, L.L. and Beranek, L.L., *Noise and Vibration Control Engineering*, John Wiley, New York, 1996.)

The sound pressure in a plane wave is given by

$$p\left(k_m, x\right) = p(0)e^{-jk_m x} = p(0)e^{-j\beta x}e^{-\alpha x} \tag{4.22}$$

where $p(0)$ is the amplitude of the wave at $x = 0$. The last term represents the wave attenuation with increasing x.

The internal structure of the porous materials varies and depends how the particular material is manufactured. There are many theoretical models for the sound propagation through materials based on the flow resistance, porosity, and geometry of the solid part and other structural characteristics of the material. Delany and Bazley measured many materials and developed empirical formulas that were further modified by Miki [9]. Their empirical formula is

$$k_m = \frac{\omega}{c}\left[1+0.109\left(\frac{\rho_0 f}{R_1}\right)^{-0.618} - j0.16\left(\frac{\rho_0 f}{R_1}\right)^{-0.618}\right] \qquad (4.23)$$

$$Z_m = \rho_0 c\left[1+0.070\left(\frac{\rho_0 f}{R_1}\right) - 0.632 - j0.107\left(\frac{\rho_0 f}{R_1}\right)^{-0.632}\right] \qquad (4.24)$$

As before, c is the speed of sound outside the material, ρ_0 the air density, R_1 the flow resistivity, and f the frequency.

Another useful constant of the material is its characteristic wave impedance. The complex-specific acoustic wave (characteristic) impedance is $Z_m = R_m - jX_m$. Let us consider a very thick layer of porous material. The sound pressure and the particle velocity inside this layer are

$$p(k_m, x) = Ae^{-jk_m x} + Be^{jk_m x} \qquad (4.25)$$

$$u(k_m, x) = \frac{1}{z_m}\left[Ae^{-jk_m x} - Be^{jk_m x}\right] \qquad (4.26)$$

The constants A and B are the amplitudes of the pressure that travels in the positive and negative directions of the x axis.

The sound pressure p_2 and the particle velocity u_2 at the back side of the material distant d from its front are (see Figure 4.3)

$$p_2(k_m, d) = Ae^{-jk_m d} + Be^{jk_m d} \qquad (4.27)$$

$$u_2(k_m, d) = \frac{1}{z_m}Ae^{-jk_m d} - Be^{jk_m d} \qquad (4.28)$$

These equations can be solved for A and B so that we can calculate p_1 and u_1 at the face of the material for $x = 0$. The impedance Z_1 is

$$Z_1 = \frac{p_1}{u_1} = Z_m\left[\frac{Z_2 \cot(k_m d) + Z_m}{Z_2 + Z_m \cot(k_m d)}\right] \qquad (4.29)$$

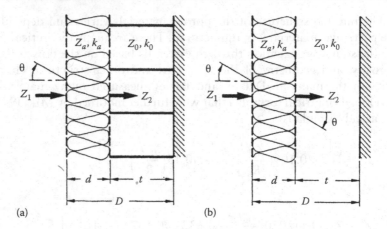

Figure 4.3 Combination of a bulk-reacting absorber layer with (a) an air gap such that sound can only travel perpendicular to the hard wall and (b) an air gap such that sound can travel parallel to the hard wall. (After Ver, L.L. and Beranek, L.L., *Noise and Vibration Control Engineering*, John Wiley, New York, 1996.)

If the material is backed by a rigid wall $(Z_2 = \infty)$, the input impedance Z_1 is

$$Z_1 = Z_m \cot(k_m d) \tag{4.30}$$

This equation permits the calculation of the characteristic acoustic impedance of the material Z_m from measurement of the impedance Z_1 after k_m has been calculated from the empirical Equations 4.23 and 4.24.

4.3.5 Porous materials in front of a rigid wall and effects of air gap

The effects of rigid wall backing mentioned in the previous section will be further analyzed because there are several alternatives [3]. The materials can be placed either directly on a rigid wall or there might be an air gap between the absorber and the rigid wall. The latter arrangement is very frequent for ceilings where a porous tile is suspended from a supporting structure. This arrangement provides an opportunity to place engineering installations (cables, air ducts) between the actual and visual ceiling.

We will now analyze the sound absorption and reflection from these installations. Let us first consider a plane sound wave that is incident normally on a wall that has a specific acoustic impedance Z_1 so that the reflection factor R and the absorption coefficients α are (see Figure 4.3a and b)

$$R = \frac{Z_1 - Z_0}{Z_1 + Z_0}$$

$$\alpha = \frac{4\,\mathrm{Re}\left(Z_1\right)Z_0}{\left[\mathrm{Re}\left(Z_1\right) + Z_0\right]^2 + \mathrm{Im}\left(Z_1\right)^2}$$

(4.31)

where $Z_0 = \rho_0 c$ is the characteristic impedance of air for plane waves. The quantities ρ_0 and c are the density and the speed of sound, respectively, in air. These are the general equations for any interface with the impedance Z_1.

Next, we will consider the case of a porous material of thickness d placed on a wall that perfectly reflects sound. The acoustic impedance is (see Equation 4.30)

$$Z_1 = Z_m \coth\left(jk_m d\right) = -jZ_m \cot\left(k_m d\right)$$

(4.32)

The absorption and the reflection can be calculated from Equation 4.31 as functions of Z_1. If d is much smaller than the wavelength in the material, Z_1 is large and the absorption coefficient is small. Thin layers of the porous material do not absorb low frequencies (unless they are very thick, which is very impractical).

This situation becomes better for higher frequencies and a maximum of absorption occurs when the material thickness approaches quarter wavelength at a relatively small d. The impedance Z_2 of the air gap (Figure 4.3) is

$$Z_2 = -jZ_0 \cot\left(k_0 t\right)$$

(4.33)

where the zero subscripts are related to the air and t is the thickness of the air gap. The input impedance Z_1 including the porous material is given by

$$Z_1 = \rho_0 c_0 Z_m \left(\frac{Z_2 + j\left(Z_m/\rho_0 c\right)\tan\left(k_m d\right)}{\left(Z_m/\rho_0 c\right) + jZ_2 \tan\left(k_m d\right)}\right)$$

(4.34)

The indices m are related to the porous material and d is its thickness. Both t and d are of interest as design parameters.

Let us assume that $t = (\lambda/4)$ a quarter wavelength thickness of the air layer. This will make the material backing impedance Z_2 zero. When analyzing both the reflection factor and the absorption coefficient for a variety of t and d values, we find that for small t the rigid wall is ineffective. The best absorption is achieved for the air layer a quarter wavelength thick.

Detailed analyses can be obtained using Equation 4.34 and the general equations for α and R.

Ingard [2] calculated the absorption coefficient as a function of distance from a hard wall for a rigid porous material. The results are shown in Figure 4.4. The absorption coefficient shows wider maxima with greater flow resistances but the maximum α are then smaller. The figure shows the absorption for both normal and random incidences. We also see that porous materials are not efficient at low frequencies. The comparison of the absorption in Figure 4.4 for locally and nonlocally reacting resonators shows that the locally reacting absorbers absorb better at low frequencies. The sound waves in the air space can be forced to have normal propagation, for instance, by the use of a honeycomb structure. Some fibrous materials have *organized* fiber directions so that they can be more locally reacting than materials that are composed of fibers at random.

4.3.6 Limp porous layers

A limp porous layer such as a cloth or a wire mesh has an extended absorption to lower frequencies as compared with rigid porous materials. Such a layer is characterized by its mass m per unit area, flow resistance R_f, and thickness t as shown in Figure 4.5.

There is a mass–spring resonance with the air backing. Ingard [2] provides curves for the calculated absorption as function of resonance frequency f_{res} and mass ratio $MR = mt\rho_0$ with flow resistance as parameter as shown in Figures 4.6 and 4.7.

More figures can be found in References 2,3. The mass–spring resonance frequency is

$$f_{res} \simeq \frac{1}{2\pi}\sqrt{\frac{\kappa\rho_0 c^2}{tm}} \tag{4.35}$$

where $\kappa = 1.4$ is the adiabatic compression coefficient. The calculated values show that the higher the MR, the lower is the resonant frequency, which is already much lower than for the rigid porous material for which data are shown in Figure 4.4. Figures 4.6 and 4.7 for normal and random incidence also show that a high absorption coefficient can be achieved at low frequencies. The higher the MR, the lower is the frequency of the peak absorption. More analysis, curves, and recommendations can be found in References 2,3.

4.4 SOUND ABSORPTION BY RESONATORS

The sound field in enclosures consists of modes. Their density depends on the square of the frequency, and as a result, the modes are distinctly separated at the lowest frequencies. The transmission functions in the room have

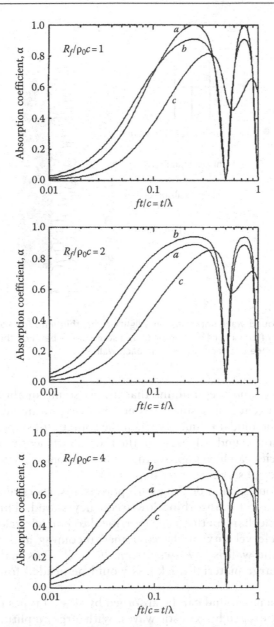

Figure 4.4 Sound absorption coefficient of a single, rigid, flow-resistive layer in front of an air space of thickness t backed by a hard wall as a function of the normalized frequency in the form of air space thickness–wavelength ratio with the normalized flow resistance of the layer, as the parameter: a—normal incidence; b—random-incidence, locally reacting air space; and c—random-incidence, nonlocally reacting air space. (After Ver, L.L. and Beranek, L.L., *Noise and Vibration Control Engineering*, John Wiley, New York, 1996.)

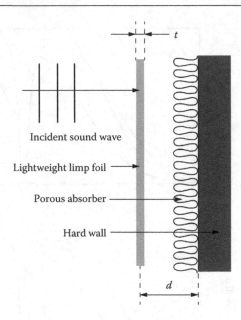

Figure 4.5 A sound wave impinging on a lightweight, thin limp foil some small distance away from a hard wall covered by a sound absorber. Foil thickness is *t* and the distance between the foil and back wall is *d*.

sharp maxima and deep minima that distort sound quality in listening. If the room is excited by a single frequency close to a mode, this frequency changes to the modal frequency when the source stops operating and only the reverberant sound still exists in the room. Sufficient wall absorption at low frequencies can help improve the modal response. Therefore, the use of low-frequency absorbers is very important.

The previous section dealing with porous absorbers revealed that they are not suitable for absorbing low-frequency sound. This is due to the absorption mechanism that is proportional to particle velocity. The maximum particle velocity of the combined incoming and reflecting wave in front of the wall is ¼ λ away from the wall. This makes it impractical since a large material thickness would be needed for absorbing low frequencies.

Low-frequency sound can be absorbed by several types of resonant systems that are excited by acoustic waves. With large amplitude at resonance, the system losses convert much acoustic energy into heat. The principal resonant systems are the Helmholtz resonators, vibrating plates and membranes, and electronically controlled absorbers. They can all operate and absorb sound at the lowest frequencies of the hearing range and have a small and practical volume.

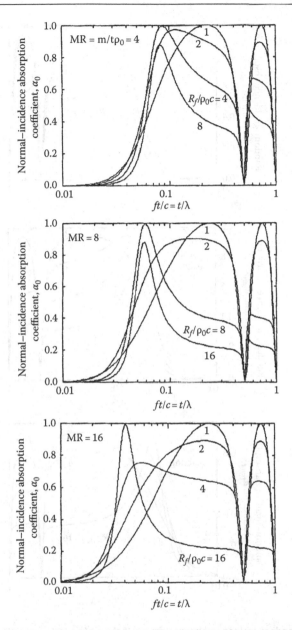

Figure 4.6 Normal-incidence absorption coefficient of a single, limp, flow-resistive layer in front of an air space of thickness t backed by a rigid wall as a function of the normalized frequency in the form of the thickness-wavelength ratio with the normalized flow resistance of the layer, as the parameter for MRs. (After Ver, L.L. and Beranek, L.L., *Noise and Vibration Control Engineering*, John Wiley, New York, 1996.)

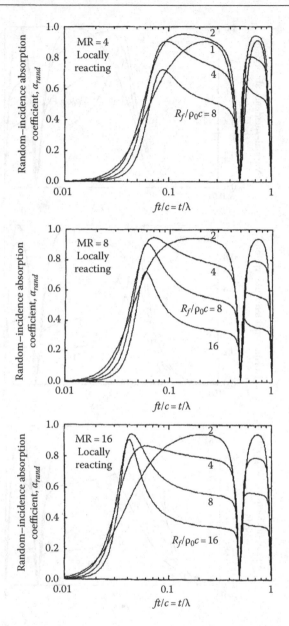

Figure 4.7 Random-incidence absorption coefficient of a single, limp, flow-resistive layer in front of a partitioned (locally reacting) air space of thickness *t* backed by a rigid wall as a function of the normalized frequency in the form of the thickness–wavelength ratio with the normalized flow resistance of the layer, as the parameter for MRs. (After Ver, L.L. and Beranek, L.L., *Noise and Vibration Control Engineering*, John Wiley, New York, 1996.)

4.4.1 Basic principles and properties of individual Helmholtz resonators

Figure 4.8 shows schematically a Helmholtz resonator. It is a *flask* that consists of a neck and closed volume. The air in the neck has a mass m:

$$m = \rho_0 l S \qquad (4.36)$$

where
   l is the length of the neck
   S is its cross section
   ρ is the air density

The compliance C of the enclosed volume $V = Sdx$ is

$$C = \frac{dx}{dF} = \frac{V}{\rho_0 c^2 S^2} \qquad (4.37)$$

where dF is the compression force. The resonance frequency ω_r is

$$\omega_r = \frac{1}{\sqrt{mC}} = c\sqrt{\frac{G}{V}} \qquad (4.38)$$

where $G = (S/l)$ is called the conductivity with $l = l_0 + 0.8D$ that is corrected for the actual length l_0 of the mass in the neck as shown in Figure 4.8.

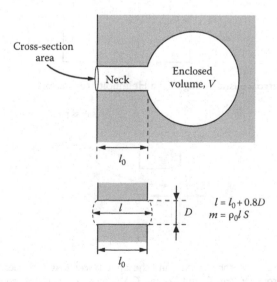

Figure 4.8 Schematic representation of a Helmholtz resonator.

4.4.1.1 Power absorbed by the resonator

The incident wave of intensity I_0 is partially absorbed. The total absorption $A = \alpha S$ is linked to the power W absorbed by

$$A = \alpha S = S \frac{I_a}{I_0} = \frac{W}{I_0} \tag{4.39}$$

In the next step, we will show the calculation of the absorbed sound power [11]. The common way is to use an acoustical analog circuit as shown in Figure 4.9.

First, let us assume that a plane wave of pressure p_i incidents on a rigid, non-absorbing wall. The sound pressure on its surface will be twice that of the free wave. With a resonator in the wall, the wave is facing the resonator radiation impedance Z_r. Figure 4.10 shows at the neck entrance a virtual massless piston.

An observer on the piston will see the input impedance Z'. The equivalent electrical circuit in Figure 4.9 shows the input voltage equivalent to $2p_i$. The wave faces the radiation impedance Z_r and the resonator impedance Z' so the resulting particle velocity u is equal to

$$u = \frac{2p_i}{(Z_r + Z')} \tag{4.40}$$

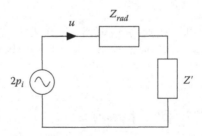

Figure 4.9 Electroacoustic analogy for a Helmholtz resonator.

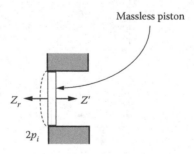

Figure 4.10 The resonator entrance and the impedances that a massless piston would see. (From Zwikker, C. and Kosten, C.W., *Sound Absorbing Materials*, Elsevier, London, U.K., 1949.)

Using the electric analogy, the absorbed power is found as

$$W_{abs} = \frac{1}{2}|u|^2 \operatorname{Re}(Z') = \frac{1}{2}\left|\frac{2p_i}{Z_r + Z'}\right|^2 \operatorname{Re}(Z') \qquad (4.41)$$

If we divide W_{abs} by the incident intensity, the absorption A of the resonator is found as

$$A = \frac{4\rho_0 c \operatorname{Re}(Z')}{|Z_r + Z'|^2} \qquad (4.42)$$

The radiation resistance seen by resonator opening in a baffle was calculated by Rayleigh [10]. Its real part is

$$R_r = \frac{2\pi\rho_0 c_0}{\lambda^2} = \frac{2\pi\rho_0}{c_0}f^2 = 0.0228 f^2 \qquad (4.43)$$

An important property of the Helmholtz resonator is its maximum absorption A_{max}. The resonators we are usually dealing with the neck cross section so large that the thickness of the boundary layer in which the principal losses of the acoustic energy are converted into heat is extremely small. Later, we will deal with perforated plates and membranes with extremely small holes so that the principal losses of acoustic energy occur there, and there is often no need to put porous materials into the resonator cavity.

The maximum energy losses occur at (or close to) resonance where the imaginary part of the denominator of Equation 4.42 is zero. Maximum power loss in the electrical circuit analyses requires $R_r = R'$ so that we obtain

$$A_{max} = \frac{\lambda_{res}^2}{2\pi} \qquad (4.44)$$

Figure 4.11 shows the dependence of the maximum absorption at resonance on the resistance that is seen at the input into the resonator normalized to the radiation resistance by the parameter μ as defined in the following.

Although the amount of absorbed energy might be reasonably large at resonance, the frequency dependence will show that the curve is usually very narrow. The frequency dependence is given by [11]

$$A = \frac{\lambda_{res}^2}{2\pi} \frac{4\mu}{\left[\mu + \left(\frac{f}{f_r}\right)^2\right]^2 + \left(\frac{f}{f_r} - \frac{f_r}{f}\right)^2 \Big/ g^2} \qquad (4.45)$$

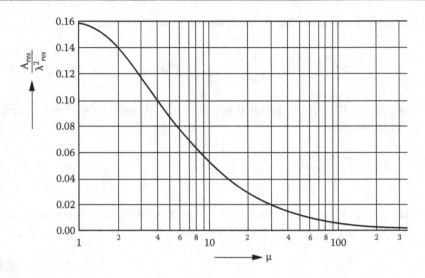

Figure 4.11 The number of m² Sabine (m² *open window*) absorbed at resonance by a single resonator, divided by the square of the wavelength at resonance, as a function of the ratio of internal resistance/radiation resistance (>1). (After Zwikker, C. and Kosten, C.W., *Sound Absorbing Materials*, Elsevier, London, U.K., 1949.)

where the quantities μ and g are

$$\mu = \frac{R}{R_{r,res}} = \frac{R\lambda_{res}^2}{2\pi\rho_0 c}$$

$$g = \frac{G}{\lambda_{res}} = \frac{S}{l\lambda_{res}}$$

(4.46)

where

    f_r is the resonance frequency
    R is the resonator resistance
    l is the neck length including the end correction

At resonance, $\mu = 1$ and $f = f_r$. Figure 4.12 shows a resonance curve and the two frequencies f_1 and f_2 at one-half magnitude of A_{max}. The greater their difference, the higher is the total absorption.

We define the absorption bandwidth in terms of the number of octaves O and using Equation 4.45 after some manipulation we obtain

$$O = \frac{f_1 - f_2}{f_r} = g\mu$$

$$f_r = \sqrt{f_1 f_2}$$

(4.47)

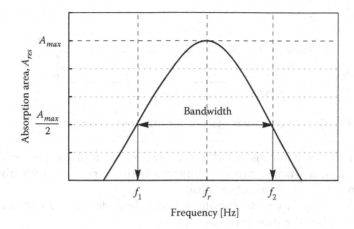

Figure 4.12 Absorption bandwidth of a Helmholtz resonator.

Combining the expressions for μ, g, and A_{res} results in

$$G = O \frac{\pi A_{res}}{2\lambda_{res}}$$

$$V = G \frac{\lambda^2_{res}}{4\pi^2}$$

(4.48)

These equations permit us to calculate the parameters for the resonator design.

4.4.2 Perforated panels

Another type of sound-absorbing construction based on the principle of Helmholtz resonator is the resonator panel. In a resonator panel, the necks are formed by a perforated plate placed at some distance from a rigid wall, either a room wall or some sturdy plate in the case of a cassette resonator system. The damping required to *tune* the system to achieve a needed bandwidth is usually achieved by placing absorbing materials into the resonator cavity or by a flow resistance fabric sheet placed on the perforated plate. There is also damping in each resonator neck along its boundary. Because of the size of the holes, the energy losses due to the viscous layers along the neck walls are typically insufficiently small.

Because the perforated panels that have the damping material inside the cavity are still most often used, their properties will be discussed first. Figure 4.13 shows the cross section of a resonator panel with an air space and hard wall backing. We are now working with an extended surface, so instead of qualifying the absorption by a total absorption $A = \alpha S$ as for

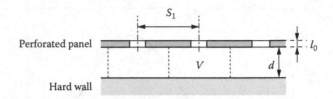

Figure 4.13 Construction parameters of perforated panel absorber.

the single resonator discussed in the previous section, we now will use the absorption coefficient α that is numerically the same as the total absorption A of one unit area (m^2) of the absorber.

If the complex z is the specific acoustic impedance of an absorber, then its absorption coefficient in terms of z is

$$\frac{z}{\rho_0 c} = r + jx \tag{4.49}$$

and

$$\alpha = \frac{4r}{\left(1+r\right)^2 + x^2} \tag{4.50}$$

Let us assume that the perforated panel consists of n single resonators per m^2 and that the area corresponding to a single resonator is S_1 (see Figures 4.13 and 4.14) and the area of the neck S_2.

The input impedance into the neck of the resonator z_2 is

$$z_2 = \frac{p}{u_2} = \frac{p}{u_1}\frac{S_2}{S_1} = z_1\frac{nS_2}{nS_1} = z_1 n S_2 \tag{4.51}$$

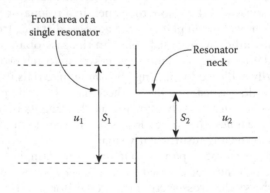

Figure 4.14 Particle velocity in front and neck of a resonator. (From Zwikker, C. and Kosten, C.W., *Sound Absorbing Materials*, Elsevier, London, U.K., 1949.)

and

$$z_1 = \frac{z_2}{S_2}\frac{1}{n} = Z_2\frac{1}{n} \tag{4.52}$$

where $nS_1 = 1$ and $Z_2 = (z_2/S_2)$ is the acoustic impedance. The frequency dependence of $z_1/\rho_0 c$ is

$$\frac{z_1}{\rho_0 c} = r + j\sqrt{\frac{1}{V'G'}}\left(\frac{f}{f_{res}} - \frac{f_{res}}{f}\right) \tag{4.53}$$

where $V' = nV \equiv d$ is the volume per m². Numerically, it is equal to the distance of the plate from the rigid wall d and $G' = nG = nS_2/l$ is equal to the total conductivity per m². The resonant frequency f_r is

$$f_r = \frac{c}{2\pi}\sqrt{\frac{G'}{V'}} = \frac{c}{2\pi}\sqrt{\frac{nS_2}{ld}} \tag{4.54}$$

where l includes the end correction so that for a circular hole of diameter D, the length $l = l_0 + 0.8D$.

The absorption of the perforated panels depends on the real part of the specific acoustic impedance. Similarly with the single resonator, the maximum absorption at resonance is

$$\alpha_{res=} = \frac{4\mu'}{(1+\mu')^2} \tag{4.55}$$

with

$$\mu' = \frac{R}{n\rho_0 c} \tag{4.56}$$

where R is the resistance of one hole. This resistance is the pressure difference across the length of the hole divided by the particle velocity of the air in the hole. More explanations and calculations of R can be found in a later section. Similarly to the case of a single resonator, we will now consider the bandwidth of the resonator panel. The bandwidth O is

$$O = \frac{\lambda_{res}^2}{2}1.442(1+\mu')g' \tag{4.57}$$

where

$$g' = \frac{G'}{\lambda_{res}} = \frac{nS_2}{l\lambda_{res}} \tag{4.58}$$

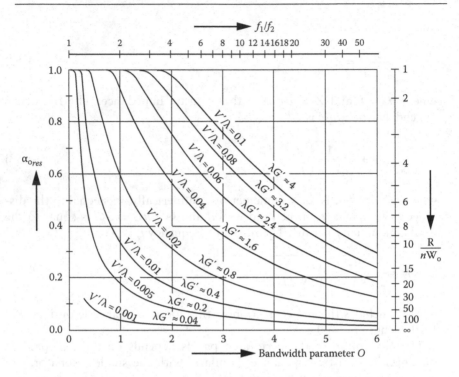

Figure 4.15 Design chart for perforated panels. (After Zwikker, C. and Kosten, C.W., *Sound Absorbing Materials*, Elsevier, London, U.K., 1949.)

Figure 4.15 constructed by Kosten [11] represents an assembly of variables and their relationship that permits to follow their trend when designing perforated absorber panels by computer. For instance, we can see that a very efficient absorber with high maximum absorption and wide band of absorption requires a rather large volume, which can be only achieved by a sufficient distance of the perforated plate from the back wall. This can be important for sufficiently damping low-frequency modes.

Figure 4.16 shows the effect on the resonator damping magnitude by placing porous materials inside the resonator cavity. We can see that the closer the materials are placed to the perforated plate the higher damping is achieved. This is important for the efficiency of the absorber construction.

4.4.3 Microperforated sound absorbers

The optimal performance of both the single resonators and absorbers with perforated plates depends on their acoustical flow resistance. The conversion of acoustic energy into heat occurs on the neck of the resonator or in the hole of the perforated plate in a thin surface layer around its perimeter, and because this damping is not sufficient, usually porous layers are

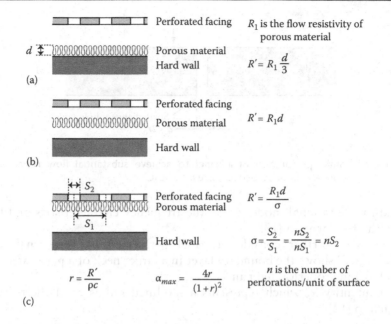

R_1 is the flow resistivity of porous material

(a) Perforated facing / Porous material / Hard wall

$$R' = R_1 \frac{d}{3}$$

(b) Perforated facing / Porous material / Hard wall

$$R' = R_1 d$$

(c) Perforated facing / Porous material / Hard wall

$$R' = \frac{R_1 d}{\sigma}$$

$$\sigma = \frac{S_2}{S_1} = \frac{nS_2}{nS_1} = nS_2$$

$$r = \frac{R'}{\rho c} \qquad \alpha_{max} = \frac{4r}{(1+r)^2}$$

n is the number of perforations/unit of surface

Figure 4.16 Effect of flow-resistive material placement on damping. (a) Porous material next to hard wall, (b) porous material centered in air space, (c) porous material next to perforated sheet.

placed in the resonator cavity to adjust the absorption to the required frequency response.

With the availability of laser technology, it is feasible to manufacture thin plates and foils from materials such as glass, plastics, metals, and similar materials with holes much smaller than 1 mm. The small holes have dimensions similar to the boundary layer and provide sufficient damping so that the sheet or foil can be used as absorber without additional damping.

Such microperforated absorbers can provide damping over a frequency range of several octaves. They are particularly advantageous for use in small rooms since they can absorb sound at low audio frequencies and provide damping of low frequency modes. Microperforation allows the design of translucent sound absorbers such as perforated glass plates and foils.

Also, since microperforation does not require conventional porous absorbers, it can provide sound absorption in special environments where porous absorbers would not be usable. They may provide sound absorption needed in complicated rooms with trusses, historical structures, etc.

Microperforated sheets and membranes made from rigid and particularly flexible (plastics) materials are very efficient absorbers also due to their vibrational modes. The bending modes of these materials depend on their mechanical properties as well as on their dimensions, mounting, and distance to a wall. The perforations contribute to the vibrational damping

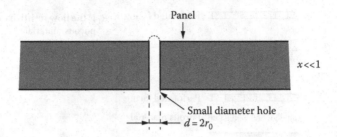

Figure 4.17 Small perforation of a panel to achieve substantial flow resistance for conversion of acoustical wave into heat.

and the vibrational modes affect the frequency characteristics and bandwidth of absorption [12].

We will further consider a panel perforated by very small holes. Figure 4.17 shows the boundary layer in a larger neck of a perforated panel and in a microperforated panel.

A quantity x, which represents a modified radius r_0 of the orifice, is equal to [12]

$$x = \frac{r_0}{\sqrt{\eta/\rho_0\omega}} = 0.65r_0\sqrt{f} \tag{4.59}$$

where

r_0 is the orifice radius in mm
$\eta = 1.8 \cdot 10^{-4}$ g/cm/s is the dynamic viscosity of air
ρ_0 is the air density
ω is the circular frequency

Crandall [13] formulated the impedance Z_1 for a hole of length t in mm and $x \ll 1$. To satisfy this condition, the diameter of the perforation must be smaller than 0.5 mm:

$$Z_1 = 8\frac{\eta t}{r_0^2} + \frac{4}{3}j\omega\rho_0 t \quad \text{for} \quad x = 1 \tag{4.60}$$

Figure 4.18b shows the equivalent electrical circuit of the microperforated panel absorber shown in Figure 4.18a. R represents the resistance and M the mass of the neck. Using $r = (R/\rho_0 c)$ and $m = (M/\rho_0 c_0)$, we obtain for [14]

$$r = \frac{0.147t}{d^2 p}\left(\sqrt{1 + 100d^2 f/32} + 1.768\sqrt{f}\frac{d^2}{t}\right) \tag{4.61}$$

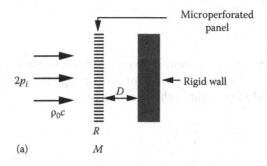

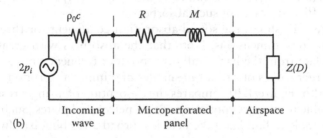

Figure 4.18 A single-layer microperforated panel resonator in front wall. R and M are symbolic to represent flow resistance and mass of air in holes [14]. (a) Microperforated panel in front of wall, (b) electroacoustic analogy for the system shown in (a).

and

$$\omega m = \frac{1.847 f t}{p}\left(1 + \frac{1}{\sqrt{9 + 50 d^2 f}} + 0.85 \frac{d}{t}\right) \tag{4.62}$$

where

d is the diameter of the perforations
t is the thickness of the plate

The symbol p is the percentage perforation, defined as the ratio of the perforation area to the plate surface area. The absorption coefficient α_N for normal incidence is given by [14]

$$\alpha_N = \frac{4r}{(1+r)^2 + \left[\omega m - \cot\left(\dfrac{\omega D}{c}\right)^2\right]^2} \tag{4.63}$$

where D is the distance of the perforated sheet from the back wall. The maximum absorption obtained for the second bracket in the denominator is equal to zero:

$$\alpha_{max} = \frac{4r}{(1+r)^2} \tag{4.64}$$

The bandwidth of absorption is given by two limiting frequencies f_1 and f_2 for which the absorption coefficient has dropped to one-half of its maximum value. Maa calculated their ratio B as

$$B = \frac{f_2}{f_1} = \frac{\pi}{\cot^{-1}(1+r)} - 1 \tag{4.65}$$

More detailed calculations can be found in Maa [14] as well as instructions for the efficient design of such absorbers.

Figure 4.19 shows the sound-absorption coefficient for three absorbers with different dimensions. Note that the absorbers with greater distance of the plate from the back wall offer broader frequency range absorption. Also, larger values of r result in smaller maximum absorption and broader bandwidth. Figure 4.20 compares the absorption of a single resonator with two double resonators consisting of two perforated plates. Such a construction is simple to build and the befit is a broad absorbing bandwidth.

4.4.4 Plate and membrane absorbers

As analyzed earlier in Chapter 2, the frequencies of the low-frequency modes in small rooms are well separated unless the modes are sufficiently damped. Depending on the room wall construction, some low-frequency room

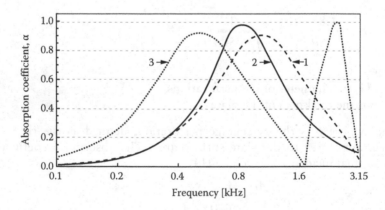

Figure 4.19 Absorption characteristics of a simple microperforated panel absorber of different design parameters: (1) $d=0.3$, $b=2$, $t=0.4$, $D=50$; (2) $d=t=0.4$, $b=3.5$, $D=100$; and (3) $d=t=0.4$, $b=4$, $D=40$. Note: d is the diameter of the perforation; b is the perforation distance; t is the thickness of the plate; D is the plate distance from the wall. (After Maa, D.Y., *Noise Control Eng. J.*, 29(3), 77, 1987.)

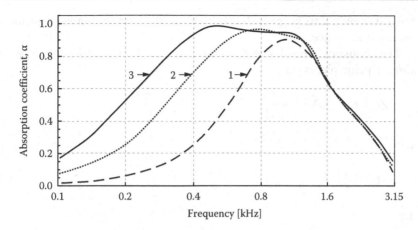

Figure 4.20 Absorption characteristics of microperforated panel constructions formed with panels of $d=0.3$, $t=0.4$, and $b=2$: (1) simple resonator, $D=50$; (2) double resonator, $D_1=D_2=50$; and (3) $D_1=50$, $D_2=100$. (After Maa, D.Y., *Noise Control Eng. J.*, 29(3), 77, 1987.)

damping may already exist, for example, cracks in walls can contribute to sound absorption. Walls that are heavy, such as brick or concrete walls, have high impedance and thus usually do not provide sufficient damping. However, the relatively thin plaster walls, mounted on studs and backed by porous damping (to provide sufficient transmission loss), may, because of their relatively low impedance, also offer damping of low-frequency modes. It is difficult to predetermine this damping by calculation and its determination usually requires elaborate measurements.

Plate and membrane absorbers often provide sound-absorptive resonant systems in addition to those offered by the Helmholtz resonators and resonator panels. Figure 4.21 shows the construction schematically. A plate or membrane is mounted on a suitable frame on a sufficiently rigid wall. The plate offers mass M, while the sealed volume formed at its back offers an air spring compliance C. This mass–spring resonator usually needs some additional damping to that provided by the losses in the membrane and the losses due to air pumping where the membrane is attached to the battens.

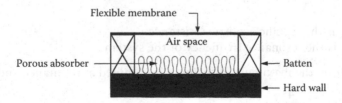

Figure 4.21 A membrane absorber.

Such damping can be provided by adding the porous material inside the enclosed air space.

The acoustical impedance Z_a of the plate, as shown earlier for systems of similar principle, is [11]

$$Z_a = R_a + j\omega M_a + \frac{1}{j\omega C_a} \tag{4.66}$$

with

$$M_a = \frac{M}{S^2} = \frac{m}{S} \tag{4.67}$$

and

$$C_a = \frac{Sd}{\rho_0 c^2} = \frac{c_a}{S} \tag{4.68}$$

where
    M is the mass of the plate
    m is its mass per unit area in kg/m^2
    S is the area of the plate in m^2
    C_a is the acoustic compliance of the cavity
    d is the depth of the air space in m
    c_a is the specific acoustic compliance of the cavity

The resonance frequency is

$$f_r = \frac{c}{2\pi}\sqrt{\frac{\rho_0}{md}} \simeq \frac{60}{\sqrt{md}} \tag{4.69}$$

The damping of the resonator depends on R_a. The absorption coefficient for normal incidence is

$$\alpha = \frac{4R_s\rho_0 c}{\left(R_s + \rho_0 c\right)^2 + \left[m\omega_r\left(\dfrac{\omega}{\omega_r} - \dfrac{\omega_r}{\omega}\right)\right]^2} \tag{4.70}$$

where
    R_s is the specific acoustic resistance
    ω_r is the resonance frequency of the system

As before, the maximum absorption is obtained at resonance and is

$$\alpha_{max} = \frac{4r}{(1+r)^2} \tag{4.71}$$

where $r = (R_s/\rho_0 c)$ so that the absorption coefficient for any frequency is

$$\alpha = \frac{\alpha_{max}}{1 + \left[\sqrt{\dfrac{m}{c_a}} \dfrac{1}{1+r} \left(\dfrac{\omega}{\omega_r} - \dfrac{\omega_r}{\omega} \right) \right]^2} \tag{4.72}$$

where c_a and m are defined in Equations 4.67 and 4.68.

The three graphs in Figure 4.22 show the absorption coefficient for three values of $r = 0.5, 1, 5$. The labeling of the curves is the ratio of the cavity compliance and the plate surface mass c_a/m. Note that the value of the coefficient $\sqrt{m/c_a}$ in Equation 4.72 affects the absorption bandwidth. The bottom figure is the most important. It shows that a large absorber volume provides a broad curve (0.5). The total absorbed energy is high due to the large bandwidth. The acoustic resistance is important for the design. Here, experimental research and absorption measurements with different porous material are necessary to secure reliable absorption data. Finally note that most loudspeakers in the room will also act as resonance absorbers because of the diaphragm mass which is supported by damped mechanical and air springs.

4.5 ABSORPTION BY COMPLIANT POROUS MATERIALS

Materials such as glass wool and mineral wool are probably the ones that first spring to mind for technical sound absorption at medium and high frequencies. The mechanical and mechanoacoustical properties such as flow resistance porosity were discussed in previous sections. The sound absorption of these materials, usually used in layers, can be predicted with technical accuracy, in spite of these materials not being perfectly homogeneous. In these fibrous materials, the conversion of the acoustic energy into heat is based on friction between the vibrating air and the fibers because of viscous losses in the air and heat transfer to the fiber *skeleton*. In principle, the skeleton is rigid although its fibers can, to some extent, covibrate with the sound waves. Ideally, the face of the absorber sheets is open so that sound waves can enter directly into the porous materials and propagate inside it.

In mineral and glass wool absorbers, the skeleton is, in principle, rigid. When these materials are used as springs, it is the air inside the material that supplies the compliance. From the viewpoint of health, these materials are, however, not ideal. The sheet surface sheds fibers, particularly when handled, which being very fine can cause respiratory problems. Mineral and glass fibers are also problematic since they are difficult to dispose of except in landfills.

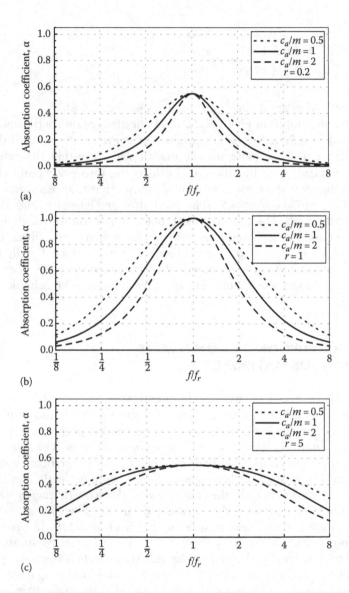

Figure 4.22 Sound absorption α of a simple mass–spring system as a function of frequency f and resistance: (a) weak damping (r=0.2), (b) optimal damping (r=1), and (c) strong damping (r=5). (After Fuchs, H.V., and Zha.X., Transparent absorption layer in the hall in the German congress (in German), *Bauphysik* 16, part 3, pp. 69–80, 1994.)

For these and for esthetic reasons, much work has been invested in designing replacement materials that have better architectural and acoustical properties and that can be better formed while not being hazardous and also more easily disposed of. Plastic foams such as polyurethane foams are suitable for these reasons.

Plastic (and rubber) foams can be manufactured with both open and closed pores. In open-pore materials, there may be a surface skin being a result of the manufacturing process. If a surface skin is present, it will seriously interfere with the sound absorption. The smaller the pores, the higher will be the flow resistivity and the more important the compliance of the plastic skeleton. Although the size and shape of the pores are irregular, on a large scale, the materials are fairly homogeneous. Generally, the materials are industrially developed using a trial and measurement technique, although if the tube measurement method is used, many samples may need to be measured for precision. Because the compliant skeleton plays a large role in the function of the materials, they can be designed to cover a broad range from those having acoustic properties similar to mineral and glass wool to those that are more similar to membrane absorbers in their acoustic behavior. One way to enhance the sound absorption of plastic foam is the use of surface corrugation. Because of the increased surface area due to the corrugation, the sound absorption of the foam can be substantially increased in the mid- and high frequency ranges.

Pure closed-pore materials such as polystyrene foams that have closed pores and high mechanical rigidity and are virtually nonabsorptive and uninteresting for acoustical purposes except as sound diffusers or possibly sheets for membrane absorbers. Sound absorption inside these materials cannot be caused by friction between the air and the skeleton but only by the deformation of the plastic structure due to the alternating pressure.

Among elastic materials, foam rubber was investigated early for its acoustical properties and most soft plastic foams have properties similar to rubber, but today, polyurethane foams are commonly used because they can be manufactured to have small but open pores. Unless the pores are small and open, there will be little sound absorption. As with fibrous materials, sound absorption tends to increase with sheet thickness. Note that a cover that has sufficient flow resistance will seriously reduce the sound absorption at high frequencies of any porous absorber as will impervious plastic films. On the other hand, a mass cover may create a mass–spring resonance that will result in high sound absorption over some low frequency range.

Figure 4.23 shows the typical sound-absorption coefficient behavior that can be achieved with engineered foams. Some data for foam materials can be found in References 3 and 11. The measurement results in Figure 4.23 refer to a CMSG combustion modifier foam based on spray polyurethane foam [15].

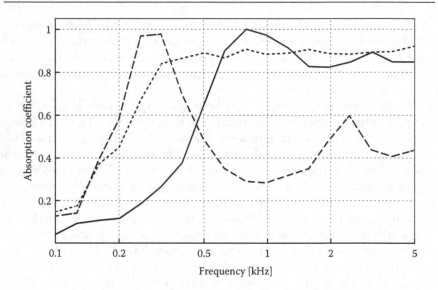

Figure 4.23 Sound absorption of a combustion modified granulated foam (CMSG). This CMSG foam is based on an open cell flexible polyurethane foam of the poly-ether type, some cells are closed, and it is loaded with melamine. The foam typically has 1.5 cells/mm and density of 43 kg/m³. Solid line: 24 mm thick layer of CMSG. Dashed line: 2 × 24 mm thick layers of CMSG foam with a bonded 35 g/m Mylar™ film coating on the top layer. Dotted line: sandwich design using 2 × 24 mm thick layers of CMSG foam with a bonded 35 g/m Mylar™ film coating on the lower layer. (From Parkinson, J.P., Acoustic absorber design, MEng thesis, Department of Mechanical Engineering, University of Canterbury, Christchurch, New Zealand, 1999.)

Because of the high sound absorption at low frequencies of semi–open pore plastic foams, they can be engineered to be optimal for damping the frequencies in the range of 50–200 Hz and they are much used in bass traps, both in homes, studios, and control rooms. In industry, they are common as sound absorbers for automotive and other transport use. Automobiles and buses use sound-absorptive foam in their engine compartment for noise control. For this use, the surface skin mentioned earlier is retained or the foam is covered by some coating on the surface. Many such products use reconstituted or chip foam made by granulating old foam. By using the foams in layers combining open volume compliance and extra inserted limp mass elements, the interested user can self-design the materials so that they fit the application. Furniture can be said to contain advanced absorbers since it in many cases uses plastic foams as cushions behind textiles and other covers. Note that any cover that has sufficient flow resistance will seriously reduce the sound absorption of any porous absorber.

Acoustical properties similar to those of plastic foam can also be had with conventional fiber materials that are covered with sintered layers that

Table 4.1 Examples of typical measured diffuse field sound absorption coefficients for various materials and constructions, as well as some single item data showing measured absorption areas

Sound-absorption coefficient α	Octave-band center frequencies						
	125 Hz	250 Hz	500 Hz	1 kHz	2 kHz	4 kHz	
Walls							
Concrete block, grout-filled, unpainted	0.01	0.02	0.03	0.04	0.05	0.07	
Concrete block, grout-filled, and painted	0.01	0.01	0.02	0.02	0.02	0.03	
Poured concrete, unpainted	0.01	0.02	0.04	0.06	0.08	0.10	
Normal, unpainted concrete block	0.36	0.44	0.31	0.29	0.34	0.21	
Painted concrete block	0.10	0.05	0.07	0.09	0.08	0.05	
6 mm heavy plate glass	0.18	0.06	0.04	0.03	0.02	0.02	
2.4 mm regular window glass	0.36	0.25	0.18	0.12	0.07	0.04	
12 mm plasterboard, wood studs spaced 0.4 m	0.29	0.10	0.05	0.04	0.12	0.09	
Plasterboard, 1.6 × 0.8 panels, 62 mm apart, studs 0.4 m. (voids filled with glass fiber)	0.55	0.14	0.08	0.04	0.10	0.10	
As earlier, but panels 128 mm apart	0.28	0.12	0.10	0.07	0.10	0.10	
Marble and glazed brick	0.01	0.01	0.01	0.01	0.02	0.02	
Painted plaster on concrete	0.01	0.01	0.01	0.03	0.04	0.05	
Plaster on concrete block or 25 mm on metal studs	0.12	0.09	0.07	0.05	0.05	0.04	
Plaster 16 mm on metal studs	0.14	0.10	0.06	0.05	0.04	0.03	
6 mm wood over air space	0.42	0.21	0.10	0.08	0.06	0.06	
9–10 mm wood over air space	0.28	0.22	0.17	0.09	0.08	0.06	
25 mm wood over air space	0.19	0.14	0.09	0.06	0.06	0.05	
Venetian blinds, open	0.06	0.05	0.07	0.15	0.13	0.17	
Velour drapes, 280 g/m²	0.03	0.04	0.11	0.17	0.24	0.35	

(continued)

Table 4.1 (continued) Examples of typical measured diffuse field sound absorption coefficients for various materials and constructions, as well as some single item data showing measured absorption areas

Sound-absorption coefficient α	Octave-band center frequencies						
	125 Hz	250 Hz	500 Hz	1 kHz	2 kHz	4 kHz
Fabric curtains, weight 400 g/m², hung 1/2 length	0.07	0.31	0.49	0.75	0.70	0.60
Ditto, but weight 720 g/m²	0.14	0.68	0.35	0.83	0.49	0.76
Shredded wood panels, 50 mm, on concrete	0.15	0.26	0.62	0.94	0.64	0.92
12 mm wood panel with holes 5 mm for 11% open area over void	0.37	0.41	0.63	0.85	0.96	0.92
Ditto with 50 mm glass fiber in void	0.40	0.90	0.80	0.50	0.40	0.30
Floors						
Terrazzo	0.01	0.01	0.02	0.02	0.02	0.02
Glazed marble	0.01	0.01	0.01	0.01	0.02	0.02
Linoleum, vinyl, or neoprene tile on concrete	0.02	0.03	0.03	0.03	0.03	0.02
Wood on joists	0.10	0.07	0.06	0.06	0.06	0.06
Wood on concrete	0.04	0.07	0.06	0.06	0.06	0.05
Carpet on concrete	0.02	0.06	0.14	0.37	0.60	0.65
Carpet on expanded (sponge) neoprene	0.08	0.24	0.57	0.69	0.71	0.73
Carpet on felt under pad	0.08	0.27	0.39	0.34	0.48	0.63
Indoor–outdoor thin carpet	0.10	0.05	0.10	0.20	0.45	0.65
Ceilings (also check floors and walls for similar material)						
Suspended 12 mm plasterboard	0.29	0.10	0.05	0.04	0.07	0.09

	Octave-band center frequencies					
	125 Hz	250 Hz	500 Hz	1 kHz	2 kHz	4 kHz
Ditto with steel suspension system frame	0.15	0.10	0.05	0.04	0.07	0.09
Plaster on lath suspended with framing	0.14	0.10	0.06	0.05	0.04	0.03
Glass fiber acoustic tile suspended	0.76	0.93	0.83	0.99	0.99	0.94
Wood fiber acoustic tile suspended	0.59	0.51	0.53	0.71	0.88	0.74
Acoustic tile, 50 mm on drywall, suspended	0.08	0.29	0.75	0.98	0.93	0.76
Ditto, air space between tile and drywall	0.38	0.60	0.78	0.80	0.70	0.75
Glass fiber, weight 360 g/m², suspended	0.65	0.71	0.82	0.86	0.76	0.62
Ditto, but weight 1120 g/m²	0.38	0.23	0.17	0.15	0.09	0.06
Luminous ceiling, typical with openings	0.07	0.11	0.20	0.32	0.60	0.85
Hanging absorbers, 450 mm, between centers	0.07	0.20	0.40	0.52	0.60	0.67
As earlier, but 159 mm between centers	0.10	0.29	0.62	0.72	0.93	0.98
Openings						
Open entrances off lobbies and corridors, 0.50–1.00						
Air supply and return grilles, 0.15–0.50						

From Reference 16

Single items (data in m²S)	125 Hz	250 Hz	500 Hz	1 kHz	2 kHz	4 kHz
Man in suit, standing (6 m² pp)	0.15	0.25	0.60	0.95	1.15	1.15
Man in suit, sitting (6 m² pp)	0.15	0.25	0.55	0.80	0.90	0.90
Woman in summer dress (6 m² pp)	0.05	0.10	0.25	0.40	0.60	0.75
Wooden chair	0.01	0.02	0.01	0.04	0	0
Cloth upholstered chair	0.15	0.25	0.30	0.35	0.40	0.40

Source: Kleiner, M., Klepper, D.L., and Torres, R.R. Worship Space Acoustics. J Ross Publishing, Plantation, FL, 2010.

have suitable flow resistance and mass. Another foamlike material is acoustic plaster, which is common in the building industry since it can be sprayed on surfaces such as curved ceilings.

4.6 EXAMPLES OF MATERIALS FOR ABSORPTION

Table 4.1 provides some sound-absorption data from measurements. These materials are typical for walls, floors, and ceilings of small rooms. These data are useful for the initial room design and for making estimates of reverberation time. However, as discussed earlier, the actual absorption depends on the sound field and therefore on room geometry and the absorber location. Nevertheless, this information is useful for consideration in material selection and for comparing the effects of the use of various materials for acoustical purposes.

REFERENCES

1. Mechel, F.P. (1989) *Schall Absorbers (in German)*. S. Hirzer Verlag, Stuttgart, Germany.
2. Ingard, K.U. (1994) *Notes on Sound Absorption Technology*, Version 94-02. Noise Control Foundation, Poughkeepsie, NY.
3. Ver, I.L. and Beranek, L.L. (1996), *Noise and Vibration Control Engineering*, John Wiley, New York.
4. Kuttruff, H. (2000) *Room Acoustics*. Elsevier Science Publishers, Essex, U.K.
5. ASTM Standards C384-98 and C423-02. ISO 354:1988, Acoustics—Measurement of sound absorption in a reverberation room, ISO, International Organisation for Standardisation, Geneva, Switzerland, 1988.
6. Schroeder, M. (1962) Correlation functions in rooms. *J. Acoust. Soc. Am.*, 34(12), 1819.
7. Attenborough, K. (1983) Acoustical characteristics of rigid absorbents and granular media. *J. Acoust. Soc. Am.*, 83(3), 785–799.
8. Allard, J.F. (1993) *Propagation of Sound in Porous Media*. Elsevier Applied Science, London, U.K.
9. Miki, Y. (1990) Acoustical properties of porous materials—Modifications of Delany-Bazley models. *J. Acoust. Soc. Jpn. (E)*, 11, 19–24.
10. Rayleigh, J.W.S. (1896) *The Theory of Sound*, 2nd edn. Dover, Cambridge, U.K., p. 319.
11. Zwikker, C. and Kosten, C.W. (1949) *Sound Absorbing Materials*. Elsevier, London, U.K.
12. Fuchs, H.V. (1995) Einsatz Mikro-Perforierten Platten (in German) The application of microperforated plates as sound absorbers with inherent damping. *Acoustica*, 81, 107–116.
13. Crandall, I.B. (1954) *Theory of Vibrating Systems and Sound*. D. Van Nostrand Company Inc., New York.

14. Maa, D.Y. (1987) Microperforated panel wideband absorbers. *Noise Control Eng. J.*, 29(3), 77–84.

15. Parkinson, J.P. (1999) Acoustic absorber design. Master of Engineering thesis, Department of Mechanical Engineering, University of Canterbury, Christchurch, New Zealand.

16. Fasold, W., Sonntag, E., and Winkler, H. (1987) *Bau- und Raumakustik (in German)*. VEB Verlag für Bauwesen, Berlin, Germany.

17. Fuchs, H.V. and Zha.X. (1994) Transparent absorption layer in the hall in the German congress (in German). *Bauphysik* 16, part 3, pp. 69–80.

18. Kleiner, M., Klepper, D.L., and Torres, R.R. (2010) *Worship Space Acoustics*. J Ross Publishing, Plantation, FL.

19. ISO 9614-3:2002. ISO, International Organisation for Standardisation, Geneva, Switzerland.

Chapter 5

Diffusion

5.1 INTRODUCTION

The reflection of sound from the absorption materials or room walls that we have been dealing with so far has been assumed specular. This means that if the size of the material is sufficiently large, the angle of reflection of a plane wave is the same as the angel of incidence. The sound at a listening location in a room consists of both direct sound that travels from the source and sound reflected from the walls and objects in the room. The purpose of the absorbers is to adjust the intensity and the spectrum of the sound reflection so that optimal listening conditions are achieved. However, the specular reflections, particularly in small rooms, cannot always completely satisfy this requirement.

In small rooms, the time difference between the arrival of the direct and the reflected sound at the listening point is very small, usually only 10–20 ms. Their interference results in sound coloration, particularly if the reflected sound is strong. To avoid coloration early reflected sound must be 20 dB weaker than the direct sound as discussed in Chapter 8. However, such attenuation by some absorbing material may result in excessive reduction of the overall acoustic energy in the room.

This situation can be corrected by using structures that diffuse the incident sound rather than absorb it. This means that the incident sound is reflected into many different directions. This will maintain the overall energy in the room but reduces the energy that is directly reflected to the listening point. Sound diffusion can be achieved by a variety of objects in the room, for instance, by furniture. However, optimum placement and diffusing properties are difficult to achieve by a single reflector.

Special structures have been developed to achieve optimum sound diffusion. This chapter will present an overview of some types, properties, design, and measurement within the scope of this book. A summary on diffusion can be found in Reference 1.

Most diffusers for use in room acoustics have been developed for a frequency range of about 0.5–4 kHz to be of usable size and cost in addition to working in the frequency range where hearing is the most sensitive. Substantial room correction is, however, often needed at low frequencies in small rooms due to the well-separated strong resonant modes that are often insufficiently causing the frequency response to be very nonuniform. These adverse effects are usually treated by the use of resonant absorbers as described in Chapter 4.

In addition to causing sound scattering or diffusion, the diffusers also have effects on sound absorption, modal density, etc., that will be covered in the following sections.

5.2 BASIC PRINCIPLES OF SOUND DIFFUSERS

The inventor of modern, numerically determined diffusers was Manfred Schroeder who in 1975 published two articles on diffuse sound reflection [2,3]. He described and analyzed a line array of narrow, approximately one-quarter-wavelength deep resonators that would scatter an incident plane sound wave over an angle of 180° over a wide frequency range. Other papers on the topic were later published by Berkhout [4,5]. The principle of Schroeder's basic idea is shown schematically in Figure 5.1a. The two wells have the same width but different depth and, therefore, the sound that is reflected from their bottoms have different phases at their tops. The two waves interfere and the resulting wave propagates under a different angle than the incident wave arrived on the two wells.

A similar idea, developed much later and shown in Figure 5.1b, shows two narrow strips of acoustic materials that have different impedances and therefore affect the phases of the reflected waves. For certain frequencies, maxima or minima of the interfering reflected sound occur as described later in this chapter.

An example of a Schroeder-type diffuser is shown in Figure 5.2 and consists of N wells of different depths, separated by thin, hard, rigid walls. The width W of each well is smaller than $W = \lambda_m/2$ where λ_m is the wavelength

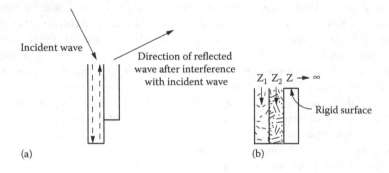

(a)   (b)

Figure 5.1 Principles of sound diffusion. (a) Two interfering tubes with phase shift. (b) Phase shift by different input impedance.

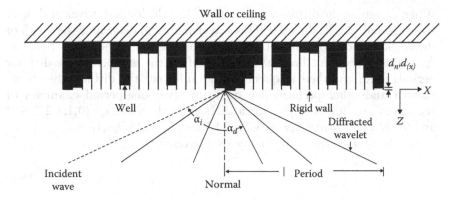

Wall or ceiling

$d_n, d_{(x)}$

Well

Rigid wall

Diffracted
wavelet

Incident
wave

Normal

Period

Figure 5.2 Scaled cross section of a 1D QRD described by Schroeder [5], consisting of two periods, with 17 wells per period ($N=17$). The width W of each of the wells, which are separated by thin and rigid walls, is equal to $0.137\lambda_0$, where λ_0 is the design wavelength, d_n is the depth of the nth well, and $d(x)$ is the depth at the lateral position x. A wave front (dashed line) making an angle of incidence α_i of $-65°$ with the surface normal and five diffracted wavelets (solid lines) occurring at $\alpha_d = -54.2°$, $-22.4°$, $2.7°$, $28.5°$, and $65°$ is shown. (From D'Antonio, P. and Konnert, J.H., *J. Audio Eng. Soc.*, 32(4), 228, 1984.)

of the maximum frequency for which the diffuser is designed. The wells must be sufficiently narrow to prevent the existence of cross modes at frequencies higher than f_m. The figure shows two repetitions of periods of N wells each.

We will now calculate the required depths of the wells. Schroeder used a mathematical sequence called quadratic residue (QR). Other sequences such as pseudorandom or primitive root can also be used to determine the well depths. The diffuser performance depends on the sequence used. Characteristic for useful sequences is that their autocorrelation has a single maximum and is close to zero elsewhere. We will describe the properties of the diffusers called quadratic-residue diffusers (QRDs). As Schroeder has shown, the Fourier transform of the diffused sound is constant, which means that the diffused energy is uniform in direction.

We can calculate the sequence of the well depths by evaluating $n^2 \bmod N$, where N is an odd prime number that is selected according to the requirements on the diffuser performance and n is an integer from zero to infinity. A QR sequence is generated by

$$s_n = n^2 \bmod N \tag{5.1}$$

where
s_n is the sequence number of the nth well
$\bmod N$ is the least nonnegative remainder
n describes the sequence of the wells

When designing the diffuser, we have to select the longest wavelength λ_0 at which sound is to be diffused. This wavelength will determine the depth of the deepest resonator. The total width of one period of the diffuser is NW. As will be shown further, the selection of N, the number of the diffuser periods used, and their frequency range influence the diffuser directivity pattern and other performance qualities. Two one-period examples of sequences are, for example, for a diffuser with $N = 7$: $s_n = \{0,1,4,2,2,4,1\}$ and for $N = 17$: $s_n = \{0,1,4,9,16,8,2,15,13,13,15,2,8,16,9,4,1\}$.

The depth d_n of the individual wells is given by

$$d_n = n_{\text{mod}N}^2 \frac{\lambda_0}{2N} \tag{5.2}$$

where
    n is the number of the well
    λ_0 is the wavelength of the lowest frequency of the diffuser
    N is the number of the wells in one diffuser period

The maximum depth of some wells can be approximately one-quarter-wavelength of the lowest design frequency. That is about 0.75 m for 100 Hz but about 15 cm for 500 Hz. Therefore, the QRDs are practically not usable for low frequencies unless using active elements in the wells, as described later.

Figure 5.3 shows how one can construct the directivity pattern of the diffused (reflected) sound by two incident wave rays (wavelets) A and B and the continuation as a diffracted wave of their travel as C and D. Should the wave from the two rays be reinforced, the path difference BC and GF must

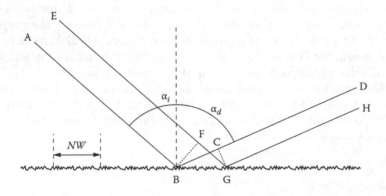

Figure 5.3 Construction showing incident (A and E) and diffracted (D and H) wavelets from the surface of a periodic reflection phase grating with a repeat distance NW. The angle of incidence is α_i and the angle of diffraction α_d. (From D'Antonio, P. and Konnert, J.H., *J. Audio Eng. Soc.*, 32(4), 228, 1984.)

be equal to $m\lambda$, where m is a positive or negative integer and $L = NW$. This condition can be from Figure 5.3 expressed as

$$m\lambda = L\big(\sin(\alpha_d) - \sin(\alpha_i)\big) \qquad (5.3)$$

where the index i is linked to the incident waves and index d to the reflected waves. To obtain the angles α_d for the directions of maximal diffusion, we solve the earlier equation for $\sin(\alpha_d)$:

$$\sin(\alpha_d) = \frac{m\lambda}{NW} + \sin(\alpha_i) \qquad (5.4)$$

Because the sine function cannot be greater than one, the solution is valid only for the numerical values of the quantities in the first term of the right-hand side of the earlier equation that will satisfy this condition.

The reflection factor $r(x)$ at the plane $z = 0$ is equal $r(x) = e^{-2\pi i \frac{2d(x)}{\lambda}}$ where $d(x)$ is the depth of the well at the position x. The scattering amplitude $a(s)$ and s are given by [6]

$$a(s) = \frac{1}{L} \int_0^L e^{-2\pi i \frac{2d(x)}{\lambda}} e^{isx}\, dx$$

$$\qquad (5.5)$$

$$s = 2\pi \frac{\sin(\alpha_d)}{\lambda} - 2\pi \frac{\sin(\alpha_i)}{\lambda}$$

where L is the total length of the diffuser. In the actual diffuser, the phase is not changing continuously but stepwise from well to well so that the upper integral is replaced by a series:

$$|a(s)| = \frac{1}{NX} \sqrt{\left[\sum_{k=1}^{NX} \cos\left(sx - 4\pi \frac{d_k(x)}{\lambda}\right)\right]^2 + \left[\sum_{k=1}^{NX} \sin\left(sx - 4\pi \frac{d_k(x)}{\lambda}\right)\right]^2}$$

$$\qquad (5.6)$$

where $\Delta x = W/NS$ is the summation interval, NS is the number of samples of depth per well and $1/NX = \Delta x/L$ relates the other quantities with NX that is the total number of sample points.

We will now show some examples of sound energy scattering from QRDs that were designed using the earlier equations [6].

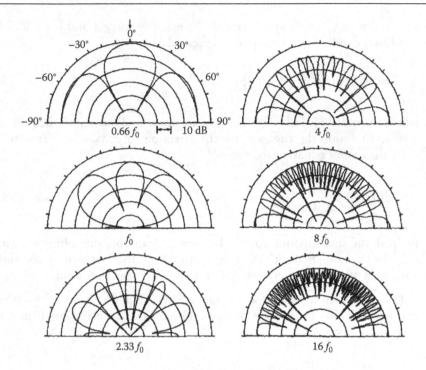

Figure 5.4 Diffracted patterns by a two-period-wide QRD. (From D'Antonio, P. and Konnert, J.H., J. Audio Eng. Soc., 32(4), 228, 1984.)

Figure 5.4 shows the diffraction patterns for $N = 53$, $f_0 = 1130$ Hz, and the direction of sound incidence that is normal to the diffuser plane. The diffuser consists of two periods. We see that the number of lobes increases with frequency, but their dependence on the reflection angle is constant.

Figure 5.5 shows the effect of increasing the number of periods for a diffuser with $N = 17$ and $f_0 = 1130$ Hz. Figure 5.5a shows the scattering from this diffuser with two periods. Figure 5.5b shows the effect of the period increase to 25. The diffracted energy is now concentrated in a few directions. Figure 5.5c shows the effect of increasing the number wells from 17 to 89.

These figures show the trends of the well and period number selection. Obviously, the acoustic goal of using a diffuser in a room is to spread the incident energy as much as possible and, therefore, to select the well number in a period as large as the room size allows is preferable.

Recently, the prediction and evaluation of the scattering by QRDs were analyzed by D'Antonio [1], and Cox and Lam [42]. They used the Helmholtz–Kirchhoff integral equation and the boundary integral method that were compared to the Fraunhofer solution, which is simple

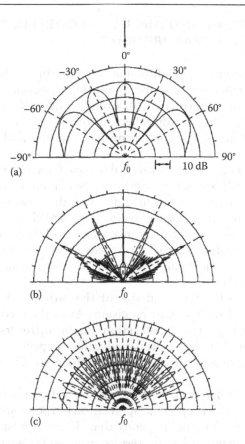

Figure 5.5 (a) Scattering intensity pattern for the QRD shown in Figure 5.2. The diffraction directions are represented as dashed lines and the scattering from the finite diffuser occurs over broad lobes. The maximum intensity has been normalized. (b) The number of periods has been increased from 2 to 25, concentrating energy in the diffraction directions. (c) The number of wells per period has been increased from 17 to 89, thereby increasing the number of lobes by a factor of 5. Arrows indicate incident and specular reflection directions. (From D'Antonio, P. and Konnert, J.H., J. Audio Eng. Soc., 32(4), 228, 1984.)

and most commonly used. Although there are differences between the results, the calculations that use the latter method are much faster than the other methods. The Fraunhofer solution, however, seems to be adequate considering the improvement of the listening quality that is often reported when even simple diffusers are used [1].

Schroeder's ideas on the QRD design and its application started research into other diffuser shapes and potential improvements. Some are shown in a later section.

5.3 SCATTERING AND DIFFUSION COEFFICIENTS
AND THEIR MEASUREMENT

The words scattering and diffusion describe in essence the same phenomenon, the reflection of sound in multiple directions as opposed to the specular reflection (mirroring). In order to quantify the scattering, the properties of the diffuser or reflecting object are characterized by two different coefficients: the coefficient of scattering and the coefficient of diffusion.

The Audio Engineering Society developed an information document AES-4id-2001 [7] and later a two-part ISO standard was completed [8]. The AES information document deals with the measurement method for obtaining a random-incidence scattering coefficient in a reverberation room. This is based on the Mommertz and Vorlaender technique [9]. The ISO standard deals with the measurement of the directional diffusion coefficient in a free field and follows the AES document developed by Cox and D'Antonio [1].

Depending on the type and size of the diffuser, the two coefficients can be measured with a good precision. Also, these coefficients are usually predicted from the construction of the diffusers. The differences between the predicted and measured values depend both on the prediction and measurement method. Both of these are analyzed in numerous publications.

The practical effectiveness of the diffusers depends on their type, construction, acoustical properties, application purpose, acoustical properties of the room, etc. The publications that describe specific diffuser application usually report that the results were *good*, whatever that means. Extensive psychoacoustic information to evaluate the effects of the listening improvement is, however, still not available and more research needs to be done.

5.3.1 Scattering coefficient

One of the metrics to characterize a diffuser is called the scattering coefficient. When a sound wave is incident on a diffuser, the reflected sound is scattered in many directions. The scattering coefficient s is defined as the ratio of nonspecularly reflected energy to the total reflected energy. The main purpose of its measurement is to provide data that will show the separation of specular energy from diffused energy. The scattering coefficient s carries no information on the directivity of the scattered energy. The measurement of the scattering coefficient can be executed in several ways. We will present the basic measurement arrangements that have been tested and standardized.

The scattering coefficient can be obtained from the measured polar distribution of sound scattered by the diffuser. The diffuser is irradiated by a small source located at a far distance (if possible) from the diffuser. The scattered sound is measured by microphones placed on a semicircle in 5° angular intervals [1,8,10] in free space or anechoic room. The sound is measured by some impulse response technique, often maximum-length sequence (MLS) [11]. Ideally, both the source and the receivers are located in the far field of the diffuser [2]. On the other hand, diffusers often find their application in small rooms, and therefore, near-field data could be useful.

In most small rooms, there are usually some reflecting objects of about 1 m typical dimension that scatter the sound. For such scatterers, it may be quite complicated to determine the diffracted sound. The boundary element method (BEM) predictions [1,13] show that there is a strong dependence on distance.

Another way to do the measurements is to use the near-field holography method developed by Kleiner et al. [14] that can project the near-field measurements into the far field or some other desired distance and direction. In another case, some reflectors are concave and concentrate the sound in the near field, which makes the measurements generally quite complicated [1]. Such concave surfaces are often found in the ceilings of rooms.

The scattering coefficient is useful for prediction of the sound field in a room. If the reverberant room technique is used, it is often not possible to measure complete size diffusers and often only models of the frequency scaled diffusers can be used. In addition, the measurements can be quite elaborate should all possible source positions be used.

The details of the scattered polar response measurement technique are well described in the book by Cox and D'Antonio [1]. The 1D diffuser to be measured is placed in the center of a semicircle. A source that irradiates the diffuser is put at a large distance from the diffuser at a selected angle. The microphones are placed on a circle at 5° angle intervals. (Some select even smaller intervals.) The impulse response is measured and, by means of the Fourier transform, provides the frequency response of the reflected sound, usually expressed in ⅓ octave bands. A 2D diffuser requires measurement to be done over arches in a perpendicular direction [13]. The scattered polar response is investigated over the application frequency range. If the measurement is done over all angles of incidence, the amount of data is obviously considerable. Additionally, the measurement has to be done in a large room or an anechoic room to eliminate the effects of wall reflections. Figure 2 in [15] shows three bandpass-filtered pulses obtained for different orientations of the measured sample. The highly correlated early part of the pulses is from the specular reflections, while the obviously uncorrelated second part contains the information on the scattered sound.

The specular reflection coefficient can be obtained from [15]

$$E_{spec} = (1 - \alpha)(1 - s) \equiv (1 - \alpha_{spec})$$

$$E_{total} = (1 - \alpha)$$

(5.7)

where
α is the absorption coefficient of the diffuser
s is the scattering coefficient
α_{spec} is the apparent specular absorption coefficient that describes the energy dispersed from specular reflection directions

The total reflected energy is E_{total} and E_{spec} is the specularly reflected energy from a surface into the diffused sound field. The scattering reflection coefficient s is obtained from the earlier equations and is

$$s = \frac{\alpha_{spec} - \alpha}{1 - \alpha} = 1 - \frac{E_{spec}}{E_{total}}$$

(5.8)

The energy E_{spec} is obtained by phase-locked averaging of the pulses for different sample positions. This principle is used for several measurement arrangements.

5.3.2 Effects of distance

Most diffusers are constructed to be used in medium and large rooms, such as concert halls, but diffusers are also designed for small rooms such as listening and control rooms. The smaller is the room the greater is the sensitivity of diffuser data on distance. Particularly when the sound distribution in a room is modeled, appropriate data have to be used. The cited articles recommend the measurements to be done in the far field, which is at distances where the sound pressure decreases at a rate 6 dB per distance doubling for 3D geometries. The distance problems can be avoided because the radiated power can also be measured at close distance using the sound intensity technique. This technique has been developed to measure the sound power radiated by complicated sources. The required microphones as well as the two channel analyzers are readily available. However, the audio industry is using the intensity technique rather rarely. To avoid the problems with the measurement distances, the sizes of the diffusers are often scaled down and the measurements are conducted at higher frequencies.

5.3.3 Random-incidence scattering coefficient

This measurement is described in the ISO standard [8]. The scattered polar response provides a detailed directivity pattern for the sound scattered by the diffuser as a function of the selected angles of incidence. Quite often, such detailed information, which requires a great effort to be obtained, is not needed. The random-incidence scattering coefficient measures the ratio of the nonspecularly (diffuse) reflected sound energy to the totally reflected energy. It is schematically shown in Figure 5.6. This scattering coefficient is measured as a function of the angle of incidence and usually in ⅓ octave bands. Because this angularly averaged coefficient in many cases does not depend too much on the angle of sound incidence, it is sufficient to measure this coefficient only for a few or only even one incidence angle.

However, this coefficient does not permit us to distinguish between the dispersion and only the redirection of sound waves because it depends only on the total energy that deviates from the specular direction. As shown later, the diffraction coefficient provides better information than the scattering coefficient. More analyses are presented with references and examples in [15]. There is no doubt that the search for dispersion criteria and the practicality of their measurement will continue for a long time particularly after their psychoacoustic effects will be more known.

In this measurement, the diffuser is placed on a turntable in a reverberation room and rotated in angle increments of 5°. The sound source and the microphone have fixed positions. Each time the diffuser is moved, the reverberation time is measured. The random-incidence absorption coefficient α_{spec} is obtained from

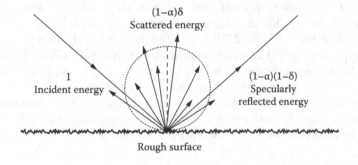

Figure 5.6 Types of scattering from a rough surface. (Adapted from Mommertz, E. and Vorlaender, M., Measurement of scattering coefficients of surfaces in the reverberation chamber and in the free field, *15th International Congress on Acoustics*, Trondheim, Norway, June 26–30, 1995.)

$$\alpha_{spec} = \frac{0.161V}{S}\left(\frac{1}{T_{60,2}} - \frac{1}{T_{60,1}}\right)$$ (5.9)

where
　　V is the room volume
　　S is the surface area of the sample
　　$T_{60,1}$ is the reverberation time without the sample and $T_{60,2}$ with
　　　the sample

The reverberation time is measured according to ISO standard 354 on Measurement of sound absorption in a reverberation room. The Vorländer paper [15] and the ISO 17497-1 standard deal with the details of the sampling technique to conduct the measurements and evaluate the scattering coefficient.

5.3.4 Diffusion coefficient

The purpose of the diffusion coefficient is to characterize the uniformity of the reflected sound. It is based on the data obtained from the measurement of the scattering coefficient that provides data to define a single dependence of this coefficient on frequency. It averages small variations of the scattering coefficient data. The decision on the definition of this coefficient was preceded by a discussion of the nature of the statistical data to be used [10,11,13,16–19]. The final decision was to define this coefficient in terms of an autocorrelation function that is used to measure the similarity properties of a time signal. In our case, the autocorrelation function, instead of time similarity, measures the spatial similarity of the signals scattered into various directions. It is based on the data obtained from the measurement of the scattering coefficient for a single irradiation source position at the angle ψ. This coefficient was proposed and its measurement formulated in the AES-4id-2001 (r2007) information document mentioned. Subsequently, in 2012, an international standard ISO 17497-2 was issued. The diffusion coefficient d_ψ is based on the ⅓ octave sound pressure levels of the polar response of the diffuser as shown in Figure 5.7.

The diffusion coefficient, which is the autocorrelation function of the pressure squared, is calculated from any the quantities, so that, for instance, reverberation can be predicted with better accuracy:

$$d_\psi = \frac{\left(\sum_{i=1}^{n} 10^{L_i/10}\right)^2 - \sum_{i=1}^{n}\left(10^{L_i/10}\right)^2}{(n-1)\sum_{i=1}^{n}\left(10^{L_i/10}\right)^2}$$ (5.10)

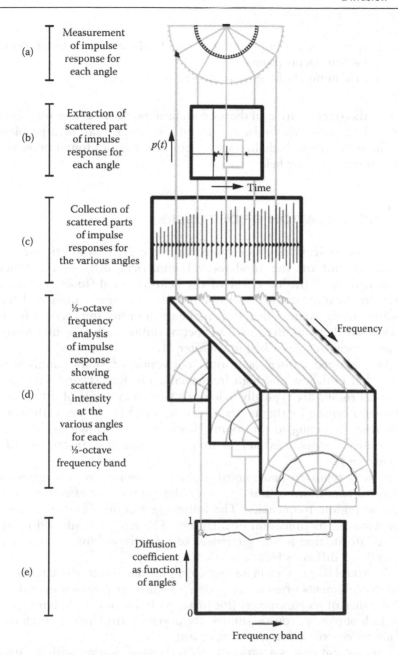

(a) Measurement of impulse response for each angle

(b) Extraction of scattered part of impulse response for each angle

$p(t)$

Time

(c) Collection of scattered parts of impulse responses for the various angles

(d) ⅓-octave frequency analysis of impulse response showing scattered intensity at the various angles for each ⅓-octave frequency band

Frequency

(e) Diffusion coefficient as function of angles

Frequency band

Figure 5.7 Process to extract the diffusion coefficient from the impulse response: (a) measurement geometry; (b) time response, h_i; (c) concatenated time responses; (d) ⅓ octave band polar responses; and (e) diffusion coefficient as a function of frequency. (After AES-4id-2001-AES (r2007) Recommendation on measurement of diffusers, *Audio Eng. Soc.*, New York.)

where
> L_i represents a set of sound pressure levels at the scattering positions measurements points
>
> n is the number of the measurement points

A very thorough analysis of the measurement problems can be found in [18]. As with the other methods, the size of the samples can be a problem for mechanical reasons including the practical difficulties found in measuring the sensors in the far field.

5.4 MODIFICATION OF DIFFUSERS

The previous sections concentrated on the principles of effective sound diffusion and on the fundamental numerical diffuser construction. Although the diffusion effect of the Schroeder's diffusers is good, the main application of such diffuser is rather in large rooms and for the midfrequency range. Their sizes make them rather impractical for small rooms but they are often used in special rooms such as sound recording control rooms as described in Chapter 11.

As explained earlier, acoustical correction of small rooms may be needed at both low and high frequencies. The lowest modal frequencies in small rooms are typically below the frequency range of diffusers. The problems caused by the modes can be lessened by adding sufficient low-frequency damping to the room. However, as is shown in this section, there are attempts to extend the frequency range of diffusers towards low frequencies.

As it is reported and published, the diffusers can also *improve* the listening conditions in small rooms although they are effective primarily only at middle frequencies. The following sections summarize some of the work on modification of diffusers. The reader is advised to obtain more information in the references of this chapter for the use of these specialized diffusers [38].

We would like to terminate this short introduction by the statement that the improvements often cannot cover the diffuser function over the complete range of its parameters (for instance, frequency) [40], but in view of the lack of psychoacoustic studies, the deviation from *perfect* performance is not necessarily subjectively important.

This section covers a variety of diffuser modification with a variety of goals. Most of them are aimed to reduce the maximum depth of the diffuser while maintaining the lowest frequency of their frequency range. Because of the volume of publications on the topic, this presentation has the format of rather an extended listing.

5.4.1 Optimization of profiled diffusers

In 1995, Cox published an important paper on diffuser optimization [20]. Cox concluded that the BEM is a practical and computationally feasible method to evaluate any diffuser modifications. He investigated the effects of the fins that characterize Schroeder diffusers. Figure 5.8a shows a diffuser as proposed by Schroeder, while Figure 5.8b shows the stepped diffuser in which the fins were removed. The depths of the Schroeder diffuser wells were varied as were the depths of the stepped diffusers. In addition, a randomly determined depth sequence was examined.

The important result was that the diffusers without fins diffused sound better than the diffusers with fins. The improved diffusivity of optimized stepped diffuser was experimentally verified. Such stepped diffusers can be manufactured at lower-cost diffusers with fins and are preferred by architects and many users for their visual properties.

The work by Cox was amended by D'Antonio who evaluated the performance of optimized diffusers [21]. The study used the results of the DISC project [22,23]. The study is wide, covering a variety of diffuser shapes and size as well as angles of incidence of incident sound. Some methods to speed up the calculations are also shown. The study also showed that the perfection of a given diffuser may not be optimal over the total range of the design parameters.

In 1995, Feldman published a paper on the construction of a diffuser that would nullify the specular reflection and create a cone of silence [24]. His goal was to achieve a cone of variable width in which the specular reflection would be suppressed.

D'Antonio and Konnert published a paper on directional scattering coefficient [25] that deals with the sound diffusion from a variety of acoustical surfaces, for instance, some absorber, edges, and similar, among others, which is very helpful for the design of listening rooms.

Tomaz analyzed the possibility to reduce the depth of diffuser wells by terminating the wells close their bottom with a Helmholtz resonator or a membrane as shown in Figure 5.9 [12]. The paper contains the design

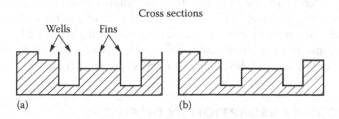

Figure 5.8 The diffusers tested: (a) Schroeder-type diffuser, and (b) stepped diffuser without fins. (From Tomaz, R., *Innovate Diffusers*, University of Salford, Salford, U.K., 2004.)

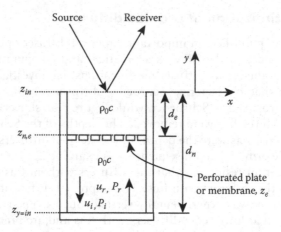

Figure 5.9 Drawing of a diffuser well with a plate or membrane element fixed at some depth. (From Tomaz, R., *Innovate Diffusers*, University of Salford, Salford, U.K., 2004.)

calculations and measurements of the dispersion by the more shallow diffusers. The diffusers are capable of extended low-frequency diffusion but the performance at the medium and high frequencies is worse. Because the diffuser is shallower, it can be useful in spite of the obviously higher cost.

As mentioned earlier, the sound dispersion can be achieved by phase shifts of reflected waves from neighboring materials. Cox and Angus suggested the use of ternary and quadriphase sequence [26]. They show a binary amplitude diffuser, consisting of hard reflecting patches {1} alternating with absorbing materials {0}. Their $N = 7$ sequence was {1 1 1 0 01 0}. A ternary sequence is {1 1 0 1 0 0 –1} where –1 is a well a quarter of a wavelength deep to provide a reflection coefficient of –1. The absorption of the patches must to be small. The paper provides a thorough analysis of the patch sequences and predicts their diffusion. However, the diffusers shown have a low diffusion coefficient. Similar research was done by Cox and D'Antonio in which they concentrated on the absorption coefficient of the diffuser components [26].

Very interesting research was done by Hughes et al. on the construction of volumetric diffusers [27]. A system of periodic arrays of cylinders of a variety of arrangements was investigated both theoretically (scattering predictions) and experimentally. This work provides a good foundation for further creativity and development of this type of diffusers.

5.5 SOUND ABSORPTION BY DIFFUSERS

The principal investigated property of diffusers is their ability to scatter sound. Perhaps the most important of the other properties is the sound absorption that can be quite high, also at frequencies out of the diffusion frequency range [39].

The reason for the sound absorption of diffusers, in particular those that have wells, is not obvious. Kuttruff attempted to explain the high sound absorption particularly at low frequencies [28]. The Schroeder-style diffusers consist of wells or tubes (in the case of 2D diffusers) with varying depths. The wells or tubes are narrow, placed one next to the other. Kuttruff examined QRDs and primitive root diffusers. He assumed that the sound pressure at the front of the wells is constant. Any pressure differences at the well entrances will cause air to flow between them to equalize the pressure. These equalizing flows cause sound absorption due to viscous losses. Such absorption does not happen if the resonators have equal depth.

Kuttruff developed mathematical models for the frequency dependence of the sound absorption. Figure 5.10 shows the absorption coefficient of a QRD with $N = 11$ and $f_r = 285$ Hz. The dotted line represents the absorption of wells with equal length and this case the losses are small. This shows that the losses are caused by the air flow.

Very important theoretical and experimental data are shown Fujiwara et al. [29–31]. He developed a theory of sound propagation and measurement technique of the particle velocity inside and in the vicinity of the resonators. Figure 5.11 shows the relative amplitude of particle velocity distribution for a certain well arrangement and a frequency 330 Hz [29]. As expected, the wave that approaches the well bottom has a wave front perpendicular to the well walls. However, we can see serious interference around the tops of the wells, of the waves that are radiated from the wells due to their phase

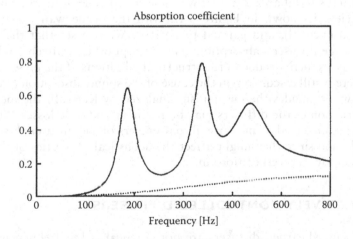

Figure 5.10 Absorption coefficient of a QRD with $N = 11$, $f_r = 285$ Hz. The dotted line shows the absorption coefficient of a structure consisting of wells with uniform length (=average length). (From Kuttruff, H., *Appl. Acoust.*, 42, 215, 1944.)

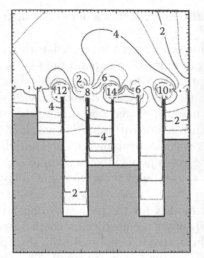 

Figure 5.11 Comparison between the predicted (left) and the measured (right) particle velocity distributions for well arrangement of [1,2,5,3,3,5,2] and a frequency of 330 Hz. (From Fujiwara, K., Nakai, K., and Torihara, H., Visualization of the sound field around a Schroeder diffuser, *Appl. Acoust.*, 60, 225, 2000.)

shifts and differing amplitudes. These phenomena result in energy losses that manifest themselves as sound absorption.

Fujiwara presents the measurements of the absorption coefficient for a Schroeder 2D QRD with prime number 7 in a reverberation chamber [29]. The 6 cm wide wells were separated by 3 mm thick plywood walls. A total of 48 units (total area 222 m²) were measured. The measured absorption coefficient is shown in Figure 5.12 with the inside wall unpainted (I), painted once (II), and painted twice (III). We can see that the painting affects the diffuser's absorption. The absorption is relatively high at low frequencies, perhaps due to the structural vibrations of the plywood.

There is still discussion on the cause of the sound absorption of diffusers with wells. Besides the mechanism analyzed by Kuttruff, the mechanical construction of the diffusers may be resonant and have losses. Therefore, it is recommended to measure all parameters of the diffusers before their installation since they might affect the acoustical properties of the room, particularly its reverberation time.

5.6 ACTIVELY CONTROLLED DIFFUSERS

As discussed earlier, diffusers are not practical at low frequencies due to their size. One of the attempts to maintain the resonant acoustical properties in a smaller device consists of replacing a well by an electronically

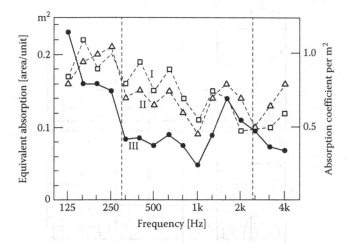

Figure 5.12 Equivalent absorption area and absorption coefficients of a 48-unit assembly of Schroeder diffusers measured in a reverberation chamber. (From Fujiwara, K., Nakai, K., and Torihara, H., Visualization of the sound field around a Schroeder diffuser, Appl. Acoust., 60, 225, 2000.)

controlled loudspeaker to provide, at the loudspeaker diaphragm surface, the resonator input impedance $z = -j \cot (kd)$. The electronic resonators are suitable for 2D diffusers because the loudspeaker dimension is approximately equal to the cross-section size of a typical diffuser well [12]. In principle, a low-frequency diffuser could be constructed completely from a wall of loudspeakers that would radiate the frequencies electronically randomly phased.

The design and performance of a multiple element hybrid diffuser is described in [32]. Figure 5.13 shows schematically (a) passive and (b) active (hybrid) diffuser, while (c) shows an engineering mock-up. The reference analyzes the principles, function, and implementation of the circuitry for the loudspeaker (that functions as a virtual well extension) as shown in Figure 5.14. The performance of a passive and active diffuser at 500 and 1000 Hz is shown on Figure 5.15. The performance difference between the passive and active diffuser is smaller than about 2 dB. Considering the complexity of the filter design problems with the stability, finite sizes, and shape of the loudspeaker, this is an excellent performance.

Another paper [33], dealing with the design, construction, and measurement of active diffusers, published by the same authors as paper [32] contains further data on the performance of hybrid diffusers. Another paper on active diffuser technology was published by Naqvi [34] and Angus [38].

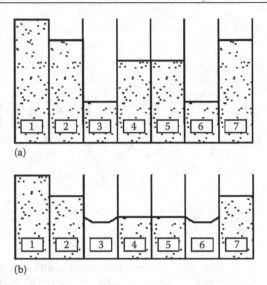

(a)

(b)

Figure 5.13 Experimental diffuser prototypes: (a) passive diffuser, and (b) diffuser with active elements. (From Avis, M.R. et al., *J. Audio Eng. Soc.*, 53(11), 1047, 2005.)

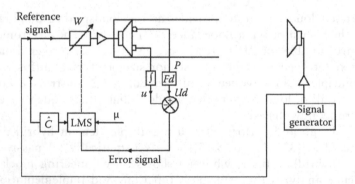

Figure 5.14 Schematic of active control of acoustic impedance. (From Avis, M.R. et al., *J. Audio Eng. Soc.*, 53(11), 1047, 2005.)

5.7 VOLUME DIFFUSERS

As sound is incident on objects in the room, furniture, lighting fixtures, artwork, humans, etc., its energy is to some extent absorbed and to some extent reflected. Which frequencies are absorbed and how sound is reflected will depend on the acoustic nature in a wide sense of the object, which is its impedance and geometry and whether the object is away from or tight against a wall in the room.

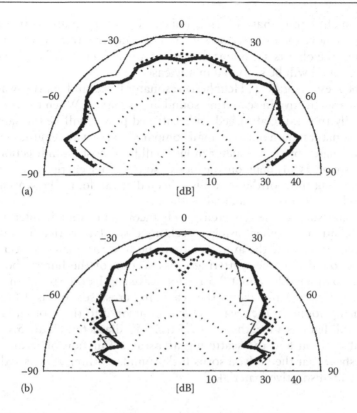

Figure 5.15 Measured scattered polar responses for active and passive diffusers. (a) 500 Hz. (b) 1000 Hz. Thick solid line: passive Schroeder diffuser; dashed line: active Schroeder diffuser; thin solid line: plane surface. (After Xiao, L., Cox, T., and Avis, M., Active diffusers: Some prototypes and 2 D measurements. *J. Sound Vib.*, 285, 321, 2004.)

Small rooms are likely to have relatively more cluttered volume than large rooms. It is obviously not possible to specify the absorption and scattering properties of all the types of objects that can be met with in a listening environment. The properties of humans as sound scatterers and absorbers are of course of particular interest when dealing with a listening environment.

The idea of a frequency-dependent scattering cross section is often used when considering the acoustic effect of free objects. The scattering properties of an object are usually characterized by its scattering cross-section area S_{SCS} [m²], which is defined as the area that a scattering object appears to have as it receives and scatters the sound power from an incident plane wave:

$$S_{SCS} = \frac{P_{scat}}{I_{inc}} \tag{5.11}$$

One might think that covering objects by a sound-absorptive material would remove their scattering ability. This only functions when the dimensions of the objects are about the size of a wavelength or larger. For small items, sound will be scattered in any case.

Open resonators (e.g., Helmholtz resonator flasks and quarter-wavelength tubes) are effective in scattering sound at resonance. When the resonator is critically tuned, the absorbed and scattered power will be the same. Since the resonator mouth must be small compared to the wavelength, the reradiation of energy by the resonator mouth will be almost omnidirectional [35].

As sound is incident on the body, some of its energy is absorbed by clothing but most of the sound is reflected at random. The scattering will depend on the torso, head, and clothing.

By measuring the coherent and incoherent sound intensity in a reverberation chamber, using a technique similar to that in [9], it was possible to measure the average human body scattering cross section [36]. It was found that the scattering cross section of the human body could be approximated by about 0.5 m² for wavelengths the size of the body or smaller. Figure 5.16a shows the scattering cross section as a function of frequency found in the measurements compared to those of an ideal hard ellipsoid that has a length about 5.5 times its diameter. Figure 5.16b shows the absorption and the scattering areas as a function of frequency. The data show that the body absorbs little sound; clothing acts as a thin layer of sound-absorptive material.

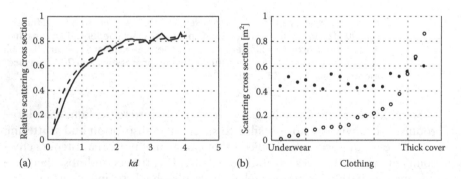

(a) kd (b)   Clothing

Figure 5.16 Scattering by the body. (a) The solid line shows the averaged scattering cross section for six humans. The dashed line shows the scattering cross section for the hard ellipsoid in air as a function of frequency. The maximum value is that at infinite frequency for a solid hard ellipsoid in air, with a total length about 5.5 diameters (d). Frequency is given in units of kd, where $k = 2\pi/\lambda$ is the wave number. (b) Total scattering cross section S_{SCS} (filled circles) and absorption area A (circles) for a single human wearing different amounts of clothing. Note that S_{SCS} is about constant as the amount of clothing is increased, whereas A increases significantly. (After Conti, S.G. et al., *Appl. Phys. Lett.*, 84(5), 819, 2004.)

Randomly positioned rigid objects in the room will effectively scatter sound if they are large compared to wavelength. The common use of vanes and wings as diffusers in reverberation chambers is an example of the usefulness of this technique.

The effectiveness of nonabsorptive *volume* scatterers can be estimated by statistical methods [37]. Assume that the volume V contains N randomly placed scatterers and that they have a scattering cross section S_{SCS} and that the bounding area of the room is S. One can show that

$$\frac{w - w_g}{w} = \frac{1}{1 + \left(\alpha S / 4 N S_{SCS} \right)} \tag{5.12}$$

where
w is the total energy density in the room
w_g is the energy density of the specularly reflected part
α is the average sound absorption of the surfaces of the room

The equation shows that when the number of scattering objects is large (but their volume small compared to V), most of the energy in the diffuse field of the room will be scattered energy. It also shows that the scatterers can be about as effective as the sound absorption by the walls in reducing specularly reflected energy.

REFERENCES

1. Cox, T.J. and D'Antonio, P. (2004) *Acoustic Absorbers and Diffusers*. Spon Press, London, U.K.
2. Schroeder, M.R. (1975) Diffuse sound reflections by maximum length sequences. *J. Acoust. Soc. Am.*, 57, 149–150.
3. Schroeder, M.R. (1979) Binaural dissimilarity and optimum ceilings for concert halls: More lateral sound diffusions. *J. Acoust. Soc. Am.*, 65(4), 968–963.
4. Berkhout, A.J., van Wulfften Palthe, D.W., and de Vries, D. (1979) Theory of optimal plane diffusers. *J. Acoust. Soc. Am.*, 65(5), 1334–1336.
5. Schroeder, M.R. (1979) Response to "Theory of Optimal Plane Diffusors". *J. Acoust. Soc. Am.*, 65(5), 1336–1337.
6. D'Antonio, P. and Konnert, J.H. (1984) The reflection of phase grating diffuser: Design, theory and application. *J. Audio Eng. Soc.*, 32(4), 228–238.
7. AES-4id-2001-AES (r2007) Recommendation on measurement of diffusers. *Audio Eng. Soc.*, New York.
8. ISO Standards 17497-1, Part 1: Measurement of the Random-incidence Scattering Coefficient in a Reverberation Room, 2004, 17497-2, Part 2: Measurement of the Directional Diffusion Coefficient in a Free Field, 2012. ISO, International Organisation for Standardisation, Geneva, Switzerland.

9. Mommertz, E. and Vorlaender, M. (1995) Measurement of scattering coefficients of surfaces in the reverberation chamber and in the free field, *Fifteenth
   International Congress on Acoustics*, Trondheim, Norway, June 26–30, 1995.
10. Cox, T.J., Dalenback, B.I., D'Antonio, P.D., Embrechts, J.J., Jeon, J.Y.,
    Mommertz, E., and Vorlaender, M. (2006) A tutorial on scattering and
    diffusion coefficients for room acoustic surfaces. *Acta Acust.*, 92, 1–15.
11. Vanderkooy, J. (1994) Aspects of MLS measuring system. *J. Audio Eng. Soc.*
    42(4), 219–231.
12. Tomaz, R. (2004) *Innovate Diffusers*. University of Salford, Salford, U.K.
13. D'Antonio, P. and Konnert, J. (1991) Directional scattering coefficient:
    Experimental determination, *The 91st Audio Engineering Society Convention*,
    New York, paper 3117.
14. Kleiner, M., Gustafsson, H., and Backman, J. (1997) Measurement of
    directional scattering coefficients using near-field acoustic holography and
    spatial transformation of sound fields. *J. Audio Eng. Soc.*, 45(5), 331–345.
15. Vorlaender, M. and Mommertz, E. (2000) Definition and measurement of ran-
    dom-incidence scattering coefficients. *Appl. Acoust.*, 60, 187–199.
16. Cox, T.J. (1995) Optimization of profiled diffusers, *J. Acoust. Soc. Am.*,
    97, 2928–2941.
17. Cox, T.J. (1996) Designing curved diffusers for performance spaces. *J. Audio
    Eng. Soc.*, 44, 354–364.
18. D'Antonio, P. (1993) The disc project: Directional scattering coefficient
    determination and auralization of virtual environments, *Proc-Noise Con
    93*, May 1993, pp. 259–264.
19. Takahashi, D. (1995) Development of optimum acoustic diffusers. *J. Acoust.
    Soc. Jpn.*, 16, 51–58.
20. Cox, T. (1995) Optimization of profiled diffusers. *J. Acoust. Soc. Am.*,
    97, 2928–2936.
21. D'Antonio, P. Performance evaluation of optimized diffuser. *J. Acoust. Soc.
    Am.*, 97(5), May 1995.
22. Konnert, J. and D'Antonio, P., (1994) The disc project: Theoretical simula-
    tion of directional scattering properties of architectural acoustics surfaces, in
    Proceedings of Wallace Clement Sabine Symposium, Boston, MA.
23. D'Antonio, P., Konnert, J., and Kovitz, P. (1993) The disc project: Auralization
    using directional scattering coefficient, *The 95th Audio Engineering Society
    Convention*, New York, October 1993.
24. Feldman, E. (1995) A reflection grating that nullifies the spectacular reflection:
    A cone of silence. *J. Acoust. Soc. Am.*, 98(1), 623–634.
25. D'Antonio, P. and Konnert, J. (1992) The direction scattering coefficient:
    Experimental determination. *J. Audio Eng. Soc.*, 40(12), 992–1017.
26. Cox, T. and D'Antonio, P., (1999) Optimized planar and curved diffusers.
    107th Convention of the Audio Engineering Society, New York, paper 5062.
27. Hughes, R., Angus, J. et al. (2010) Volumetric diffusers: Pseudorandom cylinder
    arrays on a periodic lattice. *J. Acoust. Soc. Am.*, 128(5), 2847–2855.
28. Kuttruff, H. (1944) Sound absorption by pseudostochastic (Schroeder
    diffusers). *Appl. Acoust.*, 42, 215–231.
29. Fujiwara, K., Nakai, K., and Torihara, H. (2000) Visualization of the sound
    field around a Schroder diffuser. *Appl. Acoust.*, 60, 225–235.

30. Fujiwara, K. (1995) A study on the sound absorption of a quadratic-residue type diffuser. *Acustica*, 81, 370–378.
31. Fujiwara, K. and Miyajima, T. (1992) Absorption characteristics of practically constructed Schroeder diffuser of quadratic-residue type. *Appl. Acoust.*, 35, 149–152.
32. Avis, M.R., Xiao, L., and Cox, T. (2005) Stability and sensitivity analyses for diffusers with single and multiple active elements. *J. Audio Eng. Soc.*, 53(11), 1047–1060.
33. Xiao, L., Cox, T., and Avis, M. (2004) Active diffusers: Some prototypes and 2 D measurements. *J. Sound Vib.*, 285, 321–339.
34. Naqvi, A. and Rumsey, F. (2005) The active listening room, a novel approach to early reflection manipulation in critical listening rooms. *J. Audio Eng. Soc.*, 53(5), 385–401.
35. Ingard, K.U. (1950) Scattering and absorption by acoustic resonators. PhD thesis, Department of Physics, MIT, Cambridge, MA.
36. Conti, S.G., Roux, P., Demer, D.A., and Rosny, J. (2004) Measurement of the scattering and absorption cross sections of the human body. *Appl. Phys. Lett.*, 84(5), 819–821.
37. Kuttruff, H. (2000) *Room Acoustics*. CRC Press, Boca Raton, FL.
38. Angus, J. (1995) Sound diffusers using reactive absorption grating, *The 98th Audio Engineering Society Convention*, Paris, France, paper 3953.

Chapter 6

The ear

6.1 SOUND AT THE EARS

6.1.1 Introduction

Humans can hear sound over a wide frequency and amplitude range. We can hear sound with pressure as low as 10 µPa and can endure sound pressure as high as 100 Pa, a very large dynamic range for any transducer system. A normally hearing person hears over a frequency range of about 20 Hz to 20 kHz.

For listening, we use a set of subsystems: the ears, the auditory nerves, and the brain. Functionally, the ears can be regarded as consisting of three parts, the outer, middle ear, and inner ear as shown in Figure 6.1. Together these form a system for the conversion of sound pressure oscillations to associated nerve signals in the auditory nerve. In addition, there is feedback from the neural system that affects the mechanical system of the middle and inner ear.

6.1.2 Head-related coordinate system

The sound at our ears will depend on the sound field of the room that in turn depends on the sound source, the room boundaries, and the objects (including the self and other persons) in the room. When discussing sound incident on a listener, it is practical to use the head-related coordinate system shown in Figure 6.2. In that system, the elevation angle is δ and the lateral angle is φ. The interaural axis passes through the ear canal openings and the head is centered in this axis. The angle γ toward the median plane is also important in binaural hearing but is equal to φ only in the horizontal plane. The radial distance is r.

The sound at the ears also depends on the angle of the incoming sound waves toward the head and the body. At low frequencies, the difference between the sounds at the two ears will be slight, but because of the time delay and the path length and the diffraction by the head, the differences will be large at high frequencies. For medium and high frequencies, the head

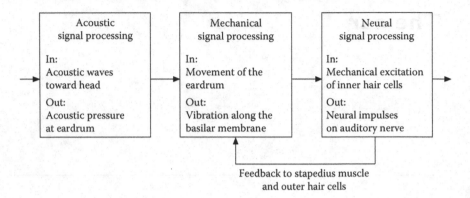

Feedback to stapedius muscle
and outer hair cells

Figure 6.1 The signal processing blocks of the ear.

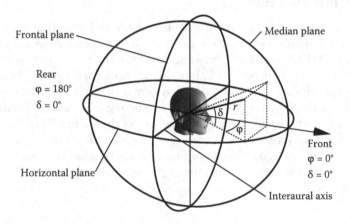

Figure 6.2 The head-related coordinate system.

acts as a reflector for one ear and a barrier for the other. The various resonances of the concha and ear canal are excited differently depending on the
angle of incidence of the sound at the ear much like an electronic antenna.

6.1.3 Head-related transfer functions

The head-related transfer functions (HRTFs) are important in describing the properties of sound at the ears as a function of the incident sound
field. Although sound is scattered by the torso, particularly its shoulders,
the head and pinnae are usually considered to provide the most salient
cues for sound localization. The HRTF is defined as the complex transfer
function $H(\omega)$ between the sound field at the ear p_{ear} and the sound field at
the position of the head in its absence p_{abs}:

$$\text{HRTF} = H(\omega) = \frac{p_{ear}(\omega)}{p_{abs}(\omega)} \qquad (6.1)$$

The HRTFs depend on the angles of arrival of the sound waves relative to the head and each angle pair, δ is a unique set of the listener's own HRTFs, and listening to those of others—using, for example, simulated sound fields in auralization, as described in Chapter 14—upsets the externalization and localization of the sounds each person is used to listen. A pair of HRTFs, HL $(r, \varphi, \delta, \omega)$ and HR $(r, \varphi, \delta, \omega)$, is used to characterize the influence of the listener's head and torso on sound waves that arrive at the listener's ears.

In practice, HRTFs are measured using a small sound source that provides a smooth spherical wave field at a radius r of 1.5–2 m from the listener's head and with miniature microphones in the ear canals or close to the ear canal openings. It can be shown that the wave field curvature at such distance and larger from a small source has little influence on measured HRTFs. Figure 6.3 shows an example of a typical setup for the measurement of HRTFs in an anechoic chamber.

Figure 6.3 Measurement of HRTFs for a manikin head in an anechoic chamber for various angles of sound field incidence. (Photo by Mendel Kleiner.)

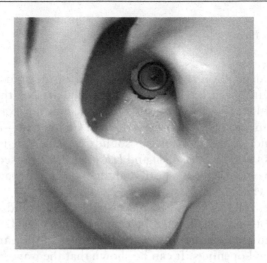

Figure 6.4 A small microphone inserted at the ear canal opening of an ear replica to sense the sound pressure in the concha. (Photo by Mendel Kleiner.)

Use of HRTFs is made complicated by a lack of agreement on the reference point at or in the ear for the measurement microphone. One good place for measurement is at the entrance of the ear canal [1]. Figure 6.4 shows a pressure-sensing microphone inserted at the ear canal entrance.

The term HRTF is often used synonymously with the head-related impulse responses (HRIRs) to which they are related over the Fourier transform. The complex transfer function $H(\omega)$ has magnitude $|H(\omega)|$ and phase $\angle H(\omega)$. The derivative of the phase with respect to frequency is the time delay τ of the signal in that frequency range:

$$\tau_g = -\frac{d\angle H(\omega)}{d\omega} \tag{6.2}$$

The frequency response function (FRF) (usually called the *frequency response*) is found from the magnitude of the HRTF and shows the relative levels of the various frequency components in an intuitive way. Since $\omega = 2\pi f$, we have

$$\text{FRF}(f) = 20\log\left(\left|H\left(2\pi f\right)\right|\right) + C \tag{6.3}$$

Here, C is a constant chosen so that the FRF is 0 dB at some suitable frequency, often 1 kHz. Figure 6.5 shows an example of the FRF for frontal sound incidence as well as for diffuse field incidence for one ear of a certain manikin head. The frequency responses are far from linear so the linear frequency response of headphones (or earbuds or in-ear phones) is therefore

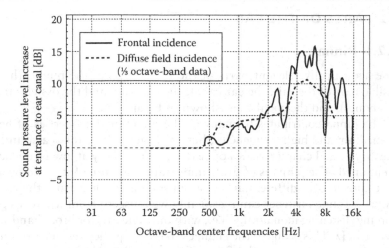

Figure 6.5 The FRF of the HRTF for an ear on a manikin head measured in an anechoic chamber for frontal sound field incidence (narrowband data) and in a reverberation chamber for diffuse field incidence (⅓-octave-band data). No torso was used with the head for this measurement; if used, it would have increased levels in the 250 Hz range. (From Kleiner, M., *Acustica*, 41, 183, 1978.)

not a good design target. Many commercially successful headphones use an equalization curve similar to that of the diffuse field shown in the figure.

6.1.4 Interaural level and time differences

The interaural level difference (ILD) and the interaural time difference (ITD) are two important quantities derived from the binaural HRTFs and are computed as follows:

$$ILD(f) = FRF_L(f) - FRF_R(f) \tag{6.4}$$

$$ITD(f) = \tau_L(f) - \tau_R(f) \tag{6.5}$$

The ILDs are strongly frequency dependent over 0.5 kHz because of the reflection and diffraction of the incoming sound waves by the head. At frequencies above about 2 kHz, the wavelength of sound starts to become small compared to the dimensions of the head and then the ILDs contribute considerably to our ability to locate sources in the horizontal plane.

At frequencies below 1 kHz, the ITDs are the main cue for our ability to localize sound. The ITDs are important at high frequencies as well since in that range hearing analyzes the time delay in the envelopes of the incoming signals.

6.2 PARTS OF THE EAR

6.2.1 Pinnae

The acoustically relevant parts of the ear consist of the pinna (the outer visible part of the ear), the ear canal, the eardrum and the ossicles (middle ear bones), and the cochlea, as shown in Figure 6.6.

The pinna contributes to many of the directional properties of hearing at medium and high frequencies, but the head and the torso also influence the directional characteristics of hearing, particularly at low frequencies as mentioned. The pinna is of great importance because of its shadow action that allows us to differentiate between sounds incident from the rear and front. A trained listener can reasonably well discern from where a signal is coming, listening monaurally (i.e., with only one ear) for broadband sounds over 1 kHz. This is due to the changes in the frequency response of the incident sound that are a result of the reflection and the diffraction by the head and torso. For sound incoming in the median plane, it is mainly the pinnae and the torso that cause the FRF variations so the localization is less accurate in this plane. A person's pinnae are not symmetrical so this may give additional cues to the direction of hearing.

The influence of the head on the FRFs is negligible at low frequencies as shown by the example in Figure 6.7 that show FRFs for various angles in the horizontal plane.

The influence of the pinnae is best seen in median plane response curves where the FRFs have been normalized to show only the influence of the angle. Figure 6.8 shows examples of such curves. The pinnae are seen to not influence the response much except over approximately 2 kHz.

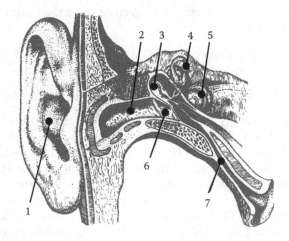

Figure 6.6 Anatomy of the ear: 1, pinna; 2, ear canal; 3, ossicles; 4, semicircular canals; 5, cochlea; 6, tympanic membrane; and 7, Eustachian tube.

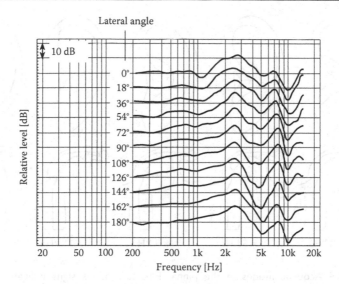

Figure 6.7 Examples of measured frequency responses of HRTFs as a function of the lateral angle φ in Figure 6.2. (After Ando, Y., *Concert Hall Acoustics*, Springer, Softcover reprint of the original, 1st edn, 1985 edn., 2011.)

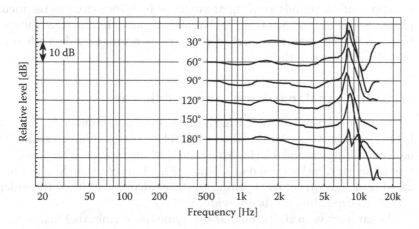

Figure 6.8 FRFs for seven different angles δ of sound incidence in the median plane (φ=0). Each curve is normalized to the response at 500 Hz for the respective angle. The median plane angle is δ in Figure 6.2. (From Kleiner, M., *Acustica*, 41, 183, 1978.)

At frequencies above 2 kHz, the level difference in sound from the rear and front directions can exceed 10 dB.

Various resonances in the pinna are excited at frequencies typically above 3 kHz and are—together with the shadow cast by the pinna—responsible for the frequency response behavior. The geometry of the resonance modes

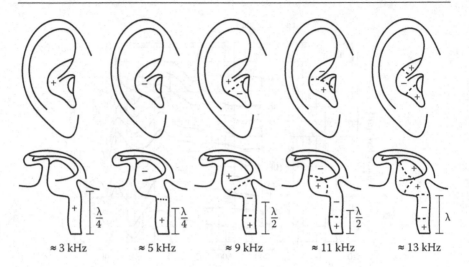

Figure 6.9 Acoustic modes in the pinna. Plus and minus signs indicate 180° phase difference. (After Shaw, E.A.G. and Teranishi, R., *J. Acoust. Soc. Am.*, 44(1), 240, 1968.)

is important for sounds arriving at angles of incidence close to the median plane. Some examples of the behavior of these modes are shown in Figure 6.9. The excitation level of the various resonances depends on the angle of incidence of the sound.

6.2.2 Ear canal, eardrum, and middle ear

The ear canal has an average length of 25 mm and diameter of 7 mm and behaves similarly to a hard-walled tube. The canal will have various resonances starting around 4 kHz. An example of the frequency response function of an ear canal is shown in Figure 6.10. The increase in SPL by the ear canal resonance (from the opening to the eardrum) can be of the order of 10–20 dB depending on the individual.

The eardrum is an elastic membrane (tympanic membrane) that separates the outer and middle ear sections. The static pressure difference between the internal and external sides of the eardrum is equalized through the Eustachian tube that forms an opening to the mouth.

The middle ear contains the ossicles, the three bones, malleus, incus, and stapes. The ossicles act as levers to transfer maximum power from the eardrum to the inner ear. The movement of the ossicles is quite linear, but at high levels, nonlinear second-order distortion is generated.

At SPLs above approximately 70 dB, the stapedius muscle in the middle ear can change its mechanical properties and give approximately 20 dB of reduction in sound transmission to act as a form of hearing protection.

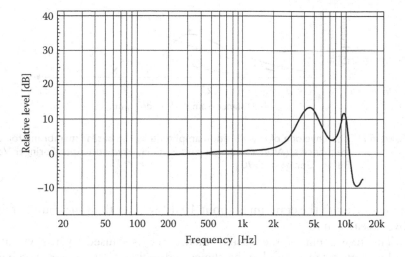

Figure 6.10 An example of the measured level difference between sound pressures at the eardrum and the entrance of the ear canal. (After Blauert, J., *Spatial Hearing*, The MIT Press, Cambridge, MA, 1996.)

This effect is called the acoustic reflex (stapedius reflex). However, because of the comparatively long reaction time required, more than 0.1 s, there is little protection against short transient sounds such as gunshots or even bursting balloons. For continuous strong sound, the reflex is passivated and no longer provides protection. The change in the mechanical properties of the stapedius muscle means that the frequency response of hearing will be time variable and possibly also that there is an optimum listening level for music.

6.2.3 Cochlea

The stapes attaches to the oval window of the cochlea, which together with the semicircular canals forms the inner ear as shown in Figure 6.6. The cochlea contains three canals: the cochlear duct and the tympanic and vestibular canals that are joined at the top end of the cochlea. The long basilar membrane, about 40 mm, separates the tympanic canal from the other canals in the cochlea. The canals contain a watery liquid that is put in motion by the vibration of the stapes for sound that arrives at the eardrum and will cause basilar membrane movement. Internally generated body sounds excite both of the cochlea canals so that body sounds are effectively short-circuited and not heard (*common-mode* rejection). Most of the vibration at the basilar membrane is due to the sound waves that cause vibration of the oval window. The bone-conducted sound that

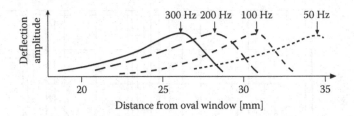

Figure 6.11 The envelope of the vibration amplitude of the basilar membrane for some frequencies. (After Zwicker, E. and Fastl, H., *Psychoacoustics*, Springer-Verlag, Berlin, Germany, 1990.)

enters through the cranium is about 40 dB weaker than the sound arriving by the ear canal.

The movement of the basilar membrane is caused by the vibrations induced at the oval window by the sound incident on the ear. The basilar membrane can be thought of as consisting of highly damped resonant sections. The sound pressure difference between the fluids in the two canals will set the membrane in motion. The resulting motion extends over a large length as illustrated in Figure 6.11, even with single-frequency excitation. The motion of the basilar membrane is, however, not frequency selective enough to explain by itself the frequency resolution of human hearing. Further signal processing takes place as the nerve signals propagate to the brain and in the brain.

The neuromechanical system block of the cochlea, shown in Figure 6.1, acts as a differential adaptive time–frequency filter bank set.

6.2.4 Hair cells

Most common sounds will have a frequency range that covers many of the filter bands. Masking experiments have shown that for engineering purposes, the auditory range can be considered as covered by 24 adjacent parallel *critical band* filters, as further discussed in Chapter 7. It is important to remember that hearing adapts the frequency band centers to the signal and the 24 bands are only an engineering approach.

The critical band filters correspond to a width of about 2 mm on the basilar membrane that holds the inner and outer hair cells. There are approximately 12,000 outer hair cells, each holding some 100–200 stereocilia (hairs). The inner hair cells number about 3500 and typically hold 50 stereocilia each. The movement of the stereocilia triggers the inner hair cells, which then send their binary signals to the various processing centers along the auditory nerve to the brain [4]. A common model for the function of the processing done by the inner hair cells is that shown in Figure 6.12 [5].

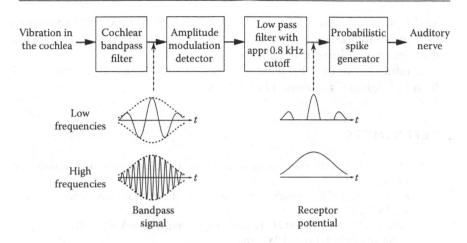

Figure 6.12 Functional, massively simplified nerve-firing model, as often used in technical applications. (From Blauert, J., *Spatial Hearing*, The MIT Press, Cambridge, MA, 1996.)

Each inner hair cell connects to more than 10 primary auditory nerve cells that emit pulses when excited by the inner hair cell. The pulses have a length of less than 1 μs and these cells cannot emit at higher rates than about 1 kHz. This means that for high frequencies, the only indication of frequency is the location of movement on the basilar membrane. The density of pulses will then follow the envelope of the vibration signal. At low frequencies, the pulses will be phase-locked to the vibration since for these frequencies, the pulse emission can take place at the rate of the vibration. For sounds that contain overtones, a large part of the basilar membrane hair cells will be active in the *analog-to-digital conversion* (ADC) of vibration to nerve signals. Pitch can thus be determined in several ways: by the place of the fundamental, by the place of the partials (overtones), and by the place difference between partials [6].

The outer hair cells play an important role as *effector* cells that are mechanically active and receive feedback from the neural system. This results in an active mechanism that amplifies the response of the ear, particularly close to the threshold of hearing. The system acts as a cochlear amplifier. One can note the active process by the phenomenon of otoacoustic emissions. If the ear is stimulated by a click, it will quickly respond by sending out sound. This forms the basis of a simple noninvasive test for hearing defects. The ear may also sometimes generate the sensation of sound without exterior stimuli, called tinnitus.

The time resolution and integration time of the cochlea's ADC will depend on the cochlea mechanics and on the feedback system mentioned, as well as on purely neural properties. Since periodic pressure fluctuations are heard as tones even down to 20 Hz and at least four cycles are necessary to form an impression of pitch, it is likely that the integration time at low frequencies is at least of the order of 200 ms [6,7].

Loudness and the threshold of hearing are discussed in Chapter 7. Binaural hearing is discussed in Chapter 8.

REFERENCES

1. Møller, H. (1992) Fundamentals of binaural technology. *Appl. Acoust.*, 36, 171–218.
2. Kleiner, M. (1978) Problems in the design and use of dummy-heads. *Acustica*, 41, 183–193.
3. Ando, Y. (2011) *Concert Hall Acoustics*. Springer, New York; Softcover reprint of the original, 1st edn, 1985 edn.
4. Zwicker, E. and Fastl, H. (1990) *Psychoacoustics*. Springer-Verlag, Berlin, Germany.
5. Blauert, J. (1996) *Spatial Hearing*. The MIT Press, Cambridge, MA.
6. Krokstad, A. (2012) *Reflections on Sound*. Music and Communication. NTNU Norwegian University of Science and Technology, Faculty of Information Technology, Mathematics and Electrical Engineering, Acoustics Research Centre, Trondheim, Norway.
7. Mohlin, P. (2011) The just audible tonality of short exponential and Gaussian pure tone bursts. *J. Acoust. Soc. Am.*, 29(6), 3827–3836.
8. Shaw, E.A.G. and Teranishi, R. (1968) Sound pressure generated in an external-ear replica and real human ears by a nearby point source. *J. Acoust. Soc. Am.*, 44(1), 240–249.

Chapter 7

Psychoacoustics

7.1 HUMAN RESPONSE TO SOUND

We are surrounded by sounds created by nature, machinery, and communication equipment, among others. These sounds cover a large frequency range, from infrasonic frequencies in the case of nature (e.g., wind, turbulence, and sea waves) to ultrasonic frequencies produced by animals and communication systems.

Psychoacoustics is the study of human response to sound. The relationship between signal stimulus and sensation—such as between signal magnitude and sensation strength—is the core of psychoacoustics.

The quality of recording and reproducing sound is of great interest because of its importance, for example, in entertainment. An outline of the factors involved in the judgment of the sound quality of a loudspeaker being listened to in a room is shown in Figure 7.1. The quality estimation and judgment will also depend on other competing information that is arriving from other sensory organs such as vision and on previous knowledge and emotive status [1,2].

An expensive brand-name label on a piece of audio equipment will make the equipment sound better if visible or otherwise known [3]. Another example is as follows: "It has been remarked that if one selects his own (loudspeaker) components, builds his own enclosure, and is convinced he has made a wise choice of design, then his own loudspeaker sounds better to him than does anyone else's loudspeaker" [4].

7.2 PROPERTIES OF SOUND SOURCES AND SIGNALS

7.2.1 Meaning

Sound waves carry information. In listening to sounds, we are affected by their meaning and also by nonaural information such as previous emotional status, whether the sound source can be seen, and cognition.

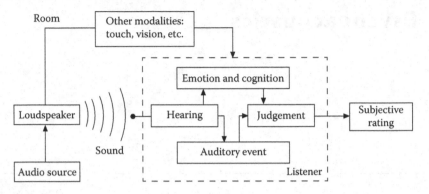

Figure 7.1 The psychophysical system in judging the sound quality by listening to a loud-speaker in a room.

Most research into psychoacoustics has been done using signals that are simple and good for use in the laboratory but lack *ecological validity*, that is, are at least reminiscent of real-world sounds met in everyday life. One scientist writes, "...under many circumstances, the perception of dynamic, ecologically valid stimuli is not predicted well by the results of many traditional stimuli experiments using static stimuli" [5,6].

An important property of sounds that have ecological validity is that they can carry information and thus meaning. The two most common examples are speech and music, but also, simpler signals such as warning sounds can carry information. The information is carried by the modulation of a carrier signal as discussed in a later section. Modulation inherently means that the signal has some frequency bandwidth.

7.2.2 Periodic and nonperiodic signals

Signals can be roughly categorized into aperiodic (nonoscillating), periodic, and quasiperiodic signals. The onset of a flue organ pipe or flute tone is created by resonance that is excited by energy provided by turbulence, that is, noise that is one form of nonperiodic signal.

A low-frequency organ pipe tone may take several hundred milliseconds to grow to its steady-state oscillation. The fundamental and the overtones will each have their own growth and decay characteristic because of the nonlinearity in the sound generation process. Because of its dependence on the turbulence supplied energy, its generation will stop instantaneously when the air fed to the flue is stopped and the acoustic energy in the air column will be attenuated quickly because of sound radiation.

A bass guitar signal, however, grows quickly because of the energy stored in the plucked string but decays comparatively slowly.

Because of the noise and limited duration of musical tones (particularly tones of wind and string instruments), they are strictly never perfectly periodic. A quasiperiodic signal is periodic over some short time but does not have the same period at some other short time, as, for example, a glissando. From the viewpoint of hearing however, such a swept signal would be heard as a sine (i.e., periodic) but with varying pitch.

In a harmonic sound the overtones have frequencies that are exact multiples of the fundamental's frequency. Piano tones are never perfectly harmonic since the propagation speed of vibration in the string varies with frequency due to the bending stiffness of the string. Drum sounds are also not harmonic because of the bending stiffness of the drum membranes and the coupling to the air inside the drum.

7.2.3 Different forms of modulation

Figure 7.2 shows the properties of some of the test signals often used for architectural acoustics and psychoacoustic research. The time variation of the amplitude of an oscillating signal is called modulation. To carry information (e.g., warning, voice, and music) and grab attention, signals need to be modulated. Amplitude, frequency, phase, etc., can all be modulated. Very slow modulation will draw little attention since the sound will be similar to a stationary sound over the attention span. Fast modulation will give the sound a rough and subjectively unpleasant character because of the presence of many tones within a critical band because of modulation. Beats between two tones are an example of modulation.

In music, amplitude modulation is known as tremolo. Naturally modulated signals are seldom modulated very quickly; the highest modulation frequencies in such musical signals reach only about 30 Hz [8]. In a frequency-modulated signal, the amplitude is constant so the envelope is constant but the spectrum is wide.

The two types of tone bursts show extreme forms of modulation: rectangular and Gaussian. The rectangular modulation resulting in the tone burst of four cycles is so short that the signal is barely heard as a tone but rather as a click. The Gaussian-modulated burst on the other hand will not have any clicks but is less well defined in time and has a spectrum quite similar to that of the rectangular tone burst. Sometimes, a raised cosine is used for modulation in place of a Gaussian envelope.

The spectrum of a sound is related to its time history by the Fourier transform. The more transient is the burst, the wider is its spectrum. Most sounds have a limited bandwidth as well as limited duration. A tone burst that has a half-length of 5 ms will have a spectral width of about 200 Hz. The product between signal bandwidth and temporal resolution is approximately constant.

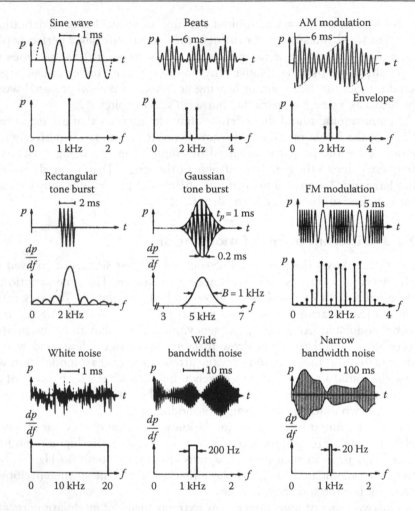

Figure 7.2 An overview of properties of some test signals used in engineering and for scientific research. Continuous sounds such as the sine tone, with beats or AM or FM modulation has line spectra. Transient sounds such as short tone bursts have continuous spectra as has noise. (After Zwicker, E. and Fastl, H., *Psychoacoustics*, Springer-Verlag, Berlin, Germany, 1990.)

Transient signals have continuous spectra, while periodic signals have discrete spectra, that is, one will be able to note one or more tones of different frequencies. A pure tone has a simple, single line spectrum, but most tones will show some spectrum width because of modulation, for example, due to musical instrument characteristics. To observe a tone's bandwidth correctly, the tone must be long enough for the analyzing instrument to resolve its spectral content. Usually, hearing analyzes tones with different time resolutions at different frequencies. This will put a limit to the temporal resolution of modulation that one can expect from hearing.

From the viewpoint of signal processing, hearing can be thought of as a short-time real-time frequency analyzer featuring correlation, pattern recognition, and memory as hinted at in Figure 8.2.

Separation of signals in time and frequency is essential to hearing, but hearing does not act as a linear sensor and analyzer combination in the Fourier sense. Due to the nonlinear sensing of the movement of the basilar membrane by the hair cells, the digital characteristics of the nerve signals, and the cross coupling and analysis by various centers and the brain, we listen in a more advanced way than we can measure using current physical devices and software.

7.2.4 Voice and music as test signals

Voice and music differ physically from common psychoacoustic test signals by their wide frequency ranges and modulation patterns. The harmonics may extend beyond the high-frequency limit of hearing. Usually, the harmonics are weaker than the fundamental.

Many musical instruments feature sounds that have fast growth and decay characteristics (turn-on and turn-off transients). In electrical engineering, the rise time of a signal is usually defined as the time it takes for the signal envelope to rise from 10% to 90% of its steady-state value. The decay time is often defined similarly or by a reverberation time. The growth characteristics are very important for the detection of the type of musical instrument that is listened to. Many instruments have similar continuous overtone spectra and timbre, but the growth characteristics are different that allow us to identify them. Figure 7.3 shows the behavior of the envelope level for some groups of instruments.

Table 7.1 shows some measured rise and decay time characteristics [9,10]. Note that these are average values. Fundamentals and harmonics will develop differently. Woodwind and brass instruments typically have about the same growth and decay times, whereas string instruments and drums tend to have much longer decay than growth times because of their larger moving mass and lower damping. Figure 7.4 shows the time evolution of the tonal pattern of a piano tone [9].

From the viewpoint of signal processing and acoustics, music and speech can be considered as wide bandwidth signals modulated by low-frequency signals. The modulation describes the amplitude variation of the multi-frequency signal's envelope at the signal's various carrier frequencies. The modulation spectra will be different at different frequencies of the carrier spectrum. The reverberation in rooms will cause the modulation depth to

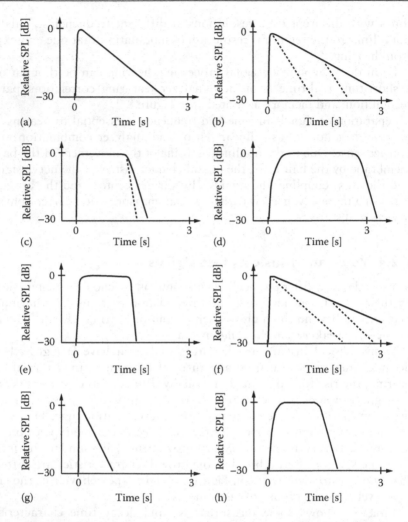

Figure 7.3 The growth and decay characteristics of the sound of some musical instruments: (a) plucked string instruments; (b) struck string instruments; (c) bowed string instruments; (d) flue organ pipe; (e) air, mechanical, and lip reed instruments; (f) percussion instruments of definite pitch; (g) drums; and (h) voice vowel sounds. Dotted lines indicate decay with added damping by musician.

diminish, and features of the signal will increasingly be blanked out as reverberation time becomes longer. (Noise will also lead to reduced modulation but is not considered here.)

Hearing, because of its time- and frequency-domain masking properties, does not treat modulation the way a simple linear physical instrument would. In particular, our hearing is much more sensitive to the rise

Table 7.1 Rise and decay times of some musical instruments based on data available in References 9,10

Instrument	Rise time [ms]	Decay time [s]
Voice	20–60	
Pipe organ	10–200	
Bass drum		4–15
Harpsichord	10–75	
Double bass	100–120	3–10 (adjustable)
Cello	60–100	1–8 (adjustable)
Violin	40–300	0.1–3 (adjustable)
Piano	10–30	0.6–3 (adjustable)
Trumpet	10–30	
Trombone	20–70	
Tuba	25–60	
Oboe	10–40	
Bassoon	20–60	
Clarinet	15–50	
Bass guitar	7–20	

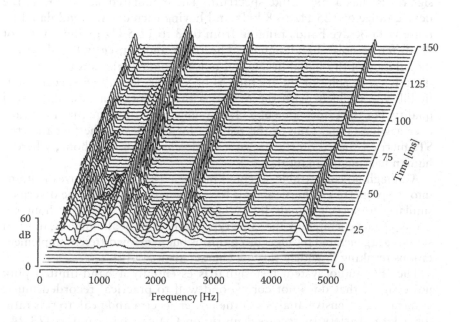

Figure 7.4 Time evolution of the tonal spectrum for a grand piano C$_6$ note being played. (From Meyer, J., *Acoustics and the Performance of Music*, Springer Science and Business Media, New York, 2009.)

of modulation envelope amplitude than to its fall. This is to be expected since hearing is developed not only for communication but also to warn us danger.

In contrast to musical instruments that usually produce harmonic or transient sounds, voice, to a large extent, generates transient or random sounds. When we speak or sing, the air from the lungs is modulated by the vocal cords. Additionally, the tongue and mouth form an acoustic filter that changes the overtone content of the vocalized sound. The lips and teeth form a barrier to the air from the lungs so the flow is cut off that causes puffs and vortices that generate transients and noise and forms consonants. The modulation results in a modulation frequency of speech that is centered on 5 Hz. Natural speech is, in many cases, an excellent test signal for audio systems since we are so trained in listening to speech. However, it is possible to fool hearing; speech signals that have been compressed using perceptual coding may still sound reasonably good in spite of the considerable information reduction.

It has been shown that—from the viewpoint of measuring the speech intelligibility in a room—speech may be well emulated by a sum of modulated noise band signals. This approach is used in the determination of the *speech transmission index* (STI) [11]. The STI is based on the evaluation of the deterioration of modulation of a wideband noise carrier signal that has a speechlike spectrum. The modulated signal covers the octave range of 125 Hz to 8 kHz and having each carrier modulated by noise in ⅓-octave bands ranging from 0.63 to 12.5 Hz giving a total of 98 different noise band signals. The room and source will reduce the modulation of the signal because of noise, room reverberation, and voice directivity. In spite of the usefulness of STI for describing speech intelligibility, a problem with the metric is that it does not describe sound quality. For the same value of STI, the subjective sound quality may vary over a wide range. One of the reasons for this lies in the fact that the STI metric is based on the use of a single-channel microphone, whereas human hearing is binaural.

Although no comparable accepted metric exists for music investigations into the properties of music signals have shown that these can be described similarly to speech signals [8]. Whereas speech is a relatively homogeneous signal generated by the same type mechanism in all humans, musical sounds are generated by a few mechanisms but in many different implementations resulting in widely varying characteristics [9,12].

The IEC audio system test signal is essentially a band-limited pink noise signal that does not correspond well to practical recorded music spectra. An extensive analysis of the power spectra and peak to rms ratio for a large variety of recorded music on CD can be found in [23,24]. Figure 7.5a shows the ⅓-octave-band spectrum measure of an average

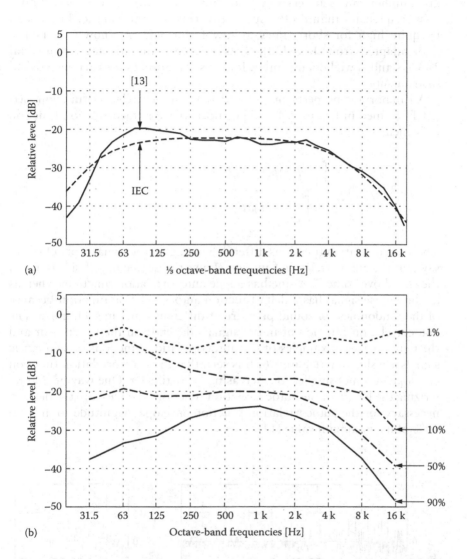

Figure 7.5 (a) Long time ⅓ octave music spectra for 13 CDs normalized to 0.5 kHz. (After Farina, A., A study of hearing damage caused by personal MP3 players, in *Proceedings of 123rd Audio Engineering Society Convention*, New York, Paper 7283, 2007.) (b) Octave-band levels exceeded for time shown in the figure. Longtime music spectra for the Telarc cannons in the Tchaikovsky, *1812* Overture TELARC CD-80041, track 1 14:42.

of 13 pieces of music recorded on CD [13]. Note that the IEC spectrum gives higher values at very low frequencies but lower values for the mid-bass frequencies than do the spectra of this recorded music. Levels can be quite high for short times at low frequencies as shown by Figure 7.5b adapted from Ref. [24]. This reference investigates commercial 24 CDs and is well documented. It also shows ranges for various musical instruments.

An important property of a signal is its autocorrelation function $\varphi(\tau)$ (ACF) defined in Equation 1.65. The normalized autocorrelation function is defined as

$$\text{ACF}(\tau) = \lim_{T \to \infty} \frac{\displaystyle\int_{-T/2}^{T/2} p(t)p(t+\tau)dt}{\displaystyle\int_{-T/2}^{T/2} p^2(t)dt} \tag{7.1}$$

The autocorrelation functions differ depending on the signal type. One can say that the autocorrelation function shows the degree of self-similarity in the signal over time T. A sine has a sine autocorrelation function, whereas a white noise signal has a delta function autocorrelation function because of the randomness of sound pressure at different times in such noise. The autocorrelation function of music signals depends on the instrument and the music played. Examples of the autocorrelation functions of two music signals are shown in Figure 7.6. It is seen that the autocorrelation function envelope becomes smaller for increasing delay times τ. One way of characterizing the length of the autocorrelation function is to measure the delay τ_e necessary for the autocorrelation function envelope magnitude to drop to 10% of its maximum value [14].

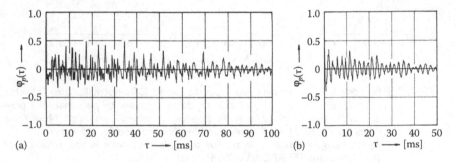

Figure 7.6 Examples of the ACF of music signals: (a) Royal Pavane, $\tau_e = 127$ ms and (b) Sinfonietta, $\tau_e = 43$ ms. (From Ando, Y., *Concert Hall Acoustics*, Springer; Softcover reprint of the original 1st edn., 1985 edn., 2011.)

7.2.5 Source center

Natural sound sources of interest in audio are musical instruments and voice. These are large so their acoustic centers are difficult to describe. Voice is radiated not only by the mouth but also by the nose, cheeks, and chest. The head and torso cause diffraction and the impulse response will be different in various directions. This causes the voice sound spectra to be angle dependent. Similarly, musical instruments tend to radiate sound from many parts, for example, for violin, from the strings, body, and f-holes. When we listen to the sound generated by a musical instrument the location of its auditory event will be blurred for this reason.

7.3 SIGNALS AND HEARING

7.3.1 Signal length and audibility

Hearing needs about five cycles of a sine to generate the perception of a tone [15]. The best way to reproduce such short tone bursts is with electrostatic loudspeakers, headphones, or earphones.

However, even if the tone stimulus gives a tone sensation, the auditory event will not have the sensation magnitude it could have since the sensation magnitude depends on the time of the excitation. The dashed curve in Figure 7.7 shows results for relative sensation strength as loudness level (LL)

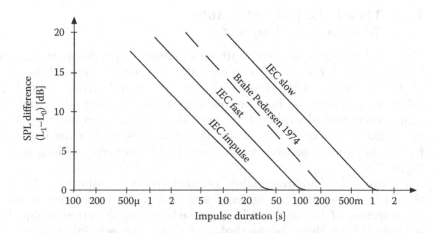

Figure 7.7 The dashed curve shows the subjectively determined level differences between 1 s exposure and shorter tone burst exposure for different tone burst lengths (at 1 kHz). The curves marked IEC slow, IEC fast, and IEC impulse correspond to the integration characteristics of sound level meters. (From Brüel & Kjær, *Acoustic Noise Measurement*, Brüel & Kjær, Nærum, Denmark, 1988.)

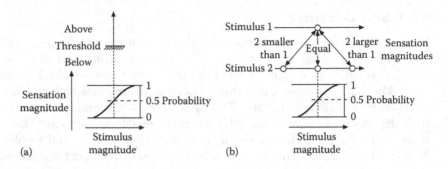

Figure 7.8 Determination of absolute threshold (a) and of equality (b).

obtained from research using 1 kHz tone bursts. A conservative estimate shows that at least a duration of 0.5 s is needed, that is, about 500 cycles at this frequency. The required duration at lower frequencies is also of this order [15].

Also shown in Figure 7.7 are the sound level meter characteristics, IEC impulse, IEC fast, and IEC slow. One notes that even the IEC impulse setting does not allow measurement of the peak values of short transient sounds. To determine the peaks, it is best to use a digital sampler or oscilloscope.

7.3.2 Threshold, just noticeable difference, and equality

The absolute threshold of sensation is defined in psychometrics as the level at which the existence of a stimulus becomes recognized. Because of the noise in our sensory systems, particularly in hearing (physiological noise such as blood flow, stomach grumbles, ambient noise, and tinnitus), there will always be some spread in the decision making. Also, the decision will usually be different if one tries to find the threshold from below (quiet) or from above (heard). This concept is illustrated in Figure 7.8a.

The principle of finding equality is related to that of finding the difference threshold. Two sounds are compared and there will be a certain probability for detection of the stimulus level of each, so the detection of equality is more difficult than the threshold detection. The principle is shown in Figure 7.8b. In the determination of these thresholds, a 50% probability level is usually chosen.

Another important idea is that of just noticeable difference (JND) or differential threshold. The JND is defined as the smallest detectable difference between a two paired sensory stimuli.

Experiments have shown that once the signal is detected and has sufficient intensity, a relative percentage change is perceived almost equally strong irrespective of intensity; this is called Weber–Fechner's law in psychology and can be written as

$$k = \frac{\Delta I}{I} \qquad (7.2)$$

where
k is the Weber constant
I is the physical intensity such as the sound intensity
ΔI is the intensity difference

It is this relationship that leads to the sensitivity characteristics of hearing and the suitability of the logarithmic level relationship to describe the strength of sounds using the decibel [dB] unit.

7.3.3 Loudness level

The two most important psychoacoustic dimensions of continuous unmodulated sounds are loudness and pitch, but for modulated sounds, there are also others such as roughness and fluctuation. For very broadband sounds, there are sharpness, timbre, etc.

The threshold of audibility (hearing threshold or minimum audible field [MAF]) is defined as the minimum perceptible free field SPL that can be detected at any frequency over the frequency range of hearing. The threshold of audibility can be determined in various ways using, for example, tones and band-limited noise.

It is difficult to measure the threshold correctly at low frequencies because of interfering low-frequency sounds and loudspeaker nonlinear distortion. As discussed previously, the tone must have minimum duration. In pure tone audiometry, a test tone in the frequency range 250 Hz to 4 kHz is used and sounded 1–3 s.

The threshold of audibility according to the ISO 226:2003 standard is shown in Figure 7.9 as the dotted curve marked MAF [26]. The curves were obtained with the listeners facing a pure tone plane wave in an anechoic chamber, and the sound levels were measured at the listeners' head position without the listener present. The listeners were young adults having normal hearing.

By the loudness level (LL) of a sound, it is meant that the SPL of a reference tone at 1 kHz is perceived as equally loud as the test sound. The LL is expressed in the unit phon. Figure 7.9 also shows some equal LL contours measured for pure tones. For these, the test persons were asked to determine when the test tone at some frequency was equally loud as the

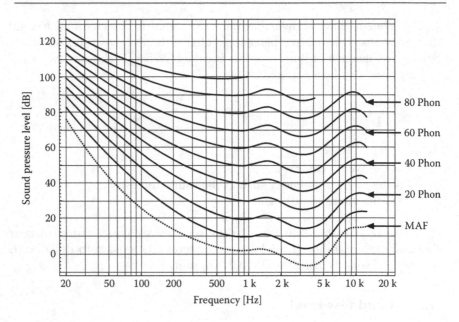

Figure 7.9 Equal loudness level contours. The curve marked MAF illustrates the minimum audible field sound pressure level for pure tones at the respective frequencies. (After ISO 226:2003, Acoustics -Normal equal-loudness-level contours, International Organization for Standardization Geneva, Switzerland.)

reference tone. Other curves would be obtained, for example, for band-limited noise or for sound coming from the side.

A trained listener has a threshold for just noticeable LL differences in SPL of about 1 dB for pure tones and may sense down to about 0.3 dB deviations in ⅓ octave bands in broadband spectra. For SPLs of about 40 dB or larger, a doubling of sound pressure at 1 kHz, for example, corresponds to an increase in LL of about 6 phon. At SPLs above 120 dB, the sound will be so strong that has a sense of feeling the sound, the threshold of feeling. The threshold of pain is often set at about 140 dB, but many persons, particularly those that have hearing impairments, may start to feel pain at much lower SPLs.

The equal LL curves are very frequency dependent at below 30 phon, particularly at low frequencies and low SPLs. This is of great significance in the design of audio equipment since it makes it virtually impossible to have correct timbre in sound reproduction unless one listens at the same LL as was present at the actual recording. Only by using an adaptive frequency response equalization system that senses the LL (and has knowledge about the original level) can one have approximately the right timbre in sound reproduction. It is impossible to achieve this control by using what is known as loudness compensation in conventional amplifiers. A sound reproduction system playing louder than another usually appears the better in comparison, having seemingly stronger base and treble response. Because of this, it is always important to use the correct presentation sound levels in listening tests.

The LL of a sound will depend also on the angle of incidence and the duration of the sound, properties that are not included in the standard. Conventional acoustic noise measurements use omnidirectional microphones sensing the sound equally well independent of the angle of incidence of the sound.

7.3.4 Loudness

An alternate way of measuring the subjective strength of a sound is to measure using a metric that has the property that an increase in loudness corresponds to a proportional increase in the metric value. Such a metric is Loudness that is expressed in units of sone. By definition, 1 sone corresponds to a 40 phon, 1 kHz, reference tone. The loudness value is equal to how many times stronger, or weaker, a sound is compared to this reference tone. Loudness can be measured subjectively using a panel of listeners judging the loudness or by analog or digital means as with LL. The approximate relationship between Loudness and LL at 1 kHz is shown by the curve in Figure 7.10.

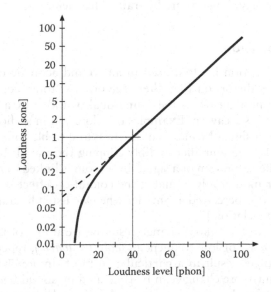

Figure 7.10 The relationship between Loudness and LL at a frequency of 1 kHz. (After Zwicker, E. and Fastl, H., *Psychoacoustics*, Springer-Verlag, Berlin, Germany, 1990.)

7.3.5 Timbre

Other subjective dimensions of sound, besides loudness, were mentioned initially. Examples of such dimensions are the following:

- Timbre—the subjective perception of spectral content (frequency and balance between various parts in the spectrum)
- Pitch—the subjectively perceived frequency of a tone
- Duration—the subjective duration of a sound
- Temporal variation in time of sounds such as modulation and fluctuation

Timbre is defined as the attribute of sound that makes two sounds having the same loudness and pitch dissimilar [7]. The timbre of a sound that has a wide spectrum depends on its relative spectral balance. It is common to describe timbre as *tonal color*. One definition of tonal color is its being that attribute by which a listener can judge that two equally loud sounds similarly presented are dissimilar [7]. By this definition, color includes timbre, other linear distortion, pitch, and added spectral components such as nonlinear distortions.

In audio equipment, various types of frequency-selective filters are used to change the timbre of sounds, such as bass and treble filters. A change of level as small as about 0.1 dB, in one critical band, may be noted in the case of wide frequency range sound by trained listeners.

7.3.6 Coloration

The term coloration is often used in audio and acoustic engineering and means audible distortion that alters the original tonal color. An example of coloration in acoustics is the sound quality added by a small reverberant room such as a cavern. Examples of coloration in audio are the change introduced in a signal by means of the bass and treble controls of an audio amplifier or by the amplifier audibly clipping the signal. In acoustics, the strength of the coloration of a signal by a room is linked to how one listens, binaurally or monaurally. Usually, the coloration effect of a room on the acoustic signal is perceived as smaller when we listen binaurally because of binaural decorrelation [7].

In contrast to the timbre change resulting from loss of high frequencies (*dull vs. brilliant*) or low frequencies (*cold vs. warm*), repetitive frequency response changes result in a particular form of timbre change called coloration. Such timbre change can be generated by sound field contributions that are correlated, somewhat time delayed, and coming from the same direction. This can be the case, for example, when one has two highly specularly reflective parallel walls dominating the sound field in a room.

The filter frequency responses are said to have comb filter properties because of the graphic appearance of the filter's spectral magnitude characteristics.

7.3.7 Pitch

Pitch is defined as the auditory attribute of sound according to which sounds can be ordered on a scale from low to high. Pitch is influenced not only by the frequency of the sound but also by the intensity, the duration, and the spectral content, and the presence of other, harmonically related, tones also influences pitch.

For hearing to detect pitch in a sine tone, the tone needs to be at least five cycles long. Five cycles of a 50 Hz signal has a length of 100 ms. Because modulated signals have periods that vary in amplitude, it will thus take longer to detect the pitch of such signal.

Pitch is expressed in units of mel. A reference sine wave, having a frequency of 1 kHz, is defined as having a pitch of 1000 mel. One measures pitch by subjective comparison to this tone. Pitch does not vary linearly with frequency, as shown by the curve in Figure 7.11. The curve shows the pitch for various frequency pure sine tones heard at a 60 phon LL.

Tones that consist of many sines, such as those produced by musical instruments, will still cause the perception of a single pitch if the frequencies of sines are in a harmonic overtone pattern.

7.3.8 Virtual pitch

If we consider a tonal spectrum that consists of a fundamental and a series of overtones, its pitch is fundamentally a result of an estimate of the pitch of the fundamental. If some of the overtones are missing, the sound will still have the same pitch but different timbre.

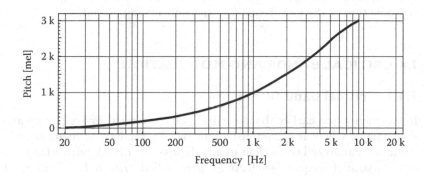

Figure 7.11 Pitch as a function of frequency (LL = 60 phon). (After Zwicker, E. and Fastl, H., *Psychoacoustics*, Springer-Verlag, Berlin, Germany, 1990.)

Band-limited noise signals also have pitch. Consider a noise that has been filtered into narrow frequency bands by a process that results in a repetitive tonal spectrum, such as a comb filter, discussed in Chapter 8. The noise will then be given a pitch that is the result of these spectral characteristics.

Now assume a sound composed of a harmonic series of tones, for example, 200 Hz plus its first ten harmonics. In listening to such a tone, one finds that the elimination of the fundamental, that is, 200 Hz, does not change the pitch, although the timbre of the sound with the missing fundamental is different from that of the full sound.

This phenomenon is called virtual pitch, sometimes also referred to as missing fundamental. The effect is of great importance in audio engineering, since it allows even a loudspeaker with poor low-frequency reproduction to have an acceptable low-frequency response for many sounds.

Most musical instruments will generate a fundamental tone together with overtones. Each of the tones will have a different envelope that may start at about the same time (as in electric guitars) or at different times (as in the case of organ tones). If the source–receiver path has poor response at some frequency, then hearing may compensate for that using information available from the overtones. This is likely to be one reason why listeners do not hear the influence of location in small rooms at low frequencies. The overtones of the instruments fill in so that the exact location relative to a dominant low-frequency mode is not essential for hearing bass. Even with dropouts at certain frequencies, the bass tone will still be perceived as present.

Commercial techniques to create bass impression without actual bass sound reproduction have been devised. In principle, these techniques generate distortion products that can be reproduced by the loudspeaker or headphone (while the original bass note cannot). Hearing then imitates the bass note using the sound having a missing fundamental to generate the virtual pitch, making the listener believe that there is full frequency sound reproduction of the low-frequency bass notes being played.

7.4 CRITICAL BANDS AND MODULATION

7.4.1 Critical band filters

It was mentioned earlier that hearing is characterized by frequency analysis, some of which is accomplished already by the basilar membrane. The stimulus is analyzed equivalent to an adaptive filter bank with a large number of parallel frequency-selective filters called *critical band filters*. The audio frequency range can be subdivided into 32 critical band filters, each having its own critical bandwidth.

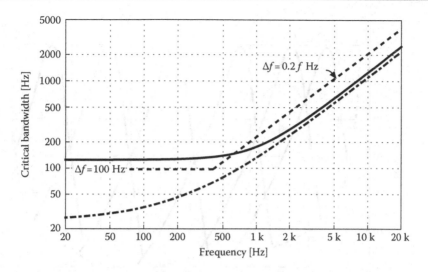

Figure 7.12 The solid curve shows the measured critical bandwidth as a function of frequency. The broken line marked 0.2*f* shows the ⅓-octave bandwidth as a function of center frequency. The dash-dotted line shows the bandwidth of ERB filters. (After Zwicker, E. and Fastl, H., *Psychoacoustics*, Springer-Verlag, Berlin, Germany, 1990.)

The critical bandwidth depends on the frequency as shown in Figure 7.12. Up to about 0.5 kHz, the critical bandwidth is fairly constant; about 100 Hz, but at higher frequencies, the bandwidth is about that of a third-octave-band wide bandpass filter. At a frequency of 1 kHz, the critical bandwidth is about 160 Hz.

There are several ways of measuring the critical bandwidth; these give slightly different results. A common way to measure the critical bandwidth is to use noise having a constant spectral density, increasing the bandwidth of the noise only, and to investigate the masking of an added pure tone. Reaching a certain bandwidth, the masking no longer increases as the bandwidth is increased. This bandwidth is the critical bandwidth. Once the bandwidth is known, one can assemble a bank of critical-band rate (CBR) filters numbered in units of Bark. Figure 7.13 shows the frequency responses of some CBR filters.

At frequencies below about 0.4 kHz, the CBR filters are about 0.1 kHz wide, and above about 1 kHz, the bandwidth may be reasonably approximated by 0.2*f*, where *f* is the filter center frequency. It is important to note that the filter function is dynamic in that hearing automatically focuses on the frequencies of interest in a sound. The CBR filters are only a practical physical implementation to provide a set of filters that cover the audio range. The auditory filters have dynamic center frequencies and do not have the flat bandpass response of the classical filters of electrical engineering.

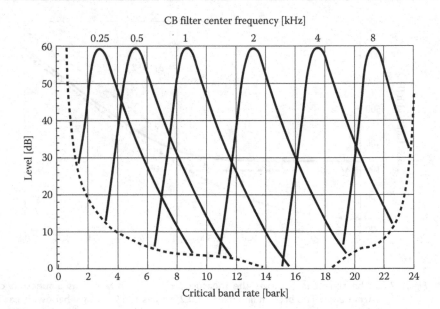

Figure 7.13 The frequency responses of some CBR filters. (From Zwicker, E. and Fastl, H., *Psychoacoustics*, Springer-Verlag, Berlin, Germany, 1990.)

Other filter functions that may be more similar to those of the ear are gammatone, ear-adequate, and wavelet-based filters. These filters are used with a different scale that is called the equivalent rectangular bandwidth (ERB) scale and is shown in Figure 7.12 [17,25]. This scale has a logarithmic behavior over a wider frequency range than the CBR filters. The width of a gammatone filter band is typically 11%–17% of the band's center frequency. Typically, a bank of such filters will use some 40 filters in contrast to the 24 of the CBR filters.

7.4.2 Start and stop transients

Consider a sine modulated by a square wave, generating tone bursts. When the modulation is fast or abrupt, hearing will detect the high-frequency content of the start and stop transients as clicks at the beginning and end of the burst.

This can be understood by studying the output response of a tone burst in a set of parallel bandpass filters, for example, a CBR filter bank. The start and stop transients appear in all the critical bands if they are due to sharp changes in the time function of the sound.

Since the out-of-band transients are attention grabbing, it is important when doing listening test to avoid fast switching and use gradual loudness increase and decrease in the test signals. Often, rise and fall times of about 20 ms or more are quite sufficient to avoid switching clicks or splatter.

7.4.3 Roughness and fluctuations

Various forms of modulation were discussed at the beginning of this chapter. Hearing is quite sensitive to amplitude or frequency modulation of a tone. The sensitivity to modulation varies with the modulation frequency and type. Modulation generates sidebands, which are frequency components that are close to the frequency of the modulated signal (usually called the carrier frequency) as discussed earlier in this chapter; see Figure 7.2. For a tone AM modulation generates fewer sidebands than does FM or phase modulation. A 100% tone modulated AM signal has only the two sideband tones and no carrier tone.

Signals that have more than one frequency component within the bandwidth of a critical band will beat, and their sum sound will fluctuate or sound rough. The term fluctuation is used for modulation frequencies below 15 Hz. The maximum sensitivity for fluctuation is found for frequencies of about 4 Hz. We can sense steady-state frequency deviations of 3 Hz or more of a single sine tone at frequencies below approximately 0.5 kHz and of about 0.2%–0.3% above this frequency.

Musicians use the fluctuation effect in tuning instruments. Using the method of beats, we listen to two tones having frequencies f_1 and f_2 simultaneously. For tones very close in frequency, we hear the combined sound as a *beating* tone. They will subjectively combine into one tone having a frequency $(f_1 + f_2)/2$, modulated by the difference in frequency between the two tones $|f_1 - f_2|$ if $|f_1 - f_2| < 15$ Hz.

For modulation frequencies in the range 15–300 Hz, the amplitude modulation subjectively changes character of the sine tone, and its modulation is more perceived as roughness. The roughness sensitivity depends on the critical bandwidth. For a sine tone at 1 kHz, the maximum roughness is heard for modulation frequencies of about 70 Hz, that is, when the sideband tone components generated by the amplitude modulation are at the frequency band edges of the critical band [1]. Once the frequencies of the sine tones are in separate critical bands, they will be heard individually.

The metric for roughness is called Roughness and is measured in units of asper. Figure 7.14 shows how Roughness for various carrier frequencies depends on modulation frequency.

We note that a sound that has been subject to comb filtering such as that shown in Figure 8.13 and even more pronounced in Figure 8.14 may have two frequency response peaks in a critical band; it generates considerable

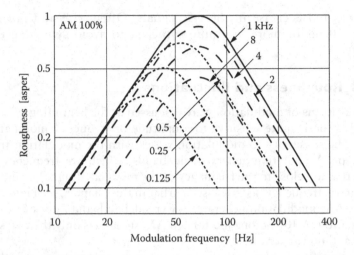

Figure 7.14 Roughness of beats (100% AM-modulated sine carrier). Frequency curves in graph show characteristics for various carrier frequencies in kHz. (After Zwicker, E. and Fastl, H., *Psychoacoustics*, Springer-Verlag, Berlin, Germany, 1990.)

roughness. When the modulated spectrum is wide such as in music or speech and the room impulse response is periodic, there will be roughness generated in many frequency bands.

7.5 MASKING

7.5.1 Spectral masking

By masking, it is meant that one sound, the masker, blocks the perception of another sound, the maskee. Masking can take place both in the time and frequency domains. The masking level is the level by which one sound needs to be increased to still be heard in the presence of a masker. Simplified, one can regard the masking level as the shift of hearing threshold in the presence of another sound, for example, a tone or a noise.

Figure 7.15 shows an example of spectral masking. It shows the shifted hearing thresholds of the maskee measured with a tone masker having a frequency of 1 kHz and different SPLs. The masking is largest when the masker and maskee are close to one another in frequency. The masker's effect is mostly active upward in frequency.

The masking effects are important in sound reproduction. Since the frequency response may lead to attenuation or amplification of low-frequency sounds relative to the midfrequency average, the actual SPL of low-frequency components is particularly important for how they will be heard.

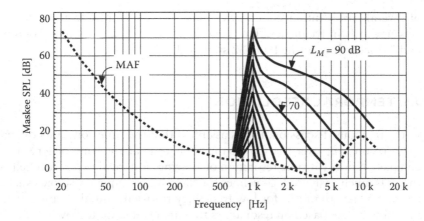

Figure 7.15 Hearing threshold shift of a test tone (maskee) due to masking. The masker is a 1 kHz tone at different SPLs L_M. Solid curves show the resulting hearing threshold of the maskee in the presence of the masker. The dashed line shows the threshold of hearing for tones (MAF). (After Zwicker, E. and Fastl, H., *Psychoacoustics*, Springer-Verlag, Berlin, Germany, 1990.)

Too much attenuation and the tone will not be heard because its SPL will be below the threshold of hearing. Too much amplification and the balance between low and mid/high frequencies will be incorrect because of the timbre change. If the SPL and frequency response on playback are set differently from at the control room, masking will be different from that in the recording studio's control room. The resulting sound will not be that intended by the producer. In practice, of course, the domestic loudspeakers are likely to have different frequency responses from those of the control room so the timbre will be different in any case.

7.5.2 Detection of signals in noise

The detection of signals in noise depends on many factors, such as cognition, binaural sound field properties, and signal properties.

In a simple case, such as the detection of a continuous sine surrounded by a band-limited noise, the noise needs to have power inside the critical band that surrounds the tone for the noise to mask the tone. Psychoacoustic experiments have shown that for a tone to be heard, its power, approximately, has to exceed that of the noise within the critical bandwidth of the tone frequency. As an approximation, one may assume that for a noise to well mask a tone, the noise needs about 3–5 dB more power in the critical band than the tone, depending on frequency [1].

In most practical situations, the masker is a noise-like signal with some bandwidth. Wide band noise will have energy over a bandwidth much wider than the critical bandwidth. It is then necessary to take the masking of several bands into account, since masking is asymmetric and primarily acts upwards in frequency. This is done in the determination of Loudness and LL according to the ISO 532 standard [27].

7.6 TEMPORAL RESOLUTION

In the previous discussion, it was assumed that the time constants of hearing are constant over the CBR bands. There is at the time of writing no complete model for time resolution of hearing. Practical experience tells us that it is much easier to hear echoes for short high-frequency sounds such as trumpet blasts than it is for low-frequency transient sound. Many musical instruments are based on resonances that are excited quite slowly (e.g., flue pipe organ tones that may have rise times of about a hundred or more milliseconds), other instruments such as bass and other guitars have rise times that are very short, sometimes only a few milliseconds. The relationship between bandwidth and time resolution in physical systems was mentioned in the beginning of the chapter. The temporal resolution becomes better as frequency is increased [18].

In modeling hearing using CBR filters or ERB filters, the signals are often treated to resemble the half-wave rectification done by the hair cells (as indicated in Figure 6.12), and then a sliding integrating window is applied to simulate the time resolution and temporal smoothing of hearing. An example of such a model is the HUTear model [19].

7.7 AUDITORY OBJECTS, RESTORATION, AND CONTINUITY

7.7.1 Perceptual restoration

Perceptual restoration can be experienced in various ways, for example, as spectral and temporal restoration or induction such as homophonic induction and illusory continuity.

By homophonic induction, it means the following effect: assume a tone that changes level from weak to strong within a short time. The sensation magnitude will then be that of the weak start tone.

Illusory continuity can be noticed as follows. Assume that a tone is subdivided into repeated tone bursts less than 0.3 s long with a 0.3 s break of silence in between. The sensation will be that of a tone with silent breaks. Now fill the silent gaps with noise, centered at the frequency of the tone, as shown in Figure 7.16. If the bursts (both tone and noise) are shorter than 0.3 s, the tone will appear to lose its breaks and instead go on continuously. The interrupts must be shorter than 0.3 s since otherwise illusory tonal

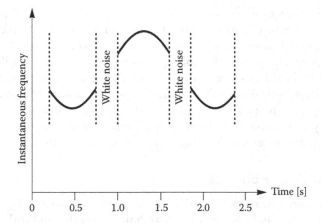

Figure 7.16 Illusory continuity of a slow frequency-modulated gliding tone. (After Bregman, A.S., *Auditory Scene Analysis: The Perceptual Organization of Sound*, Bradford, 1994.)

continuity cannot be kept. For dynamic signals such as tonal glides, the tone length can be much longer and the breaks still fill in to generate the sensation of continuity.

7.7.2 Continuity

The continuity effect also works on speech, musical notes, and melodies. It even works if two alternating noises have different bandwidths. The effect is also known as the picket fence effect and is similar to the closure/ belongingness effect in vision that is a grouping effect shown in Figure 7.17.

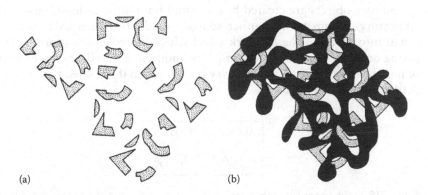

Figure 7.17 An example of the closure/belongingness effect in vision. (a) Open and (b) closed. (From Bregman, A.S., *Auditory Scene Analysis: The Perceptual Organization of Sound*, Bradford, 1994.)

The effects require the maskee (the sound that will be masked) to be known or have a recognizable pattern and that the masker and the maskee have the same assigned location. If the two sounds can be separated laterally, the sounds will be separable, that is, the noise pulses that were added to fill the silence gaps will be heard as separate events from the tone or other sound that was unmasked, the so-called *binaural release* from the illusory continuity [20,21].

Illusory continuity may be an important effect when listening to low-frequency sounds because the stimulus levels may be close to the threshold of hearing and reverberation might contribute to the effect.

7.7.3 Auditory objects

Perceptual organization is familiar from vision as shown in Figure 7.18. In audition, perceptual organization may hold spectral, temporal, and spatial meaning. If the hearing sensation has characteristic spectrum, sufficient strength, and duration, we may be able to assign it the status of an auditory object or event and perhaps even a place in space [20,21]. Of course, an echo—after a sufficient time—is a separate auditory object from the original sound event.

Grouping of sound by streaming, in analogy to the letters shown in Figure 7.18, is one way for sounds to obtain *Gestalt* (objecthood) or auditory event status. Binaural segregation is another way as are the spectral characteristics over time of sounds, such as in the case of a musical note. A short overview of hearing music in ensembles is given in Reference 28.

Auditory icons, such as specific warning sounds, may take on the character of objects, since they can be grouped into a category of their own. An example of such an icon is the buzzing or rattling of a loudspeaker when overdriven. From experience, we may know that this is a sign, an icon, of danger to the equipment.

Auditory objects are created by our mind but must—at least outside the laboratory—compete with other sensory modalities such as haptic and visual information. The McGurk effect where lip movements form a competing auditory event to that formed by audition is a well-known example of how hearing takes on a secondary role to vision if the senses are in competition for attention [22].

AI CSAITT STIOTOS

$A_I C_S A_I T_T S_T I_O T_O S$

Figure 7.18 In the upper line, the letters do not make sense, but the letters in the lower line are segregated by visual factors. (From Bregman, A.S., *Auditory Scene Analysis: The Perceptual Organization of Sound*, Bradford, 1994.)

REFERENCES

1. Zwicker, E. and Fastl, H. (1990) *Psychoacoustics*. Springer-Verlag, Berlin, Germany.
2. Blauert, J. (1996) *Spatial Hearing*. The MIT Press, Cambridge, MA.
3. Toole, F. (2008) *Sound Reproduction: The Acoustics and Psychoacoustics of Loudspeakers and Rooms*. Focal Press, New York.
4. Beranek, L.L. (1986) *Acoustics*. McGraw-Hill, New York (1954). Reprinted by the Acoustical Society of America.
5. Neuhoff, J.G. (2004) *Ecological Psychoacoustics*. Elsevier Academic Press, San Diego, CA.
6. Asutay, E. et al. (2012) Emoacoustics: A study of the physical and psychological dimensions of emotional sound design. *J. Audio Eng. Soc.*, 60(1/2), 21–28.
7. Salomons, A.M. (1995) Coloration and binaural decoloration of sound due to reflections. Dissertation Technische Universiteit Delft, Delft, the Netherlands.
8. Polack, J.-D., Alrutz, H., and Schroeder, M.R. (March 1984) The modulation transfer function of music signals and its applications to reverberation measurement. *Acta Acust. Acust.*, 54(5), 257–265.
9. Meyer, J. (2009) *Acoustics and the Performance of Music*, Springer Science and Business Media, New York.
10. Ahnert, W. (1984) Sound energy of different sound sources, in *Proceedings of 75th Audio Engineering Society Convention*, Paris, France, Preprint 2079.
11. Houtgast, T. and Steeneken, H.J.M. (1985) A review of the MTF concept in room acoustics and its use for estimating speech intelligibility in auditoria. *J. Acoust. Soc. Am.*, 77(3), 1069–1077.
12. Fletcher, N.H. and Rossing, T.D. (2010) *The Physics of Musical Instruments*. Springer, New York.
13. Farina, A. (2007) A study of hearing damage caused by personal MP3 players, in *Proceedings of 123rd Audio Engineering Society Convention*, New York, Paper 7283.
14. Ando, Y. (2011) *Concert Hall Acoustics*. Springer, Tokyo, Japan. Softcover reprint of the original, 1st edn., 1985 edn.
15. Mohlin, P. (2011) The just audible tonality of short exponential and Gaussian pure tone bursts. *J. Acoust. Soc. Am.*, 29(6), 3827–3836.
16. Brüel & Kjær. (1988) *Acoustic Noise Measurement*. Brüel & Kjær, Nærum, Denmark.
17. Lokki, T. and Karjalainen, M. (2000) An auditorily motivated analysis method for room impulse responses, in *Proceedings of Digital Audio Effects Conference (DAFx-00)*, Verona, Italy, pp. 55–60.
18. Moore, B. (ed.) *Auditory Frequency Selectivity, NATO ASI Series A, Life Sciences*, vol. 119, Plenum Press, New York.
19. Härmä, A. and Palomäki, K. (1999) HUTear—A free matlab toolbox for modeling of human auditory system, in *Proceedings of 1999 Matlab DSP Conference*, available at www.acoustics.hut.fi//software//HUTear (accessed May 2013).
20. Bregman, A.S. (1994), *Auditory Scene Analysis: The Perceptual Organization of Sound*. Bradford, Chester, NJ.

21. Bregman, A.S. (1996) *Demonstrations of Auditory Scene Analysis*, The MIT Press, Cambridge, MA.
22. McGurk, H. and MacDonald, J. (1976) Hearing lips and seeing speech. *Nature*, 264, 746–748.
23. Chapman, P.J. (1996) Programme material analysis, in *Proceedings of 100 Audio Engineering Society Convention*, Copenhagen, Denmark, Paper Number: 4277.
24. Greiner, R.A. and Eggers, J. (1989) The spectral amplitude distribution of selected compact discs. *J. Audio Eng. Soc.*, 37(4), 246–248.
25. Slaney, M. (1993) An efficient implementation of the Patterson-Holdsworth Auditory Filter Bank, Apple Computer Technical Report #35, Perception Group, Advanced Technology Group, Apple Computer, Inc., Cupertino.
26. ISO 226:2003. *Acoustics - Normal equal-loudness-level contours*. ISO, International Organisation for Standardisation, Geneva, Switzerland.
27. ISO 532:1975. *Acoustics - Method for calculating loudness level*. ISO, International Organisation for Standardisation, Geneva, Switzerland.
28. Deutsch, D. (2010) Hearing music in ensembles, *Physics Today*, American Institute of Physics, February, pp. 40–45 (accessed July 2013).

Chapter 8

Spatial hearing

8.1 AUDITORY EVENT AND SOUND SOURCE

Our subjective experience tells us that a sound should have *locatedness*, a feeling that a sound has a source from which it emanates. The true location of that source may be different from the location assigned by listening. In a reverberant room, for example, it may be very difficult to locate the source of a tonal sound. However, we can form auditory event locations that are different from the sound source location. The most obvious discrepancy is noted in the case of headphone listening. In spite of the sound sources mounted at the ears, we can still have auditory events that appear inside the head or that are externalized to the ears.

The location of a sounding object can be described by its radial distance and its azimuth and inclination angles using the system shown in Figure 6.2. However, in some cases, sounds do not have locatedness, for example, a pure tone in a sound field that excites many modes (and thus is diffuse) cannot be assigned a source position, since in such a sound field, the sound is incident from many directions at the same time. This is easy to confirm by a simple listening experiment.

Wide bandwidth sounds may be easy to assign a location to, particularly transient repetitive sounds. These are usually easier to locate than continuous sounds because of their larger high-frequency content that allows binaural hearing to use the precedence effect explained later. Sounds can however have wide frequency range and be difficult to locate anyway such as various types of random noise sound fields as discussed later in this chapter.

It is important to differ between the ideas of lateralization and localization. Sources around us are reasonably expected to have a place in the space around us. The process of finding the sound location is called localization. In headphone listening, the virtual sound sources often seem to appear on the interaural axis between our ears. Placing the sources on that line is called lateralization.

8.2 BINAURAL HEARING

Natural listening is done using both ears, called binaural hearing. For simplicity, it is common to model the human, somewhat egg-shaped, head as a sphere. On the average, the breadth of the head at the ears is 145 mm and the length 197 mm at the plane of the head. Measured interaural time delays suggest that the equivalent diameter of the average head is about 160 mm so a model using a sphere that has a radius r of 80 mm, with the ears positioned diametrically, as shown in Figure 8.1 is common. Men and women have slightly different head as well as pinnae sizes [1].

The additional travel distance difference for signals reaching the far ear is $r(\varphi+\sin(\varphi))$ where φ is in radians. For a sound arriving from an angle of 55°, the difference in arrival time will be approximately 0.45 ms. In the case of a sine having a frequency of 0.1 kHz, this corresponds to a phase difference of 0.045 periods, at 1 kHz of 0.45 periods and at 10 kHz of 4.5 periods.

Hearing uses the interaural time differences (ITDs) in the arrival of sound to the ears for determination of the angle of incidence of sound relative to the median plane. Experiments show that at frequencies below approximately 1 kHz, hearing primarily uses phase differences for localization of sounds. At frequencies above 2 kHz, it is primarily the ITDs Δt in the ear signal envelopes (ITDs) and interaural level differences (ILDs) that are important for localization. Note that for high precision, the time–bandwidth product $\Delta t \cdot \Delta f$ must be large so it is easier to find the direction of wide bandwidth sounds.

A common model for illustrating the fundamental principle of operation of the binaural system is shown in Figure 8.2.

The ILDs can be as much as 5 dB already at frequencies as low as 1 kHz, and at frequencies above 3–4 kHz, the ILDs of about 20 dB is sufficient for localization. Note that (assuming the head a sphere) there are many

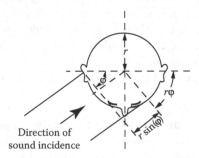

Figure 8.1 A simplified model of the human head for the calculation of the interaural path length difference.

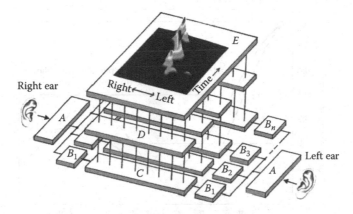

Figure 8.2 Architecture of the bottom-up part of a model of binaural signal processing. A, middle-ear modules; $B_{1...n}$, inner-ear modules, working in spectral bands; C, modules to identify and assess interaural arrival time differences in bands (ITDs); D, modules to assess ILDs in bands; E, running binaural-activity map (intensity, sidedness, time). (From Blauert, J. and Braasch, J., Binaural signal processing, Preprint *IEEE Conference on Digital Signal Processing 2011*, Corfu, Greece, 2011.)

angles around the axis between the ears that will fulfill the interaural level and time difference criteria. Human heads, however, are more egg-shaped than spherical so there will be many more height resolving cues both in the interaural level and time difference domains. It is likely that filtering by the pinna, head, and torso (i.e., by the HRTFs) provides cues that help resolve the locatedness of sounds that have similar ITDs. If the direction of the incoming sound is difficult to determine using these cues, small head movements are necessary to decide on the arrival direction. One can say that the brain uses the head as an acoustic antenna array that can be tilted and rotated for the best reception properties. Once the diameter of the head is the size of the wavelength, the ILDs become very large, more than 20 dB.

8.3 CORRELATION AND SOUND FIELD

8.3.1 Sound fields

The correlation between signals is an effective way of comparing their information content. The correlation between sound pressures at different points in a sound field can yield important information about the properties of the sound field. In our everyday life, we use hearing for direction finding, for example.

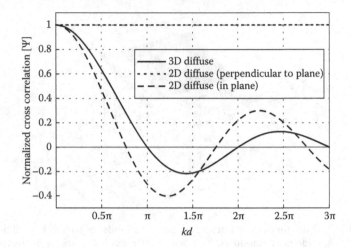

Figure 8.3 Theoretical cross-correlation coefficient between points in a sound field at distance *d* for a wave number *k*. Solid line is for 3D diffuse, dashed line is for 2D diffuse in plane, and dotted line is for 2D diffuse perpendicular to plane of diffusivity.

The correlation between the sound pressure signals at the ears depends on the spatial distribution of incoming sound at the listener. A 2D sound field can be found in a room where the walls are reflective and the ceiling and floor absorptive, such a sound field generally has a smaller point to point cross-correlation than a 3D sound field as shown in Figure 8.3.

In a simple wave field, the use of correlation helps us locate the source because we can measure the time or phase difference between the sounds at the two ears (and from that sound field angles of incidence) at various places in the sound field and from that information derive the location of the sound source.

The direction of arrival of a sound wave can however be determined not only by such binaural comparison of the two ear signals for phase or time information. Other information can be used, for example, the change in timbre as the head is moved in the sound field. This requires the sound to have a minimum duration, which makes transient sounds difficult to analyze in this way.

Many other factors, some of them nonacoustical, affect the determination of the angle of incidence of sounds as was illustrated in Figure 7.1. Our memory of sounds and the information content of sounds help us make initial guesses as to where sounds heard are coming from. The discussion in the next few sections will describe human auditory performance primarily without other sensory information or cognition.

8.3.2 Interaural correlation

It is to be expected that the correlation between the sound pressures $p_{LE}(t)$ and $p_{RE}(t)$ at the left and right ear plays an important role in sound source localization and the subjective impression of the sound field's spatial properties. The cross-correlation function (CCF) was described in Section 1.7.4. The normalized interaural cross-correlation function IACC $\psi(\tau)$ between two ears is

$$\text{IACC}(\tau) = \max_{|\tau| < 0.001} \frac{\int_0^T p_{LE}(t) p_{RE}(t + \tau) dt}{\sqrt{\int_0^T p_{LE}^2(t) dt \int_0^T p_{RE}^2(t) dt}} \tag{8.1}$$

The ITDs in a time-changing dynamic sound field are best measured using the *moving* short-time interaural cross-correlation (STIACC) function of the sound pressures $p_{LE}(t)$ and $p_{RE}(t)$ at the left and right ear, respectively. The STIACC is defined as

$$\text{STIACC}_M(t,\tau) = \max_{|\tau| < 0.001} \frac{\int_t^{t+\Delta t} p_{LE}(t - \tau/2) p_{RE}(t + \tau/2) dt}{\sqrt{\int_t^{t+\Delta t} p_{LE}^2(t) dt \int_t^{t+\Delta t} p_{RE}^2(t) dt}} \tag{8.2}$$

where $p_{LE}(t)$ and $p_{RE}(t)$ are the sound pressures at the left and the right ear, respectively [3]. The choice of the length of time $\Delta\tau$ depends on the purpose of the measurement and which measurement signal that is being used. When the cross correlation is measured using sine signals, the STIACC function will also become sinusoidal. The larger the frequency range Δf, the more sharp will the peak of the STIACC be. The STIACC is, however, often measured using octave band filters so the bandwidth Δf varies with center frequency.

The IACC metric used in room acoustics is similarly defined but uses the impulse responses at the two ears instead of the sound pressures [3]. The metric was originally suggested as a means of investigating the diffuseness of sound in concert halls. In this case, the measurement was done using orchestral music as the test signal. Care is advised when specifying, measuring, and using IACC data to determine subjective sound field quality properties.

It is important to note that IACC can be used with many different types of test signals and with bandpass-filtered impulse responses. Care is advised when using IACC data and specifying and measuring IACC. Recently, the use of the quantity 1-IACC has become common since many associate a larger number with a better property.

8.4 LOCALIZATION

8.4.1 Localization using spectrum and level

At close distance to a small sound source, its sound pressure magnitude decreases monotonously with distance r by $1/r$ since the intensity decreases as the power in the spherical wave is spread over a larger area. This results in a sound pressure level drop of about -6 dB for each distance doubling. At high frequencies, the attenuation is larger because of the sound absorption by atmosphere, resulting in spectral changes at far distance as shown in Figure 8.4.

In nature, the sound level is further reduced by scattering due to trees and atmospheric turbulence. Since most practical environments, both in and out of doors, are subject to background noise, there will be some distance beyond which the sound from the source of interest will be masked by the inevitable background noise.

Much of the psychoacoustic research relating to localization has been done under anechoic conditions, which is convenient since in an anechoic chamber the sound field may be well controlled and the background noise very low. By moving into such a space, new difficulties are introduced. We are not used to listening and determining distance to a sounding object in anechoic surroundings. The clues that could possibly be given by the interference dips due to sound reflections by the floor in normal rooms will be absent. Under normal circumstances, we will be further helped by the early reflections and reverberation generated by all room surfaces.

Factors that will influence distance perception for identical ear signals are the sound level, spectral changes, pressure gradient, sound field curvature,

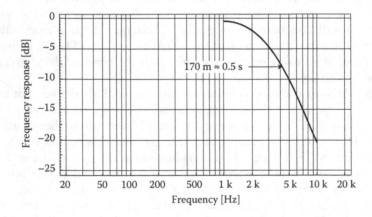

Figure 8.4 Approximate frequency response characteristic of sound transmission over 170 m (roughly 0.5 s travel time) at 20°C and 50% RH due to sound absorption of air.

and of course cognition. We can only determine the distance to continuously sounding objects (such as a loudspeaker emitting a pure tone) by our knowledge or experience of the sound source. For voice and musical sound, we can only rely on experience in judging the distance.

For sound sources at a distance smaller than about 1 m, the HRTFs will be different from those of a plane wave and it is usually possible to deduce the sound source distance by the special properties of the ILDs at close range. Results of experiments in anechoic environments for the intermediate distance range of 2–15 m show that for well-known sounds, the distance perception is quite good, that is, listeners are able to assign auditory event distances that agree with the sound source locations.

One example is shown in Figure 8.5. The lines in this graph show results from experiments on estimation of auditory events as a function of sound source distance. The sound source was positioned in the frontal direction as defined in Figure 6.2. The influence of uncommon voice character such as shouts and whispers can be seen to give considerable deviations from the distance perceived for the auditory event for normal speech. Loud speech is perceived as more distant particularly at close distance, which indicates that cognition plays an important part in distance perception as well: "no need to shout if you are close." The same in reverse applies to the results for whisper: "why whisper at a distance?"

In listening to music over loudspeakers in a stereo or multichannel sound system in a small room such as a living room or control room, it is not possible to have the true distance to the loudspeakers the same as that for the orchestra. First, an orchestra has many instruments and clearly

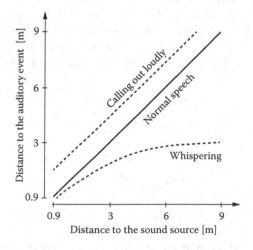

Figure 8.5 Estimated distance for some auditory events as a function of sound source distance, frontal sound incidence in an anechoic chamber. (After Blauert, J., *Spatial Hearing*, The MIT Press, Cambridge, MA, 1996.)

the loudspeakers cannot be at all places at the same time; second, because of typical recording techniques, the sound from any one instrument will be reproduced by several loudspeakers. The situation is analogous to that of looking at a photo or watching a movie. The pictured object is often implied to be much farther away than the photo or screen at which our eyes are focusing.

8.4.2 Sound sources in the median plane

The distance perception experiments discussed in the previous section were special cases of sound arriving in the median plane. Such sounds will be very similar at both ear signals because of head and pinnae symmetry. Since there are no binaural clues, the median plane localization ability is poor and must to some extent depend on memory cues. Some examples of the spectral differences as a function of angle of arrival in the median plane were shown in Figure 6.8. The timbre differences due to frequency response variations are primarily useful for judging wideband sounds. Of course, the more well known the sound is, the easier it will be to use memory to determine the angle.

Tests have shown that one can change the subjectively perceived angle of sound incidence in the median plane by varying the center frequency of bandpass-filtered noise. The result of such an experiment in an anechoic chamber is shown in Figure 8.6, however, only for one test subject. In this case, the sound was always coming from the same place but the perceived location in the median plane of a narrowband noise changed, as the center frequency of the noise changed. The localization was in the directions where the frequency response characteristics amplified the respective frequencies.

Figure 8.7 shows results from experiments where the test subject has been asked to determine the angle of incidence for sounds incident in the median plane. In these experiments, the sound was that of speech by a

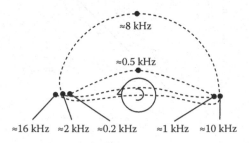

Figure 8.6 Result of experiment showing virtual median plane auditory event directions for various narrowband noise signals for one test subject. Frontal sound incidence in an anechoic chamber. (After Blauert, J., *Spatial Hearing*, The MIT Press, Cambridge, MA, 1996.)

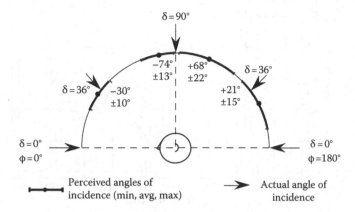

Figure 8.7 Result of experiment showing localization accuracy and blur in the median plane for continuous speech by a familiar person. Test done in an anechoic chamber. (After Blauert, J., *Spatial Hearing*, The MIT Press, Cambridge, MA, 1996.)

familiar person so the spectrum was rather wide and the signal known. In the case of music, the spectrum is likely to be wider but the recorded signal to be less familiar. The consequence of this for localization and localization blur is probably that the accuracy and precision will be even less than that for familiar speech. Later investigations have shown that the dominant frequency range for front–back resolution is the high frequency range over 8 kHz [5].

The poor localization in the median plane and the poor localization in the horizontal plane when using monaural hearing are related. In both cases, the direction finding ability is limited to what is provided by memorized sound cues, and thus localization and localization blur are poor.

Spectral changes can, however, sometimes be used to change the apparent height of the auditory event in a useful way [6]. A typical use is to make the sound of speech from loudspeakers appear at the same height as the video image in television when the loudspeakers are mounted under the screen.

8.4.3 Localization using binaural information

For a low frequency tone, hearing resolves the differences between the two ears using phase. Using the model shown in Figure 8.1, we find that an incidence angle of 45° will lead to an arrival time difference of about 0.35 ms. The period time T corresponds to a 360° phase shift. For a 100 Hz sine sound arriving, the phase difference will be about 13°, at 1 kHz about 130°, and at 10 kHz about $1300° = 3 \cdot 360° + 220°$. Clearly, this will lead to ambiguity in determination of the phase difference at higher frequencies,

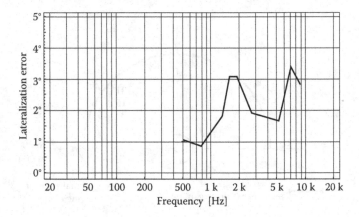

Figure 8.8 The localization blur in the horizontal plane for tonal sounds of various frequencies. (After Blauert, J., *Spatial Hearing*, The MIT Press, Cambridge, MA, 1996.)

since the same phase (plus multiples of 360°) will be present several times. Figure 8.8 shows the localization blur in the horizontal plane for frontal incidence of continuous tonal sounds of various frequencies [4]. By blur is here meant the change in position of the sound source that can be heard as a change in the position of the auditory event by 50% of the test persons.

Note that close to frontal incidence, there is no risk of ambiguity and the localization uncertainty is small even at frequencies over about 1.5 kHz. At higher angles, a different strategy is needed though. Lateralization is then achieved by a combination of ILDs and ITDs. At above 2 kHz, the head already casts a considerable shadow creating substantial ILDs that together with the ITDs give quite good performance as shown in Figure 8.9. The data shown in this figure were derived from listening tests using 100 ms long bursts of (nominally) white noise so the envelope was very well defined and the signal had a very large bandwidth. Both the length of the pulses and their spectra are likely to influence the localization and the localization blur [4].

Note that the results discussed so far in this section have concerned tones; most real-world signals—such as most voice and music sounds—are transient or at least modulated. The localization mechanism used by binaural hearing above 2 kHz is essentially similar to that of a pair of onset detector in the form of a filter bank with edge detectors [2,29]. For sounds that have even and wideband spectra, higher frequencies will have larger influence on localization than lower frequencies. The more transient the modulation, the easier it will be to localize the sound provided that the modulation frequency (or repetition rate) is sufficient.

For multiple signal sound fields, the localization is determined by the precedence effect discussed later in the chapter.

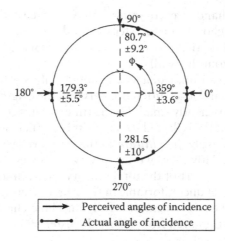

Figure 8.9 Localization accuracy and blur in the horizontal plane for 100 ms white noise pulses. Test done in an anechoic chamber. Note that in this test the loudspeaker was moveable along a circle in the horizontal plane and adjusted by the test persons to give the perceived angles of incidence. (After Blauert, J., *Spatial Hearing*, The MIT Press, Cambridge, MA, 1996.)

8.5 MULTIPLE SOUND FIELD COMPONENTS

8.5.1 Introduction

The mathematical definition of correlation was introduced in Chapter 1 and the autocorrelation properties of sounds were discussed earlier in this chapter. Due to the time–frequency analysis of hearing and the signal analysis and memory provided by the brain for sounds from the voice and musical instruments, *subjective correlation* must also include the meaning of sound. A sound that is reflected over long paths so that it is perceived as an echo may have low physical correlation to the original but large *cognitive* correlation. In a musical signal, the short-time CCF is likely to be different in the attack, steady state, and decay parts of the sound.

The CCF of two sounds where one sound is a delayed copy of the original will have a peak at the delay time. For the CCF to show the delays in a distinct way, the autocorrelation function (ACF) of the original sound must be short, that is, the ACF's τ_e must be much shorter than the delay time τ caused by the delayed sound path.

Consider, for example, the sound from a source that is reflected by the room surfaces as well as by various objects in the room. We are all walking on floors speaking but we seldom react to the sound reflected by the floor, neither for our own voice nor for the voice of the person we are having a conversation with.

On the other hand, we are immediately aware of the reverberation of a cavern or the echo of our footsteps caused by sound reflection from the facade of a building that we are walking by (provided that we are at sufficient distance from the wall).

Three reasons can immediately be thought of as a hypothesis for these effects: (1) The sound level difference between the original and the reflected sound must be sufficiently small, (2) the time difference between the arrival of the original and the reflected sound must be sufficiently large, and (3) the reflected sound must be *in addition* to the sound reflected by the floor that we usually hear. However, we are always subconsciously aware of the reflected sound as a part of the total sensory environment.

Most persons feel uncomfortable in their first meeting with the quiet and sound-absorptive environment of a fully anechoic chamber where there is little sound reflection even from below the listener.

8.5.2 Threshold for a single frontal repetition

The *reflection masked threshold* (RMT) is defined as "the amplitude threshold below which a human listener is unable to perceive the effect of single reflections, multiple reflections, and reverberation".

The threshold for detecting repetitive sounds can be investigated in many ways, for example, by (1) feeding one loudspeaker with a direct and delayed version of the same sound, (2) using two physically displaced loudspeakers along the same line from the listener's head, or (3) using two physically displaced loudspeakers, one of which is sending a direct sound and the other sending a delayed version of the sound. The two sounds may be delayed sound and have different intensity or timbre. In contrast to continuous white noise, a speech signal is modulated and cognition plays a large role in detecting the voice contents that roughly can be described as consonant–vocal–consonant (CVC) combines. These have a length of about 0.25–0.5 s.

The time delay dependency of the RMT for single early reflections, for some different sounds, is shown in the graph in Figure 8.10 [7].

Although the RMT shown in the figure is for different lateral angles of incidence, the data still show that the RMT varies considerably depending on the source signal [7]. The delay characteristic seems to be related to the effective autocorrelation length of the signal. The test clicks used had very short ACFs and were separated by quiet (only 2 clicks per second), and consequently repetitions such as those created by wall reflections became very audible. The speech (e) had longer ACF than the test clicks and thus was more masked, particularly since in running speech, there is little quiet. The classical music (a) and (b) was characterized by longer autocross-correlation functions still. Large variation in ACFs were found for some anechoically recorded music as shown in Chapter 7 [8,9]. Results for the RMT as a function of signal duration are shown

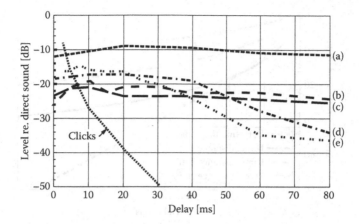

Figure 8.10 The lines show the RMTs for noticing a delayed lateral reflection added to a direct signal for various signals and angles of lateral sound incidence. The curves show as follows: (a) Handel concerto grosso φ = 30°, (b) Mozart φ = 40°, (c) pink noise φ = 65°, (d) Pizzicato violin φ = 30°, and (e) speech φ = 30°. (After Olive, S.E. and Toole, F.E., *J. Audio Eng. Soc.*, 37(7/8), 539, 1989.)

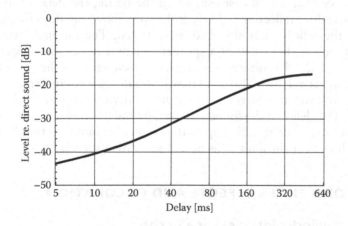

Figure 8.11 Prediction of RMT dependence on the direct sound duration. (After Buchholtz, J.M. et al., Room masking: Understanding and modelling the masking of reflections in rooms, in *Proceedings of the 110th Audio Engineering Society Convention*, Amsterdam, the Netherlands, preprint 5312, 2001.)

in Figure 8.11. The longer the duration of the signal, the less likely are reflections to be detected.

Figure 8.12 shows the threshold for detecting a delayed speech signal when both signals are incident from the frontal direction with distance to an omnidirectional sound source as reference.

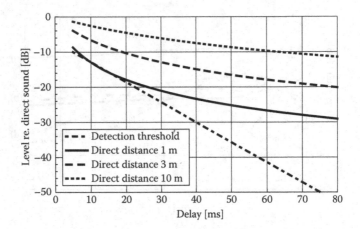

Figure 8.12 The dot-dashed line shows the detection threshold for noticing a delayed signal added to a direct signal for the case of frontal sound incidence (speech at $L_p = 70$ dB). The curves show the relative level of the delayed sound assuming an omnidirectional sound source at the direct distance shown. (After Kuttruff, H., *Room Acoustics*, 5th edn., Spon Press, London, U.K., 2009.)

We note that for all cases shown in the figure, the delayed sound by a wall or ceiling reflection is likely to be well above the detection threshold unless the reflection is absorbed or scattered. The internal level of one's own speech is about that of a speaker at 1 m distance. Since attenuations of –10 and –20 dB corresponds to sound-absorption coefficients of 0.9 and 0.99, respectively, we also note that it should be easy to bring one sound reflection down to a level where it is not noticeable compared to the direct sound. The delay of the sound reflected by the floor of one's own speech is about 10 ms when standing up so the graph also indicates that it should be difficult to hear the floor reflection of one's own speech.

8.6 COMB FILTER EFFECTS AND COLORATION

8.6.1 Periodic interference in the frequency response

Any summation of coherent signals results in positive and negative interference that leads to some frequency components being amplified and some being canceled, creating an uneven frequency response. The resulting frequency response is said to have been subject to comb filtering because of the graphic appearance of the frequency response characteristic. The principle is shown in the impulse response shown on the left in Figure 8.13a and b. In this figure, the level of the delayed signal is $L = 20 \cdot \log(g)$ where g is the gain of the delayed sound component. The detection threshold for such

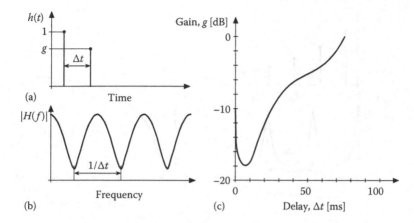

Figure 8.13 An example of the impulse response and its associated transfer function magnitudes for a comb filter resulting from an added delayed signal. (a) The impulse response $h(t) = \delta(t) - g\delta(t - \Delta t)$ and (b) the associated frequency response. (c) The gain factor g that gives just perceptible coloration of white noise. (After Kuttruff, H., *Room Acoustics*, 5th edn., Spon Press, London, U.K., 2009.)

comb filtering when the direct sound is a continuous white noise is shown on the right-hand side (Figure 8.13c) of the figure.

We note that the maximum sensitivity in this case occurs for delays of about 8 ms that corresponds to frequency response dips starting at 60 Hz and then repeating at 120 Hz intervals. Since the test sound is white noise, detection is based on the perceived coloration of the sound. For very short time delays $\Delta t < 0.1$ ms, the effect is that of a low pass filter, a timbre change. For larger delays in the range 2–20 ms, the coloration effect on white noise becomes that of a tonal character, often called repetition pitch [12].

The time window of the ear at midfrequencies is about 30 ms so sound that arrives with a $\Delta t < 30$ ms will be processed as one signal and the detection is based on spectral pattern recognition. For larger delay times, the effect of the delay is sometimes perceived as having rhythm, because of the temporal recognition. Depending on the auditory filters, the roughness of response within one filter is going to be dependent on the combination of the auditory filter frequency, its bandwidth, and the signal delay time.

Signals such as speech and sounds from musical instruments are more subject to recognition than white noise. Speech, for example, has a typical modulation frequency of about 7 Hz and the length of the wide frequency range sibilants is about 30 ms. Because of the different operation of hearing above and below 2 kHz, coloration by delayed added signals is likely to be perceived differently above and below this frequency border.

It is reasonable to assume that multiple repetitions of the same sound should follow the same basic pattern as that shown in the two previous

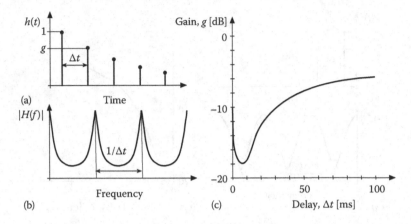

Figure 8.14 An example of the infinite impulse response and the associated transfer function magnitudes for the resulting comb filter. (a) The impulse response $h(t) = \Sigma g^n \, \delta(t - n\Delta t)$ ($n = 0$, 1, 2, 3, 4,...) and (b) the associated frequency response. (c) The gain factor g that gives just perceptible coloration of white noise. (After Kuttruff, H., *Room Acoustics*, 5th edn., Spon Press, London, U.K., 2009.)

figures. A common occurrence in a small room is that of repeated reflections of sound between walls or between the floor and ceiling. Figure 8.14 shows how repetitive reflection results in even more pronounced comb filter frequency response. We note again how a wideband noise can be given a pitch as a result of the filtering as discussed previously.

The auditory effect of the repeated reflections depends on the length of the direct sound. If the sound is longer than a substantial part of the repetitions, the effect will be that of a timbre change due to coloration. If the sound is much shorter than Δt—such as a handclap—the effect is called a flutter echo. We then hear the time structure of the repetitiveness. Coloration and flutter echo are easiest heard on signals where the repetitions are generated electronically.

8.6.2 Cone of confusion

Comb filter coloration because of sound reflection is not heard when the sound field is diffuse such as in multichannel sound reproduction or in a reverberant sound field. Binaural hearing removes comb filter effects by a mechanism called binaural unmasking or binaural decoloration that will be discussed later in this chapter. Of course, if the sound is already subject to comb filtering, for example, because of its recording or the loudspeaker's characteristic, the effect will still be heard.

The condition for comb filter coloration is often described using the idea of a cone-of-confusion that springs out of the common spherical head model

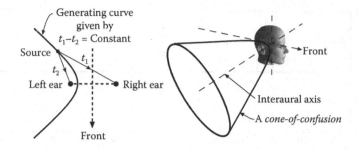

Figure 8.15 The cone-of-confusion defines source positions that result in the same binaural time difference.

shown in Figure 8.15. This cone can be thought of as the source locations from which ear signals arrive with the same binaural time separation. Repetitive sounds that cannot be resolved binaurally cause coloration. Filtering by the signals reflected by pinna, head, and torso (i.e., by the HRTFs) provides cues that help resolve such problems as will minute head movements.

8.6.3 Random reflections

Another possibility to remove coloration in binaurally inseparable sound reflection is to remove regularity and repetition in the impulse response. Randomness in a room impulse response can be improved by addition of scattered sound by objects and uneven surfaces or by the use of room designs that provide nonrepeating reflection sequences such as those composed of time delays that are randomized [13] or primes [14].

The presence of repetition and echo in a system can be analyzed using the Fourier transform of the system's impulse response or—if the impulse response is not available—by analyzing the system's cepstrum. The cepstrum is the inverse Fourier transform of the logarithm of a system's spectrum. The cepstrum will reveal if there are any artifacts due to interference because of echo such as those shown in Figures 8.13 and 8.14.

The behavior of the inverse Fourier transforms of two impulse responses that have different repetition characteristics is shown in Figure 8.16. The impulse response in (a) has extreme spectral repetitivity as shown in (c), whereas the impulse response shown in (b) has randomly placed impulses in its spectrum (d). The (d) spectrum is much smoother than the (c) spectrum so the coloration caused by filter (b) is much smaller than that by filter (a).

8.6.4 Echo

An echo is a reflected sound with sufficient magnitude and delay as to be distinguishable as a repetition of the direct sound. For sounds to be heard

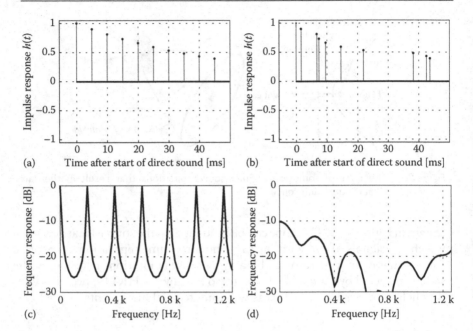

Figure 8.16 The figure shows two impulse responses and their associated spectra. (a) and (c) with regularly repeating impulses and (b) and (d) with a prime number delay sequence of impulses.

as echoes, they need to be of a transient nature such as speech or the sound of musical instruments. Echo is unusual in small rooms since a certain minimum delay relative to the primary sound is needed for an echo to be heard. If the echo has the same level as the primary sound, it will be annoying at delays of 50 ms or more for speech, a delay that corresponds to an extra acoustic path length of more than about 17 m. If the echo is 20 dB below the primary sound, it is not likely to be annoying in running speech; some test results are shown in Figure 8.17. For music, the time delay needed for echo is longer unless the musical sound generated has clearly audible transients such as the sound from maracas, trumpets, etc. In the small rooms discussed in this book, echo is only a problem under special circumstances that are discussed in Chapter 9.

8.6.5 Spatial effects of sound pairs

It is to be expected that multiple copies of a sound arrival from different angles and with short time differences will result in additional and/or different effect to those of comb filtering and coloration.

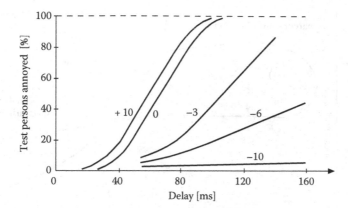

Figure 8.17 Percentage of test persons annoyed by echo in listening to speech with a direct sound in combination with a delayed sound (φ=0°) with the relative level of the echo to the direct sound as parameter. (After Haas, H., *Acustica*, I, 49, 1951 [translation appears as Haas, H., J. Audio Eng. Soc., 20(2), 146, 1972].)

When listening to a sound such as speech or music in the room, the signal spectrum is filtered by the room's frequency response that is determined by the impulse response through the Fourier transform. The source signal is heard convolved with the binaural impulse response of the source and room for the listening position. So far, only impulse responses that belong to a point source and point receiver pair were considered. However, hearing has binaural direction sensing capability and sources are directional. This makes a more advanced approach necessary where the directivity of the source and the binaural detection methods of hearing are taken into account along with the transient and spectral characteristics of the source signal.

When we hear two sounds one after the other—so that they are recognized as two separate sounds—we can assign directions of incidence for each sound separately. For correlated sounds—spatially separate sound field components that were generated by the same source at the same time—the resulting sound (with the delayed version) may appear to come from one direction or two directions or be spatially smeared, usually called diffused depending on the time gap between the two sounds.

Figure 8.18 shows results from investigation into how various levels and time delays of only a single additional sound field component from the side (φ=90°, δ=0°) affect the sound heard from the front. In the research cited, the sound was classical music radiated by two loudspeakers [15]. Effects were noted such as change of source location, coloration, feeling of spaciousness, and echo. Echo is the time-delayed repetition of sound and as shown needs a time delay to be apparent. Since time delayed sounds have

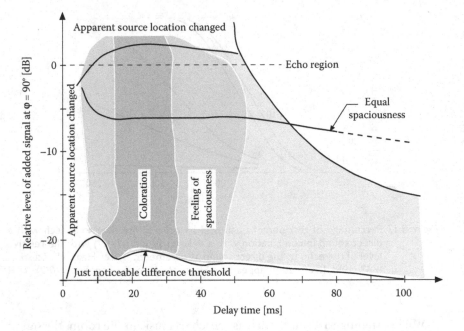

Figure 8.18 Various subjective effects of an added sound field component at φ = 90°, for various delays and levels relative to the direct sound, which has frontal incidence (classical music). (After Barron, M., *J. Sound Vib.*, 15(4), 475, 1971.)

traveled farther than the earlier arriving sound, its intensity will usually be weaker because of geometric spreading of power so its relative level will be lower. Typically, the echo limit is set to about 50 ms but will in practice depend on the arrival of the earlier, time-smeared sound. Because of the long time delay necessary for it to be heard, echo is not usually found in small rooms.

For the shorter time delays of small rooms, it is the effects of change of source location, coloration, and feeling of spaciousness that are dominant. For very short time delays $\Delta t < 0.8$ ms, the apparent source location depends on the delay and level differences between the sound at the ears. The apparent source is called a *phantom source*.

8.6.6 Influence of off-center arrival angle

Since the head and torso reflect sound, the HRTFs will depend on the angle of arrival of sound. The sound pressure level increase at the ear on the side of sound arrival may increase by about 10 dB. From the viewpoint of room acoustics, it is important to know that the sensitivity to early reflected sound is highest at angles of about 45° as shown in Figure 8.19 although the angle is not very critical.

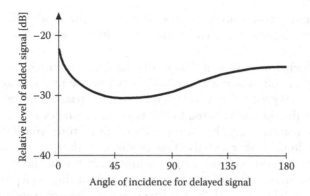

Figure 8.19 Relative threshold for noticing a signal delayed by 50 ms relative a direct speech signal as a function of horizontal angle for the delayed signal (in an anechoic environment). Both sound components incident in the horizontal plane, direct sound has frontal incidence. (After Seraphim, H.P., Über die Wahrnehmbarkeit mehrerer Rückwürfe von Sprachschall, *Acustica*, 11, 80, 1961.)

8.6.7 Phantom sources, time, and level trading

Phantom sources are virtual sources created by hearing. The most well-known type of phantom source is the one that appears between two loudspeakers driven by two equal signals in stereo sound reproduction as shown in Figure 8.20. The line between the loudspeakers is called the baseline. The listener is situated at right angle to the baseline, symmetrically between the loudspeakers.

If the signals to the loudspeakers are equally strong and arrive at the listener at the same time, the phantom source will be located straight ahead of the listener. Since the sounds heard are subject to different HRTFs than true sound from the front, the phantom source will have a different

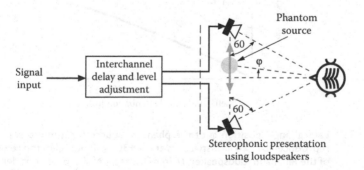

Figure 8.20 A phantom source can be generated between the loudspeakers in stereophonic playback.

timbre than a true source. Its sound will also be louder than that from a single loudspeaker driven by the same signal since two loudspeakers are operating.

Experiments have shown that in this listening situation, one can generate phantom sources on the baseline between the loudspeakers and with suitable signal processing also outside the stereo triangle (auralization) provided that the signals radiated by the two loudspeakers are correlated.

Phantom sources can be moved as a result of time and level trading of the stereo loudspeaker signals. The localization sharpness of the phantom source depends on the time structure and spectrum of the sound from the two loudspeakers. Transfer function matching of the loudspeakers is critical for precise imaging. If the loudspeakers are not accurately matched the phantom sources appear blurred. In an anechoic environment, the effect is easy to hear for clicks, but for complex sounds, the effect is sometimes confusing even there.

Using magnitude and arrival time differences between the sounds from the two loudspeakers, one can move the phantom sources along the baseline between the two loudspeakers. Figure 8.21 shows how the perceived angle of incidence depends on the relative level difference between the loudspeaker sounds. Figure 8.22 shows similarly how the perceived angle of incidence φ depends on the relative time difference [17].

It is reasonable to assume that it should be possible to sum or combine the effects due to signal intensity and signal delay to provide the desired auditory event location. Informal experiments (probably using speech) reported on the effects on the lateral angle of phantom sources in a stereo loudspeaker setup by combining signal delay and amplitude differences give results as shown in Figure 8.23 [18]. The curves in this figure are

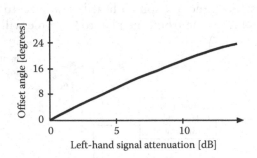

Figure 8.21 Lateral angle position φ of the phantom source for a mono sound source reproduced by two loudspeakers at $\varphi = \pm 30°$ as function of the reduced level of the left-hand loudspeaker. (After Franssen, N.V., Some considerations on the mechanism of directional hearing, Dissertation, Electrical Engineering, Mathematics and Computer Science, Technical University of Delft, Delft, the Netherlands, 1960.)

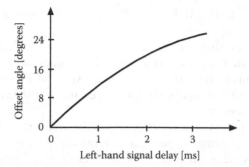

Figure 8.22 Lateral angle position φ of the phantom source for a mono sound source reproduced by two loudspeakers at φ=±30° as function of the extra time delay of the left-hand loudspeaker. (After Franssen, N.V., Some considerations on the mechanism of directional hearing, Dissertation, Electrical Engineering, Mathematics and Computer Science, Technical University of Delft, Delft, the Netherlands, 1960.)

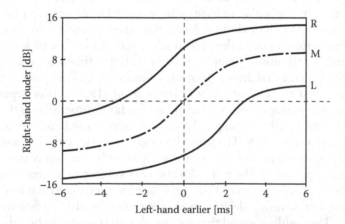

Figure 8.23 Lateral position φ of the phantom source in the case of a transient sound reproduced by two loudspeakers at φ=±30° with simultaneous variation of delay and level between loudspeakers. Dot-dashed line in center: M, phantom source in the middle between loudspeakers; Solid lines: L, phantom source in left-hand loudspeaker; R, phantom source in right-hand loudspeaker. (After Franssen, N.V., Some considerations on the mechanism of directional hearing, Dissertation, Electrical Engineering, Mathematics and Computer Science, Technical University of Delft, Delft, the Netherlands, 1960.)

redrawn from the results presented in [19,20]. In practice, however, there is large variation in phantom source distinctness and placement depending on signal character; a transient signal is located differently from a steady-state sine or a steady-state sawtooth wave. The importance of signal onset characteristics are further discussed in the next section that deals with the precedence effect. The influence of cognitive factors on phantom source blur and placement is large.

At frequencies above about 2 kHz, the sound reflection and diffraction by the head results in large level differences between the sounds at the ears for sound that arrives from the two loudspeakers. It is then difficult to move the phantom sound source using a delay for the stronger sounding loudspeaker.

The time delays and level differences between the channel signals in stereo sound recordings depend on the microphone properties and placement, as discussed in Chapter 9. Using suitable signal processing, phantom sources can even be generated to appear outside the baseline connecting the stereo loudspeakers [21].

8.6.8 Precedence effect

The summing localization discussed in the previous section is of great importance for listening to stereo sound reproduction in rooms. It is, however, only applicable to cases where the delay Δt between the arriving sounds is smaller than about 0.8 ms. For larger delays up to about 30 ms, a different effect occurs: the arrival direction of the first of several coherent sound contributions (temporally and spatially displaced) determines the subjectively perceived direction of the sound source. This *precedence effect* is often called *the law of the first wave front* [19,20]. It is dependent on the integrative properties of hearing, so for large time delays, the effect is reduced until finally for long delays the delayed sound appears as an echo as shown in Figure 8.18. The precedence effect is particularly important for localization of sound in large rooms such as auditoria and concert halls.

In one experiment, the test subjects were subject to a stereo sound field with the first arriving sound (speech) from the left-hand loudspeaker and the same but variably delayed sound arriving from the right-hand loudspeaker. The results showed that for the phantom source to be shifted to the right-hand loudspeaker, its sound had to be about 10 dB stronger than the direct sound component, as shown in Figure 8.24 [4,19,22].

When the two loudspeakers emitted the same and equally strong sound, the auditory event was placed at the loudspeaker emitting the first sound as long as the delay of the latter sound was in the interval of about 1–30 ms. For delays larger than about 40 ms, two sound sources were heard: one at the loudspeaker that emitted the undelayed signal and an echo from the loudspeaker emitting the delayed signal as indicated in the figure.

In some cases, such as when the direct sound from the source is obscured or when there are focusing surfaces, the directional impression can shift in an undesirable way. For playback of stereo and multichannel recordings, the precedence effect must be allowed to function properly. This requires the sound field generated by loudspeakers and room to be fairly

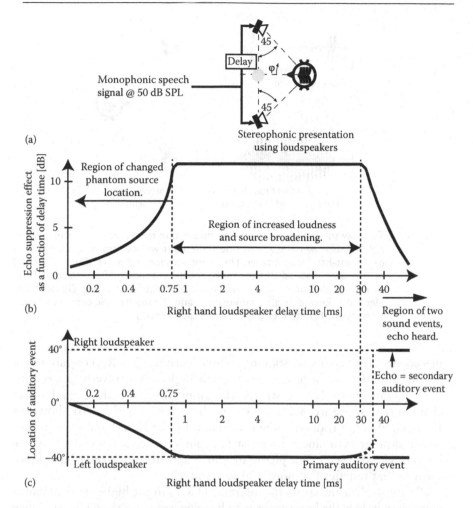

Figure 8.24 The precedence effect (note the logarithmic time scale). (a) Stereo listening arrangement, (b) Haas' data for level increase necessary to move phantom source from left to delayed right loudspeaker, and (c) auditory events heard and their locations when the two sounds are equally strong. (After Haas, H., *Acustica*, I, 49, 1951 [translation appears as Haas, H., *J. Audio Eng. Soc.*, 20(2), 146, 1972]; Madsen, E.R., *J. Audio Eng. Soc.*, 18(5), 490, 1970.)

symmetrical around the listening axis. It will be difficult to obtain good listening conditions in rooms that are L-shaped or have other asymmetries unless precautions are taken to compensate for the missing wall section by sound absorption. The limits of the principle are therefore of interest.

It is important to note that the precedence effect is strongly dependent on the onset characteristic of the sound. Click trains, noise bursts, and speech may give different results, for example, as found when trying to implement

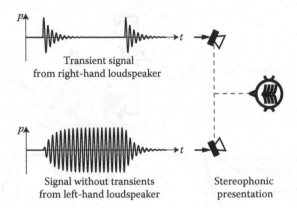

Figure 8.25 To show the Franssen effect, a transient sine tone burst is radiated by the right loudspeaker and a slow sine tone burst without transients is radiated by the left-hand loudspeaker. The phantom source appears at the right-hand loudspeaker that radiates the more transient sound. (After Franssen, N.V., Some considerations on the mechanism of directional hearing, Dissertation, Electrical Engineering, Mathematics and Computer Science, Technical University of Delft, Delft, the Netherlands, 1960.)

precedence effect characteristics in machine learning [23]. Reverberation will confuse the effect so the onset is particularly important in reverberant rooms.

The Franssen effect illustrates the importance of the signal onset characteristics. Figure 8.25 shows the listening situation. To show the Franssen effect, a transient sine is radiated by the right-hand loudspeaker and a slow growth sine tone burst without transients is radiated by the left-hand loudspeaker. The phantom source appears at the right-hand loudspeaker that radiates the more transient sound.

Obviously, hearing trusts the interaural CCFs in the higher critical bands more than it does the lower ones as to be expected since the CCFs are likely to have sharp peaks in these critical bands where the critical bandwidth is large.

It is also important to note that the precedence effect only applies to the localization of the first arriving sound of a series of repeating sounds with short time delays such as those generated by reflections of sound from source in a room. It would be incorrect to brand the precedence effect as a masking phenomenon since sound arriving up to 30 ms after the direct sound contributes to the subjectively perceived strength of the (first arriving) sound [19]. The arrival time of the delayed sound in this time slot affects the timbre of the first arriving sound.

In a small room, the reflections by the walls all contribute to the strength and timbre of the direct sound, so when a speaker's head is turned, we do not notice much of a change in level or timbre of the speech when we listen in a small room. In a large room where the majority of reflections are likely to arrive much later than 30 ms, the change in direct sound will be much more noticeable.

8.6.9 Precedence effect and Gestalt

Since the reverberation time of a small listening room is typically rather short and its level low, new Gestalt carrying sounds will compete for attention and the low level of reverberation of the room will be reduced to being perceived similar to background noise. This is in contrast to the situation in a large room since there the torrent of early reflections will start to arrive only after the direct sound and the early floor reflection have arrived.

8.6.10 Fusion and analysis time

In listening, hearing (with the help of its cognitive abilities) follows the sound event along in time and follows its Gestalt. A 3D echogram is helpful to visualize the situation. The echogram is equivalent to an impulse response but shown as level in dB. In Figure 8.26 it also shows the angle relative to the median plane for the arriving sound components. The gray sliding time frame in the 3D echogram shown indicates the time slot within which hearing will analyze a transient signal for the precedence effect. It is difficult for hearing to follow the sound reflections in space and time. After some time the reflected sound components merge and cannot be separated by hearing, and they become fused. In a large room, the earliest reflection of speech, if sufficiently strong, can be heard as echo if they arrive more than about 40 ms after the direct sound.

The time frame available for the fusion process and forward temporal masking is about 30 ms with a drop to zero for larger delays up to 500 ms [19,24,25]. In the figure, a rectangular window is used for simplicity. For lower frequencies, the fusion time frame is longer than that at 1 kHz. At high frequencies, it is shorter, almost following the product of frame time and bandwidth $\Delta t \cdot \Delta f$ being constant. (In the frequency range around 1 kHz, hearing has an effective bandwidth Δf of about 200 Hz.)

Fused sound components that are physically separated in time and space will lead to different psychoacoustic effects depending on the spatial impulse response time pattern. Figure 8.18 shows the effects that can be heard for a direct sound at 0° and a reflected sound at 90°. Examples are the change of apparent source location, coloration, and spaciousness. Since the early sound field components in a small listening room, such as that shown in Figure 11.2, come already after a few milliseconds, the primary effect of the reflected sound will be the sense of spaciousness since in the fusion of the direct sound and the early reflections, the precedence effect will be dominating.

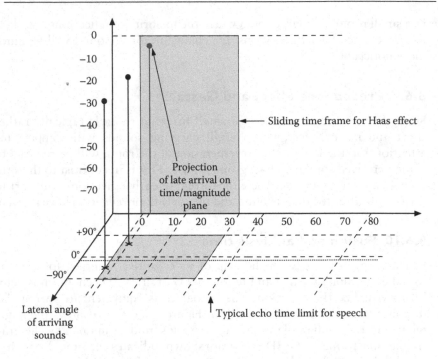

Figure 8.26 An example of a 3D echogram for a sound played back equally strong by two loudspeakers in a stereo system in an anechoic environment. The sound coming from the right-hand loudspeaker assumed slightly delayed. The gray region indicates approximate time frame of 30 ms for speech as shown in Figure 8.24b.

8.6.11 Precedence effect and reverberation

Without the precedence effect, it would be difficult for us to give location to sound sources in a home or other semi-reverberant environment. As was shown in Chapter 2, the reverberation radius, the distance at which the direct sound from a source has the same sound pressure level as that of the reverberation, is quite small, typically less than 1 m. Because of the precedence effect, the phantom sources will remain in place as long as the reverberation and early reflections are weak as indicated by Figure 8.24. The margin has its highest value for reflection delay times around 20 ms, which corresponds to path lengths of about 6 m, that is, typically a reflection order of 2 in a small room such as the one shown in Figure 11.2 [26].

In listening to stereo sound reproduction in the room, the sound field is composed of the two competing direct path signals that arrive from the two loudspeakers shown in Figure 8.20 as well as of the reflections of the loudspeaker signals by the room surfaces. The room wall reflections will help in widening the sound stage although the phantom sources will become blurred but still placed mainly between the two loudspeakers. When omni-directional loudspeakers are used in a sound reflective environment, the phantom source becomes more blurred than with directional loudspeakers because of more competing information added by the early reflections.

In a relatively reflection-free environment such as a control room designed according to *live end–dead end* or *reflection-free zone* principles (see Chapter 11), the reverberation will arrive relatively late (compared to the situation found in most domestic environments), but it will be much attenu-ated because of sound absorption due to sound absorbers and diffusers.

If the listening environment has reflecting surfaces that block the direct sound or focus early reflections, the apparent location of the sound source will be determined by the early reflections. In larger rooms that have curved reflecting surfaces or corners reflectors there may be echo). Unfortunately, much of the psychoacoustics research applicable to practical situations has been done with speech and classical music that have their respective spec-tral centers of gravity in the midfrequency range, so the results are typically not directly applicable to high- and low-frequency extremes and transient sounds.

8.6.12 Cocktail party effect

Binaural hearing leads to a much improved localization ability and sub-jective perception of spaciousness compared to that of monaural listening because of the three phenomena of localization, noise suppression, and bin-aural unmasking.

The apparent loudness of a signal is almost doubled when the signal is presented binaurally rather than monaurally [27], resulting in an improved the signal-to-noise ratio.

Localization together with the precedence effect offers noise suppression. The action of hearing is that of adaptive beam forming. Hearing focuses cognitively on the sound field features of interest. This together with the forward masking of an earlier arriving sound means that much of the rever-beration, that is, room noise, is attenuated. The same applies to conven-tional noise. This effect is sometimes called *spatial release from masking.*

Although the apparent direction of a person's voice is usually determined by the precedence effect it is important to remember that visual informa-tion in the form of lip reading is also contributing to the localization [28].

For a musical instrument that is being played in a room, the visual information will be less relevant although a cognitive element will still be involved. Since the modulation and directivity characteristics of musi-cal instruments are different from those of speech—and different between instruments—the effective signal-to-noise ratio and the clarity of sound in the room will be slightly different from case to case.

The spatial separation of the direct sound and the reverberation causes an intelligibility level difference from the viewpoint of speech and music clarity in a room. This difference drops quickly if there are several identifiable sounds surrounding the head and particularly if those are located on opposite sides of the head. For continuous speech, the decrease is less pronounced, which indicates that binaural hearing can only suppress interference from one lateral region at a time and that it can use temporal pauses in the interference to suppress a second interference from a different direction, thus contributing to the *cocktail party effect* (CPE) [29].

Other effects help binaural hearing along with CPE. For example, hearing also uses spectral differences between the concurrent voices of speakers to separate the voices. Interfering speech will also carry linguistic content so speech that is understandable will be informationally masking. In a complex environment, the CPE will also depend on the interactions between the type, number, and location of interfering speech, reverberation, and noise [30].

Speech intelligibility may be defined as the percentage proportion of speech items (syllables, words, or sentences) that can be repeated without error by listeners. In an analogous manner, one could speak of music intelligibility. Few people however are able to repeat a musical phrase, for example, but from the viewpoint of the clarity and correct sound reproduction for music listening, there is much in common between rating of the characterization and quality of speech and music in a room environment. Forward masking by reverberation due to music is smaller than that by speech since music has lower modulation frequencies [44]. Since music enjoyment does not seem to depend on clarity (many listeners prefer to sit far back in concert halls where reverberant sound usually dominates), the idea and expectation of quality in sound reproduction of music clarity seems to be highly individual.

8.6.13 Spatial impression

The listener's auditory spatial perception of a room or other space is influenced by many sound field properties. Some of those properties are energy related, for example, reverberation time, the relationships between the levels of direct and reverberant sound, and the spectra of the incoming sound components; some are spatially related, for example, the spatial distribution of the incoming sound field components; and still others are dependent on arrival time, such as echoes and other strong sound reflections and of course sound field diffusivity [31].

The subjective perception of spatial characteristics often does not map well onto the physically measured dimensions of the spatial sound field properties. For example the perceived reverberation times in auditoria often do not correspond to those obtained by using instrumentation and traditional evaluation techniques. It is necessary to be careful when using results from classical measurement, such as reverberation time, in judging room acoustic quality. A review of factors influencing the hearing of room acoustical spatial properties can be found in [32].

The detection of reflections and image shift thresholds in listening in typical domestic rooms were investigated in a study [7], the following being the most important results:

- Fundamental similarities between the detection of broadband reflections and the detection of delayed resonances.
- Delayed sounds arriving from the same direction as the direct sound are often less audible than from other directions.
- Little difference between thresholds for impulsive and continuous sounds.
- Natural reflections and reverberation have relatively little effect on the absolute threshold of an individual delayed reflection for delays shorter than 30 ms.

In research investigating the spatial aspects of reproduced sound in small rooms, it was found that test persons could reliably distinguish between timbre and spatial aspects [33]. It was further found that it is the spectral energy above about 2 kHz of the individual reflection that determines its importance for the spatial aspect.

8.7 SPATIAL ATTRIBUTES WITH REFLECTED SOUND

8.7.1 Spaciousness

Early investigations of sound reproduction showed that stereo sound reproduction was essential to achieve plausible spatiality. The recording technology used was shown to be decisive in achieving a good stereo sound stage combined with some envelopment [34,35].

In a more extensive test listeners were found to focus on the following attributes: definition of sound images, continuity of sound stage, width of sound stage, impression of distance/depth, reproduction of ambience/spaciousness/reverberation, and perspective [36].

Some spatial and timbral attributes as important for the spatial impression in the reproduction of music are shown in Table 8.1 [after 37]. Some attributes pertaining to on-site listening in concert halls and similar venues also were perceivable in reproduced sound, but there were also attributes unique to reproduced sound [37].

Table 8.1 Spatial and timbral attributes important for the spatial impression in the reproduction of music

Attribute	Description
Naturalness	How similar to a natural (i.e., not reproduced through loudspeakers) listening experience the sound as a whole sounds.
Presence	The experience of being in the same acoustical environment as the sound source, e.g., to be in the same room.
Preference	If the sound as a whole pleases you. If you think the sound as a whole sounds good.
Low-frequency content	The level of low frequencies (the bass register).
Ensemble width	The perceived width/broadness of the ensemble, from its left flank to its right flank. The angle occupied by the ensemble. The meaning of *the ensemble* is all of the individual sound sources considered together. Does not necessarily indicate the known size of the source, e.g., one knows the size of a string quartet in reality, but the task to assess is how wide the sound from the string quartet is perceived. Disregard sounds coming from the sound source's environment, e.g., reverberation, and only assess the width of the sound source.
Individual source width	The perceived width of an individual sound source (an instrument or a voice). The angle occupied by this source. Does not necessarily indicate the known size of such a source, e.g., one knows the size of a piano in reality, but the task is to assess how wide the sound from the piano is perceived. Disregard sounds coming from the sound source's environment, e.g., reverberation, and only assess the width of the sound source.
Localization	How easy it is to perceive a distinct location of the source—how easy it is to pinpoint the direction of the sound source. Its opposite is when the source's position is hard to determine—a blurred position.
Source distance	The perceived distance from the listener to the sound source.
Source envelopment	The extent to which the sound source envelops/ surrounds/exists around you. The feeling of being surrounded by the sound source. If several sound sources occur in the sound excerpt, assess the sound source perceived to be the most enveloping. Disregard sounds coming from the sound source's environment, e.g., reverberation, and only assess the sound source.
Room width	The width/angle occupied by the sounds coming from the sound source's reflections in the room (the reverberation). Disregard the direct sound from the sound source.

Table 8.1 (continued) Spatial and timbral attributes important for the spatial impression in the reproduction of music

Attribute	Description
Room size	In cases where you perceive a room/hall, this denotes the relative size of that room.
Room sound level	The level of sounds generated in the room as a result of the sound source's action, e.g., reverberation, i.e., not extraneous disturbing sounds. Disregard the direct sound from the sound source.
Room envelopment	The extent to which the sound coming from the sound source's reflections in the room (the reverberation) envelops/surrounds/exists around you, i.e., not the sound source itself. The feeling of being surrounded by the reflected sound.

Source: After Theile, G., Multichannel natural music recording based on psychoacoustic principles, in *Proceedings of Audio Engineering Society 19th International Conference*, Schloss Elmau, Germany, May 2001.

In sound reproduction it is important that the phantom sources can be freely located in auditory space and not be located at the loudspeakers. Outside of a *sweet spot* (region of best sound reproduction), spaciousness in sound reproduction can only be plausibly recreated by dedicated surround sound recording techniques or surround sound simulators that use an auralization approach.

8.7.2 Spaciousness and diffusivity

Localization of externalized single sound field components was shown to be fairly straightforward but dependent on many factors. Localization of sound field components that have identical sound levels at the ears will depend on further factors such as phase difference.

When sounds are correlated, such as a monophonic signal that is presented binaurally, the auditory event occurs inside the head, *inside head localization* (IHL). If the sounds at the ears are fully uncorrelated, such as two separate noise signals that are presented binaurally, there will be two auditory events, one at each ear.

An interesting effect can be heard when presenting a monophonic wide bandwidth noise signal in stereo (over loudspeakers or headphones) if the stereo signals are out of phase. The noise frequency components below 2 kHz are then perceived as spatially diffuse—having *spaciousness*—whereas those for higher frequencies are perceived as located between the loudspeakers (or for headphones, IHL occurs). The time difference in the low-frequency components provides phase cues that are ambiguous thus providing apparent sound field diffuseness, whereas the high-frequency sounds are analyzed by their envelopes and those will be identical at the two ears causing a located auditory event.

Similarly, when a wideband noise signal is provided over headphones to a listener and one of the headphones is fed with the signal delayed by a millisecond or more, the sound is perceived as diffuse.

What constitutes a diffuse sound field is thus different in the physical and psychoacoustic domains. In the latter, a diffuse sound field is that that provides nonlocatedness of sounds or, alternatively phrased, that provides a sound that is located over all spatial angles (or rather upper hemisphere in a concert hall that has sound-absorptive seating).

In physics on the other hand, a diffuse sound field is defined as a sound field where all angles of sound incidence have equal probability, where the sound from each spatial angle is out of phase, and where the energy density is the same everywhere.

Obviously, the two ideas of what constitutes diffuseness are different in the two sciences. A physically diffuse sound field will also be psychologically diffuse but not necessarily the reverse. From the viewpoint of listening, it is of course the psychoacoustic properties that are of importance, not the sound field properties.

8.7.3 Auditory source width and image precision

As we listen to sounds, the apparent width of the auditory event, often called the *auditory source width* (ASW), will depend on many issues. To those listening to stereo or multichannel recordings of sound, it is quite clear that the width of the array of phantom sources treated by the recording or playback is determined by not only the layout of the loudspeaker setup in the listening room and the directional properties of the loudspeakers but also on the listening room itself. The more reflections arriving from the sides of the listening room, the wider will the ASW be. However, the ASW will be frequency dependent above 0.5 kHz and a 2 kHz sound arriving at ±45° relative the frontal direction will produce maximum ASW [38,39]. This is to be expected since the masking by direct sound is the smallest for this angle of incidence of early arriving reflections [16]. The ASW also depends on the low-frequency content of the signal, more low-frequency energy increases ASW [38,40,41]. Psychoacoustic testing shows that the spatial aspects of the early reflections are primarily determined by the reflection spectrum above 2 kHz [33].

Reliable data for sound reproduction in small rooms are difficult to find. A single omnidirectional loudspeaker judiciously placed close to the corner of a room may well create as large an auditory image as a conventional stereo loudspeaker setup placed out in the room as discussed in Chapters 9 and 11.

Using digital signal processing, the ASW can be made to extend far outside the bounds set by the stereo baseline. Sound field cancelation techniques are then used to process the sound supplied to the loudspeakers (assuming an ideal stereo listening situation) so that image sources can be sensed on the extended baseline. Such signal processing is sometimes used for effect sounds in movies.

However, also a single small loudspeaker or a stereo loudspeaker pair being used to play back sound in an anechoic chamber will have an associated ASW. Phantom source *accuracy* is the offset from the intended position of the phantom source and *precision* is the relative random offset around that position. It is to be expected that frequency and arrival time irregularities in a stereo or other multichannel loudspeaker system will cause a widening of the phantom image. A good test for phantom source precision is a good stereo or binaural recording of applause by an audience in a reverberant room. In playback, the handclaps should be evenly spread out over the stereo baseline and each handclap should have a distinct position.

Some degree of ASW can be achieved by using loudspeaker types that are designed to be omnidirectional or radiating away from the listener provided the room is sufficiently large. However, as long as most of the reflected energy occurs within 30 ms of the arrival of the direct sound, the room will still sound small.

In recording a sound source, it may be possible to capture the influence of the acoustics generated by the sound reflected by surfaces close to the source such as the walls, canopy, or shell that surround the source. This will result in a source width that is related to the ASW but that can be heard separately from that generated by the listening environment at playback.

8.7.4 Envelopment

In the small home listening room, the natural reverberation often dies out quickly because the reverberation time is often as short as 0.4–0.5 s (Refs. [22–24] of Chapter 2). Additionally, the natural reverberation is masked by early reflections so that it is perceived only weakly as a contribution to a pleasant feeling of ambience.

In a concert hall, however, as the initial auditory event caused by the direct sound and early reflections fades out, the decay of the multitude of late reflections will be heard as spacious reverberation. The late reverberation in a good hall is diffuse so it has about the same intensity at all angles. The term *envelopment* (LEV) is used to describe the spaciousness effect of this type of diffuse sound field.

Some researchers split LEV into three components: *early spatial impression* (ESI), *background spatial impression* (BSI), and *continuous spatial impression* (CSI) [38]. Figure 8.27 shows how ESI and BSI are related. Some researchers consider ASW a metric for the ESI. The BSI increases as masking by the direct sound diminishes and is fully developed after about 160–200 ms but drops for later times of the reverberation process as the

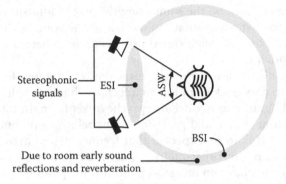

Figure 8.27 Sound field components necessary to generate CSI using loudspeakers or room reflections. CSI gives 360° surround but not necessarily an impression of sound reflections coming from above. (Adapted from Griesinger, D., Spaciousness and envelopment in musical acoustics, *Proc. 101 AES Convention*, Los Angeles, Paper 4401, 1996.)

reverberation level drops. This type of drop is however only heard at the end of a musical piece. In a complicated orchestral piece, it is not possible to focus on many items in the auditory scene so the earlier and stronger sounds can mask reverberation and the BSI will not be formed until these strong early sounds have dropped sufficiently in level.

Usually, LEV is a characteristic of reverberation that arrives from a multitude of random angles. LEV can also be created—even in an anechoic chamber or outdoors—by using multichannel playback of incoherent sound and similar techniques. However, envelopment may also be created by simple psychoacoustic tricks as was hinted at in the previous section.

Since the acoustic distance between the ears is about 16 cm—roughly half a wavelength at 1 kHz—one finds that it is quite easy to produce BSI in this frequency range because even small amounts of time delay fluctuations of about 1 ms or more between the signals arriving at the ears will lead to the formation of spaciousness. At frequencies below 0.5 kHz, it is found that loudspeakers placed diametrically on the ear axis produce maximum spaciousness. This leads to the spaciousness heard when listening to a stereo loudspeaker pair increases as the angle subtending the loudspeakers increases (side wall reflections neglected).

It stands to reason that the more diffusely reflecting the walls behind the listener, the larger will the LEV be. Reflections from the ceiling are not necessary to create large LEV although the feeling of sound coming from the upper hemisphere rather than from the horizontal plane is improved by sound having arrivals that have elevation angles of 30° or more above the horizontal plane. The reverberant sound in a concert hall may or may not provide the feeling of spatial diffuseness. For example, the sound under a deep auditorium balcony may be uncorrelated at the ears, but it will still appear as coming from the balcony opening in front of the listener.

The LEV is related to the IACC, and the lower the IACC, the more evenly distributed will LEV be around the listener. The IACC is usually measured using a binaural manikin as described in Chapter 14. Room design can be used to enhance the LEV as can using loudspeakers that radiate much sound away from the listener.

Besides its width, the auditory event may appear at different distance from the listener. Loudness is the most important factor to determine the subjective distance: weak sounds appear far while strong sounds appear close. Timbre may contribute to distance perception since low-frequency sounds become louder when sound pressure level increases as shown in Figure 7.9.

Familiarity with the sound typical for the source is important for distance perception. There is no natural distance for a pure tone, for example, since such tones do not appear in nature. Experienced listeners can determine within a few dB what the sound level should be in a concert hall using ASW, LEV, and visual and cognitive cues. Familiarity with the sound typical for the source is important for distance perception. There is no *natural* distance for a pure tone, for example, since such tones do not appear in nature.

8.7.5 Simulated diffusivity

It is to be expected that as the number of incident sound directions (and sound sources in sound reproduction) grows and as they become increasingly uncorrelated (as in late reverberation), the auditory event will become large and unfocused.

Experiments made in an anechoic chamber with a number of loudspeakers radiating uncorrelated noise have shown that even with as few as four uncorrelated noise, sound field components, spatially evenly distributed in a plane at a small elevation angle, are sufficient to provide a spatially diffuse sound field as shown in Figure 8.28. The more uncorrelated the ear signals, the larger was the diffusely perceived solid angle of incoming sound. For an even distribution of the sound field in the upper hemisphere, a requirement for good spaciousness, the IACC value should be smaller than about 0.1 [42].

These results indicate that a 5-channel surround system could provide diffuse reverberant sound, but in the standard implementation, the rear channels are compromised (for practical reasons) to have a too large opening angle, usually 110°–120° wide, that results in weak or nonexistent auditory events to the back. A 6- or 7-channel system with smaller interloudspeaker angles in the rear, similar but not equal to those of the front loudspeakers, will provide better subjective diffusivity and more precise localization of rear sound sources.

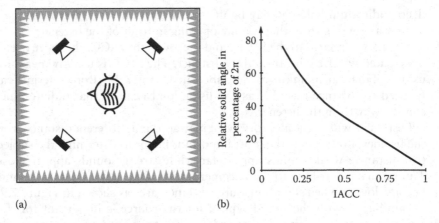

(a)   (b)

Figure 8.28 (a) Loudspeakers used for presentation of four uncorrelated signals in anechoic chamber, and (b) perceived solid angle of the sound field in the upper hemisphere as a function of IACC. (After Damaske, P., *Acustica*, 19, 199, 1967/68.)

The investigations, however, only covered the limited frequency range of 0.25–2 kHz. In view of the different signal processing by hearing of sound field components above 2 kHz from that below 2 kHz, this leaves further research to be done. The answer to the question how many subwoofers are enough in low-frequency sound reproduction is related to these missing psychoacoustic experiments.

8.8 REVERBERATION

8.8.1 Ratio of direct to reverberant sound

The spaciousness of a sound field is related to the feeling of diffuseness but is also affected by the relative level of early reflected sound to late reflected, reverberant sound, the direct-to-reverberant energy ratio (H), Figure 8.29 shows the result of experiments in which the H was changed and how that affected the spaciousness of the sound field. One unit of change on the spaciousness scale corresponds to the JND, that is, where 50% of the time the test persons were able to discern a difference.

The importance of the early part of the reverberation is also shown by the curves in Figure 8.30 that show the subjectively judged reverberation time T_{subj} in the experiment as a function of H, with different reverberation times T_{60} as parameter. The results confirm the possibility of adjusting the subjectively perceived reverberation time by manipulation of strong early reflections.

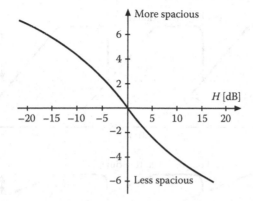

Figure 8.29 Subjectively perceived steps of spaciousness as a function of the level difference $H = 10 \log(E_{direct}/E_{reverberation})$ between direct and reverberant sound. Note that high spaciousness is obtained for small values of H. (After Kuttruff, H., *Room Acoustics*, 5th edn., Spon Press, London, U.K., 2009.)

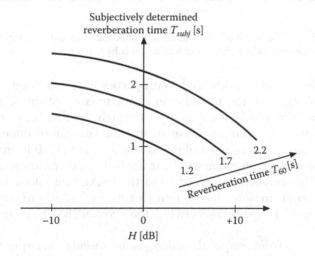

Figure 8.30 Subjectively perceived reverberation time T_{subj} as a function of H with reverberation time T_{60} as parameter. (After Kuttruff, H., *Room Acoustics*, 5th edn., Spon Press, London, U.K., 2009.)

8.8.2 Modulation and masking

The reverberation of a room changes the time envelope of the sound and gives it a fast growth and slow decay as indicated in Figure 8.31. Reverberation and early reflections reduce the modulation of the signal. It is in the initial fast growth at the start of the reverberation process that the early reflections can be noted.

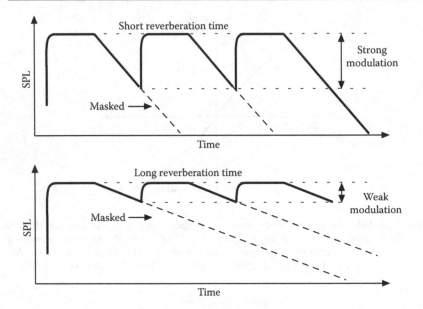

Figure 8.31 Reverberation of previous sound masks later sound. The modulation depth decreases as a result and less detail is heard.

The sliding time window and its properties were discussed in a previous section. Because of the modulation characteristics of music and speech, hearing is primarily focused on approximately the first 0.15 s of the reverberation process in *running* music or speech. This can be understood from the importance of masking as illustrated in Figure 8.31. It is only at the end of a piece of music that one will hear the full reverberation process so that we hear the reverberation recede into the background noise of the room. It is important to note though that in running music—to some extent—the difference in spatiality between the ESI and BSI helps overcome the masking.

Full release from temporal masking is not found to occur until after several hundred milliseconds after the end of the sound since the sliding time window is not rectangular as drawn previously in Figure 8.26 but slowly decaying from about 0.03 to 0.5 s as hinted at in Figure 8.24 [24,25].

Because the filter characteristics of the room feature strong resonant behavior, the sound in any reverberant room will have less modulation than the source signal. The MTF is used to describe the influence of reverberation on the signal envelope and is related to STI, discussed in Section 7.2.4. A similar music transmission index (MTI) has been defined [44]. Such a metric could be used to measure the clarity and transparency of a music room provided that it was suitably modified to take spatial room acoustic properties into account. This would require the development of a BMTI.

There are however many musical styles and performance types, and these would probably all require different BMTI values and other characteristics so it seems difficult to develop such a metric. In any case, this is an interesting field for future research into small-room acoustics and could result in much improved room acoustic metrics.

A simpler way of measuring the temporal envelope response of the room is to use tone bursts, rectangular or Gaussian shaped [45,46]. The envelope of such tone bursts will depend also on the loudspeaker driver's electroacoustic properties [47].

Under special circumstances a low reverberation level can be combined with long reverberation time generated electronically to provide an unrealistic but pleasant sounding reverberation.

8.8.3 Distance perception

The relative level difference between direct sound and early reflections to reverberation is, in addition to the sound level, an important cue to the perceived distance to the sound source. The arrival time difference between the direct sound and the reverberation will also influence distance perception. One way of measuring this difference is to use the center-of-gravity time t_g of the reverberation. The larger the room the later will the main contribution of reverberant energy be arriving. The center-of-gravity time t_g of an impulse response $h(t)$ is defined as [48]

$$t_g = \frac{\int_0^\infty t \cdot h(t)\,dt}{\int_0^\infty h(t)\,dt} \tag{8.3}$$

As indicated by the previous discussion of ASW, LEV, and the balance between reverberant and direct/early reflected sound, they all will influence the distance perception in the sound stage. Sound reflected laterally by the room surfaces (or provided by multiple loudspeakers) will increase the relative level of surround sound. However, both the reverberation time and the reverberation level and process details affect the distance perception.

It is difficult to generate a plausible impression of a low-frequency sound source being far away in a small room without a sufficiently long reverberation process. The analysis time frame discussed earlier is frequency dependent, and at low frequencies, it may be a few hundred milliseconds long. In small rooms, the growth and decay of sound may essentially already be over within that time. Our impression of far away thunder outdoors depends acoustically on the many reflections that reach us with long time delays.

In most practical listening occurring in small rooms, the listener will be sitting or standing so the height of the listener's ears above floor level will be about 1.1–1.7 m. In conversation as well as in music reproduction, the sound source will be in about the same height. Because most floors are acoustically hard at low frequencies (a thin carpet have negligible sound absorption at low frequencies), there will be reflection of the sound from the floor. Similar reasoning applies to the sound reflected by the ceiling since most ceilings in cold climates will be made of plaster, gypsum boards, and other (relatively) heavy materials. For the sake of the discussion, both floor and ceiling may be assumed acoustically fully reflective. Also the sound sources that one usually finds in small rooms may usually be assumed to be point sources. This means that there will be a possibility to estimate the nearness of sound sources using the timbre change caused by the floor and ceiling reflections.

As an example, consider the geometry of the stereo listening setup shown in Figure 8.20. In a living room the sides of the triangle may be 2-3 m long, the height of the listener's ears and the sound sources about 1.2 m. The ceiling height can further be assumed to be 2.4-2.5 m in many dwellings.

In such a room the first-order reflections affect the timbre change since the sound reflected by the floor and ceiling will occur in approximately the same cone of confusion. There will be cancelation of sound pressure at the listener for those frequencies where the distance traveled by the reflected sound is an odd multiple of one-half wavelength, that is at about 0.2 and 0.6 kHz. Because of the directivity of the sound sources, sound absorption by carpets and other floor coverings, and other similar effects, the interference dips in the frequency response will be less pronounced at frequencies above 0.5 kHz.

The 0.2 kHz interference dip occurs at about twice the frequency of the interference dip in main floor seating in concert halls [49]. While the frequency response dip in concert halls becomes increasingly important on the concert hall main seating as one moves away from the stage, the frequency response dip will be about the same in most homes. Most recording of classical music is done either using close-up or suspended microphones so the interference dips that occur in recording are likely never to be those that occur in listening in the home or the concert hall (except at balcony fronts as mentioned previously).

REFERENCES

1. Burkhard, M.D. and Sachs, R.M. (1975) Anthropometric manikin for acoustic research. *J. Acoust. Soc. Am.*, 58, 214–222.
2. Blauert, J. and Braasch, J. (2011) Binaural signal processing. Preprint *IEEE Conference on Digital Signal Processing 2011*, Corfu, Greece.

3. Ando, Y. (2012) *Architectural Acoustics: Blending Sound Sources, Sound Fields, and Listeners*. Springer, New York. Reprint of the original 1st edn. 1996.

4. Blauert, J. (1996) *Spatial Hearing*. The MIT Press, Cambridge, MA.

5. Langendijk, E.H. and Bronkhorst, A.W. (2002) Contribution of spectral cues to human sound localization. *J. Acoust. Soc. Am.*, 112(4), 1583–1596.

6. Bloom, P.J. (1976) Creating source elevation illusions by spectral manipulation, in *Proceedings of the 53rd Audio Engineering Society Convention*, Zürich, Switzerland, preprint E-1.

7. Olive, S.E. and Toole, F.E. (1989) The detection of reflections in typical rooms. *J. Audio Eng. Soc.*, 37(7/8), 539–553.

8. Ando, Y. (2011) *Concert Hall Acoustics*. Springer, Berlin, Germany. Softcover reprint of the original 1st edn. 1985 edition.

9. Ando, Y. and Gottlob, D. (1979) Effects of early multiple reflections on subjective preference judgments of music sound fields. *J. Acoust. Soc. Am.*, 65(2), 524–527.

10. Buchholtz, J.M. et al. (2001) Room masking: Understanding and modelling the masking of reflections in rooms, in *Proceedings of the 110th Audio Engineering Society Convention*, Amsterdam, the Netherlands, preprint 5312.

11. Kuttruff, H. (2009) *Room Acoustics*, 5th edn. Spon Press, London, U.K.

12. Salomons, A.M. (1995) Coloration and binaural decoloration of sound due to reflections. Dissertation, Technische Universiteit Delft, Delft, the Netherlands.

13. Knudsen, V.O. (1962) Acoustics of music rooms, in *Fourth International Congress on Acoustics*, Copenhagen, Denmark, paper M27.

14. Moorer, J.A. (1979) About this reverberation business. The MIT Press, *Comput. Music J.*, 3(2), 13–28.

15. Barron, M. (1971) The subjective effects of first reflections in concert halls. *J. Sound Vib.*, 15(4), 475–494.

16. Seraphim, H.P. (1961) Über die Wahrnehmbarkeit mehrerer Rückwürfe von Sprachschall. *Acustica*, 11, 80.

17. De Boer, K. (1941) Stereofonischegeluidsweergave. Dissertation, Applied Sciences, Technical University of Delft, Delft, the Netherlands.

18. Franssen, N.V. (1960) Some considerations on the mechanism of directional hearing. Dissertation, Electrical Engineering, Mathematics and Computer Science, Technical University of Delft, Delft, the Netherlands.

19. Haas, H. (1951) Über den Einfluß eines Einfachechos auf die Hörsamkeit von Sprache. *Acustica*, I, 49. (translation appears as Haas, H. [1972] The influence of a single echo on the audibility of speech. *J. Audio Eng. Soc.*, 20(2), 146–159.)

20. Meyer, E. and Schodder, G.R. (1952) Über den Einfluß von Schallrückwürfen auf Richtungslokalisation und Lautstärke bei Sprache. Drittes, Physikalisches Institut der Universität, Göttingen, Germany.

21. Atal, B.S. and Schroeder, M. (1966) US pat. nr. 3236949 APPARENT SOUND SOURCE TRANSLATOR.

22. Madsen, E.R. (1970) Extraction of ambiance information from ordinary recordings. *J. Audio Eng. Soc.*, 18(5), 490–496.

23. Wilson, K. (2005) Learning the precedence effect, in *2005 IEEE Workshop on Applications of Signal Processing to Audio and Acoustics*, New Paltz, NY.

24. Griesinger, D. (1996) Spaciousness and envelopment in musical acoustics, in *Proceedings of 101st Audio Engineering Society Convention*, Los Angeles, CA, Paper Number: 4401.
25. Krokstad, A. (2012) *Reflections on Sound. Music and Communication*. NTNU Norwegian University of Science and Technology, Faculty of Information Technology, Mathematics and Electrical Engineering, Acoustics Research Centre, Trondheim, Norway.
26. Lochner, J.P.A. and Burger, J.F. (1964) The influence of reflections on auditorium acoustics. *J. Sound Vib.*, 1(4), 426–454.
27. Zwicker, E. and Fastl, H. (1990) *Psychoacoustics*. Springer Verlag, Berlin, Germany.
28. McGurk, H. and MacDonald, J. (1976) Hearing lips and seeing speech. *Nature*, 264, 746–748.
29. Peissig, J. and Kollmeier, B. (1997) Directivity of binaural noise reduction in spatial multiple noise-source arrangements for normal and impaired listeners. *J. Acoust. Soc. Am.*, 101(3), 1660–1670.
30. Hawley, M.L., Litovsky, R.Y., and Culling, J.F. (2004) The benefit of binaural hearing in a cocktail party: Effect of location and type of interferer. *J. Acoust. Soc. Am.*, 115(2), 833–843.
31. Toole, F. (2008) *Sound Reproduction: The Acoustics and Psychoacoustics of Loudspeakers and Rooms*. Focal Press, Oxford, U.K.
32. Toole, F.E. (1990) Loudspeakers and rooms for stereophonic sound reproduction, in *Audio Engineering Society Eighth International Conference on the Sound of Audio*, Washington, DC.
33. Bech, S. (1998) Spatial aspects of reproduced sound in small rooms. *J. Acoust. Soc. Am.*, 103(1), 434–445.
34. Williams, M. (2004 and 2013) *Microphone Arrays for Stereo and Multichannel Sound Recording*, (Volumes 1 & 2) Editrice Il Rostro, Via Monte Generoso 6a, Milano, Italy.
35. Theile, G. (2001) Multichannel natural music recording based on psychoacoustic principles, in *Proceedings of Audio Engineering Society 19th International Conference*, Schloss Elmau, Germany, May 2001.
36. Toole, F.E. (1986) Loudspeaker measurements and their relationship to listener preferences. *J. Audio Eng. Soc.*, 34(4), 335–336.
37. Berg, J. and Rumsey, F. (2003) Systematic evaluation of perceived spatial quality, in *Proceedings of Audio Engineering Society 24th International Conference on Multichannel Audio*, Banff, Alberta, Canada.
38. Griesinger, D. (1999) Objective measures of spaciousness and envelopment, in *Audio Engineering Society Proceedings of the 16th International Conference on Spatial Sound Reproduction*, Rovaniemi, Finland.
39. Griesinger, D. (1997) Spatial impression and envelopment in small rooms, in *Proceedings of the 103rd Audio Engineering Society*, New York, Paper Number: 4638.
40. Morimoto, M. and Maekawa, Z. (1988) Effects of low frequency components on auditory spaciousness. *Acustica*, 66(4), 190–196(7).
41. Morimoto, M. and Iida, K. (1995) A practical evaluation method of auditory source width in concert halls. *J. Acoust. Soc. Jpn. (E)* 16(2), 59–69.

42. Damaske, P. (1967/68) Subjektive Untersuchung von Schallfeldern. *Acustica*, 19, 199.
43. Houtgast, T. and Steeneken, H.J.M. (1985) A review of the MTF concept in room acoustics and its use for estimating speech intelligibility in auditoria. *J. Acoust. Soc. Am.*, 77(3), 1069–1077.
44. Polack, J.-D., Alrutz, H., and Schroeder, M.R. (1984) The modulation transfer function of music signals and its applications to reverberation measurement. *Acta Acustica united with Acustica*, 54(5), 257–265.
45. Kihlman, T., Kleiner, M., and Kirszenstein, J. (1975) Problems in room acoustics. Speech intelligibility and modulation attenuation. Report 75-09, Department of Building Acoustics, Chalmers University of Technology, Gothenburg, Sweden (in Swedish).
46. Farina, A., Cibelli, G., and Bellini, A. (2001) AQT—A new objective measurement of the acoustical quality of sound reproduction in small compartments, in *Proceedings of the 110th Audio Engineering Society Convention*, Amsterdam, the Netherlands, paper 5283.
47. Kleiner, M. (2013) *Electroacoustics*. CRC Press, Boca Raton, FL.
48. Kürer, R. (1969) Zur Gewinnung von Einzahlkriterien bei Impulsmessungen in der Raumakustik. *Acustica*, 21, 370–372.
49. Schultz, T.J. and Watters, B.G. (1964) Propagation of sound across audience seating. *J. Acoust. Soc. Am.*, 36, 885–896.

Chapter 9

Sound reproduction in small rooms

9.1 SMALL SPACE

This chapter discusses the reproduction of speech and music over loud-speakers in small rooms. All the three methods common in room acoustics, that is, physical, statistical, and geometrical acoustics, are required to fully understand the acoustic behavior of the room from the viewpoint of sound reproduction. The sound reproduction system requirements together with the loudspeaker characteristics determine the desired acoustic conditions for the listening room. Since loudspeakers interact with the acoustics of the room, the chapter also reviews loudspeaker properties.

The most neutral sound reproduction is usually not the goal of sound reproduction except in the control rooms of sound recording studios. In homes, both for music appreciation and home movie theater, the goal for many listeners is rather *suspension of disbelief* a term used in presence research. For sound reproduction in homes, the ambition is to recreate the sound space intended by the recordist. For classical music, it often means that we wish to transport the listener to some real or imaginary concert hall or opera, for example, in this case, the reproduced sound is influenced by the sound character of the musical instrument, the room and microphones used for the recording, the acoustic properties of the playback room, the electroacoustic properties of the loudspeakers used, and the placement of the source and listener in the room.

For music that relies on electronic generation or amplification, the studio engineer can create acoustic worlds that do not or cannot exist in reality, such as rooms that have 30 s long reverberation times, but that still need to be rendered as intended.

In either case, the distributed recorded sound is typically intended to be listened to in a small room or cabin, or over headphones. It is difficult to express the sound space of such small room satisfactorily by physical mea-surement. Physical metrics, suitable for the description of such properties, are under development for description of acoustics of large auditoria with some success, but the acoustics of small rooms have not received the same attention. Subjective measurement of the sound stage uses ideas such as distance perception, auditory source width, and listener envelopment and is described in Chapter 8.

A central problem contributing to the difficulties met with in small contrary to large room acoustics is that the acoustic properties of the room boundaries change drastically over the audio frequency range. This requires a combined approach using separate acoustics methods for low and high frequencies as required by the properties of the sound signals and hearing.

9.2 TWO APPROACHES

9.2.1 Low- and high-frequency room acoustics

The two ideas around that most discussions of acoustics of small rooms tend to focus are early reflections and modes or roughly high and low frequencies, respectively. It is difficult to put an exact frequency limit between the two regions since room acoustics, psychoacoustics, and sound source characteristics all influence the decision. We listen in different ways to low- and high-frequency sounds. This may have to do with the properties of hearing (such as binaural properties), with the properties of voice and music, or a combination.

As humans, we are specialized in voice communication so it seems natural to assume that modulation envelope characteristics would be very important in the speech information carrying frequency range, that is, primarily above 250 Hz, whereas at lower frequencies, other strategies would be preferred by hearing. The poor sensitivity of hearing at low frequencies indicates that these—in a natural environment—were not very important.

The Schroeder frequency criterion, Equation 2.50, is sometimes suggested as a way of separating the low- and high-frequency regions of room acoustics. Its value depends on room volume and damping and is typically 0.15 to 0.3 kHz for domestic rooms. Above the Schroeder frequency, the resonances are typically so wide that there are at least three mode resonances per mode bandwidth. The Schroeder frequency has however little bearing on the low-frequency limit by psychoacoustic considerations, often set to 0.4 kHz.

The typical reverberation time in living rooms is in the range 0.4–0.5 s with a corresponding mode bandwidth of 4–5 Hz. The more resonances the more diffuse the sound field since each resonance can be associated with a set of propagation angles for the waves that combine to make up the individual mode.

At frequencies well below the Schroeder frequency, the number of modes is sparse. It is usually difficult to separate other than the first 10 modes by measurement of the room's transfer function at one receiving point only. The frequency curve becomes too complex to allow visual separation of the modes. Subwoofers are often used to reproduce frequencies in this *sparse* mode low-frequency region below about 80 Hz, below the fundamental frequencies of the human voice. Measuring amplitude and phase at several positions spaced about one-half wavelength, it is possible to separate the various modes for analysis.

The modal approach of room acoustics is best used to analyze room behavior for sounds that have long onset, steady state, and decay times such as low-frequency bass sounds from pipe organs and wind instruments. Some bass instruments such as bass guitars and pianos have strong initial transients, but also extended decay times, so both low- and high-frequency analysis methods are necessary in considering room behavior for these types of sound.

9.3 LOW FREQUENCIES

9.3.1 Modes

Modal resonances can be considered a result of waves moving in opposite directions along a closed path as explained in Chapter 2. The more rigid the surfaces and the smaller the sound absorption, the more resonant will the space be and the more noticeable will the modes' resonance and spatial characteristics be. Characteristic for resonant systems is the frequency-dependent response of the system to excitation at the resonance frequencies and the long rise and decay times.

9.3.2 Radiated power and source placement

The room can be regarded as a physical filter that alters the time history of the source signal by inflicting its impulse response on the signal. To what extent this is audible depends on the signal's properties as well as on the room, source, and receiver properties: the resonance frequencies of the modes, their damping, and the locations of the source and receiver relative to the mode eigenfunctions as given by Equation 2.39.

The importance of source location on the radiated power was discussed in Chapter 2 with relation to the modes, their eigenfunctions, and damping. If the loudspeaker is small and has high internal impedance, its influence on the damping of the modes will be small. In practice, however, many loudspeakers, because of their low internal impedance in the range close to the driver resonance frequency, add considerable damping to the low-frequency modes. A room that has many loudspeakers will usually sound better because of both the added damping to the low-frequency modes and the scattering provided by the loudspeaker boxes at medium and high frequencies. This may be an important effect in the use of full frequency range loudspeakers in surround sound systems. One of the reasons audio

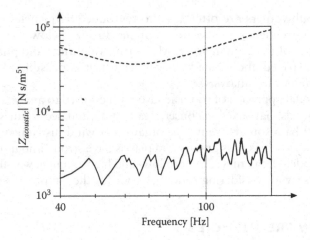

Figure 9.1 Examples of measured acoustic impedances: solid line shows the room's acoustic impedance in a room corner. Dashed line shows acoustic impedance of the small loudspeaker used in the measurement. (From Kleiner, M. and Lahti, H., Computer prediction of low frequency SPL variations in rooms as a function of loudspeaker placement, in *Proceedings of the 94th Audio Engineering Society Convention*, Berlin, Germany, 1993.)

equipment showrooms tend to have better bass reproduction than homes is the presence of many loudspeakers that together contribute the damping.

Figure 9.1 shows the measured acoustic impedances in a room corner and of a small loudspeaker [1]. The rectangular room had dimensions that were $5.45 \times 4.85 \times 3.55$ m^3 and the reverberation time was in the range 1–2 s. The loudspeaker was set on a small custom-built closed enclosure of about $0.4 \times 0.2 \times 0.1$ m^3 with an accelerometer mounted on the 6.5 in. diameter driver diaphragm to measure its velocity.

The figure shows that with about 10 such loudspeakers the room and loudspeaker impedances would be of about the same magnitude so there could be strong modal damping because of the loudspeakers.

The human voice is also affected by the acoustic properties of the environment. When speaking or singing in a suitably resonant environment, it is easy to sense the extra acoustic impedance load of the room on the voice mechanism at the mode resonance frequencies. Putting a short horn or tube in front of one's mouth one can notice the impedance change.

9.3.3 Frequency response

The response of a resonant environment can be characterized by the transfer function usually defined as the ratio between the sound pressure at the measurement point and the volume velocity of the source. The log magnitude of the transfer function can be thought of as the frequency response of the room for steady-state sound sources. For low frequencies, the transfer function of the room can be measured using reciprocity as discussed in Chapter 14. The frequency response gives a simple possibility to determine the reverberation time of low-frequency modes from their bandwidth since the –3 dB bandwidth Δf of a resonance depends on its reverberation time T_{60} as discussed in Section 2.3.4.

$$\Delta f \approx \frac{2.2}{T_{60}} \tag{9.1}$$

Using the FEM, we can determine the mode bandwidth simply and with high precision as described in Chapter 13. Using the FEM, we can also compare room properties in a way not possible when doing measurements since the air is at rest, and air temperature and room properties can be kept fully constant. It is thus of interest to use the method to analyze the properties of a simple listening room by calculating some transfer functions from loudspeaker to the listener.

Figure 9.2 shows a modeled room in which the four walls were assigned a surface impedance $Z_W = 8\varrho_0 c$, while the rest of the surfaces were assumed

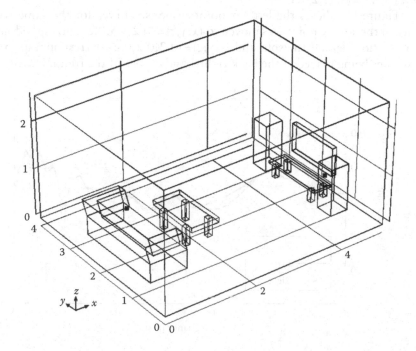

Figure 9.2 An imaginary listening room with length $x = 5$ m, width $y = 4$ m, and height $z = 2.4$ m. The wall impedances were assigned as described in the text so that the reverberation time was about 0.5 s. (Geometrical model courtesy of Comsol.)

to have no sound absorption and were assigned $Z_W = \infty$. The dimensions of the room were assumed as length $x = 5$ m, width $y = 4$ m, and height $z = 2.4$ m. The modes of such a room will have a reverberation time of about 0.5 s. The source position was assumed as the diaphragm of the right-hand loudspeaker at $(x;y;z) = (4.39;1;0.8)$. The listener was not modeled but the listening positions of the two ears of an imaginary receiver at the sofa were modeled as two points at $(x;y;z) = (0.65;2.1;1)$ (left ear) and $(x;y;z) = (0.65;1.9;1)$ (right ear). The loudspeaker was assigned a volume velocity of $Q = 1$ m³/s. Note that a real woofer loudspeaker would likely have a volume velocity that drops by about –6 dB per octave over its resonance frequency [38]. The influence of this effect is not shown here.

Figure 9.3 shows the frequency response of the room for the two *ears* of the imaginary listener. The spacing of 0.2 m between receiver points corresponds roughly to the interaural distance.

Figure 9.4 shows the left ear point response curves for the same room when the receiver point was moved 0.4 m to the left side $(x;y;z) = (0.65;2.5;1)$ or 0.4 m to the front $(x;y;z) = (1.05;1.9;1)$. For comparison, the frequency response curves for a point represent that of the left ear of a standing person at $(x;y;z) = (1.45;2.1;1.7)$.

Figure 9.5 shows the left ear point response curves for the same room when the source point was moved to $(x;y;z) = (0.2;1.5;0.2)$ corresponding to a position behind the sofa and $(x;y;z) = (0.2;0.2;0.2)$ corresponding to the woofer being placed at the back right-hand corner of the room. Having the

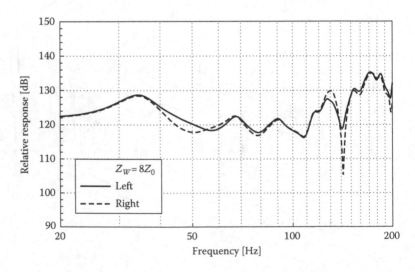

Figure 9.3 Frequency response curves calculated by the FEM for the rectangular room shown in Figure 9.2 driven by a loudspeaker with a constant volume velocity. The curves represent two receiver positions 0.2 m apart. The modal damping was adjusted so that the reverberation time was about 0.5 s at all frequencies.

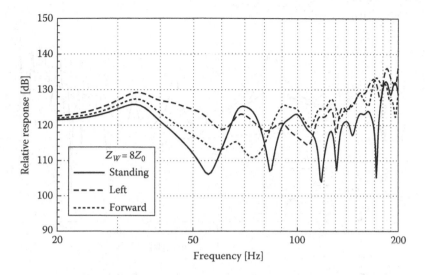

Figure 9.4 Frequency response curves calculated by the FEM for the rectangular room shown in Figure 9.2 driven by a loudspeaker with a constant volume velocity for three different receiver points representing listener standing in front of sofa and at points +0.4 m left and +0.4 m in front of sofa as described in text.

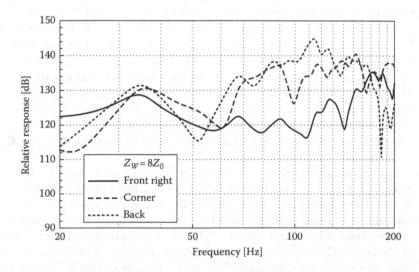

Figure 9.5 Frequency response curves calculated by the FEM for the rectangular room shown in Figure 9.2 driven by a loudspeaker with a constant volume velocity for three different source points representing the loudspeaker behind the sofa, in a corner, and at its normal front right positions as described in text.

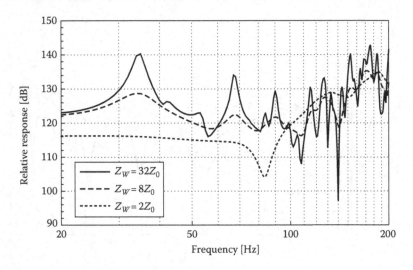

Figure 9.6 Frequency response curves calculated by the FEM for the rectangular room shown in Figure 9.2 driven by a loudspeaker with a constant volume velocity for three different wall impedances Z_W. Listener at the center of sofa and at the front right-hand loudspeaker as described in text.

loudspeaker back of the sofa in this case results in about the same coupling as the corner position. As expected, placing the loudspeaker at some distance from the corner results in the most linear frequency response.

Low-frequency sound reproduction attributes such as *boominess* and *boxiness* are caused by the uneven frequency characteristics of reverberation time, which is coupled to modal damping [2]. Figure 9.6 shows the frequency response at the left ear with the source at the diaphragm of the right-hand loudspeaker at $(x;y;z)=(4.39;1;0.8)$ when the wall impedances were set as $Z_W = 2\varrho_0 c$, $Z_W = 8\varrho_0 c$, and $Z_W = 32\varrho_0 c$, respectively, to have different modal damping. The low impedance corresponds a reverberation of about $T_{60} \approx 0.2$ s time and the high impedance to about $T_{60} \approx 1.6$ s.

The frequency response function for $Z_W = 32\varrho_0 c$ is seen to fluctuate considerably but most of the response is kept within a 20 dB wide SPL band as noted in Chapter 2. The curves for $Z_W = 8\varrho_0 c$ and $Z_W = 2\varrho_0 c$ are much smoother and more well controlled. In the midfrequency range well above the threshold of hearing, a 20 dB level range corresponds to a loudness ratio of about 1:4. However, because of the reduced sensitivity of hearing at low frequencies shown by the equal loudness contours in Figure 7.9, the loudness difference at low frequencies will be much higher. A single weak tone at a mode resonance may fluctuate between being audible and inaudible as one moves through the room because of the mode's geometric pressure distribution.

However, since most musical sounds have a fundamental and harmonics they will have similar timbre at different locations because of the tonal frequency components regenerated by the virtual pitch phenomenon, explained in Chapter 6.

9.3.4 Frequency shift

In the sparse mode range at low frequencies, a sine will force the vibration of a single mode if the sine's frequency is close to the mode's resonance frequency. When the sine stops sounding, a frequency shift effect may be audible. The energy that is stored by the room resonance will decay with the *free* natural resonance frequency of the room even if the applied tone had a slightly different frequency. Such shifts can be quite easily heard on isolated pure sine tones if the tone ends abruptly. The ear is most sensitive to frequency shifts of about 3–5 Hz [3].

For *running* music, the frequency shift (that can be thought of a slow modulation) will lead to *muddiness* at the lowest frequency in the music. Once several modes are excited by the tone, some will lead to positive while others to negative frequency shifts of the tone as it decays so a slight muddiness in low frequency sound reproduction will always be present at frequencies where the modal density is insufficient.

9.3.5 Room aspect ratios

The search for *optimum* room aspect ratios has long been a favorite topic for research [4–8]. Most work has been based on the assumption that useful results can be obtained by investigating various properties of the magnitude of the transfer function between two diagonal corners of a rectangular room. Typically, the work has been focused on the spacing, distribution, and damping of low frequency modes; a good review is given in Reference 9.

Various metrics have been suggested for expressing the goodness of these transfer functions as a function of room aspect ratios such as variance of spatial average (VSA) across seats across the frequency band of interest, mean spatial variance (MSV), and standard deviation of the modal frequency spacings.

There are numerous problems with this approach, the first being the default assumption of the room being rectangular, second the treatment of all modes being similarly important, and third, to name but a few, that the damping of all modes is similar. Axial modes are typically less damped than tangential and oblique modes, for example [10]. Since any object present in the room changes the frequency distribution, spatial properties, and damping of the modes, any work along these lines can only hint at the properties that should be strived for.

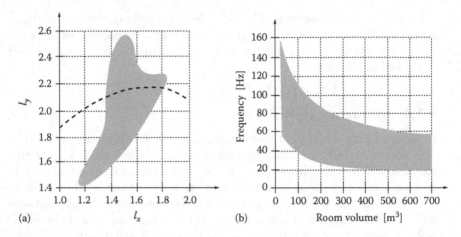

Figure 9.7 (a) Gray area shows Bolt's criterion for rectangular room dimension ratios that give the best spacing of mode resonance frequencies in rectangular rooms. Region below dashed line fulfills condition $l_{max} < 1.9V^{1/3}$ for nonextreme shape for physical diffusivity according to the ISO 354 standard [49]; see Chapter 3. (b) Gray area shows range of validity for (a). (After Bolt, R.L., *J. Acoust. Soc. Am.*, 18, 130, 1946.)

It is also well known that it is useful for the room to be asymmetric at low frequencies so that modal nulls are avoided. Because the frequency response varies with loudspeaker and listener positions, the frequency response between room corners is not useful in practice except for finding all modes.

Initially, the scientific discussion centered on the spectral distribution of the mode frequencies. When two modes have the same frequency, they are often referred to as degenerate or coincident. By choosing room dimensions carefully, one can adjust the resonances to be evenly distributed in frequency. The gray area in Figure 9.7 shows Bolt's tentative room proportion criterion assuming $l_z = 1$. The gray area in the graph in Figure 9.7a shows a region that encloses the room dimension ratios that have the smoothest frequency response at low frequencies in a small rectangular room based on a frequency spacing index. Damping must also be considered to obtain a true measure of the frequency response irregularity. The criterion indicates that a relatively broad area of $l_x{:}l_y$ values may be acceptable. Figure 9.7b shows the range of validity for Figure 9.7a as a function of room volume and frequency. Note that the lower boundary of the region of validity does not correspond to the very lowest mode frequencies in the room [4].

The Bonello criteria are based on similar thinking as Bolts with the difference of being based on mode energy rather than frequency spacing [5]. In using the Bonello criteria, one assumes that each mode has the same damping and is equally excited. The criteria are as follows: (1) within a ⅓ octave band, there should not be any coincident modes, unless compensated by several additional noncoincident modes, and (2) each higher frequency ⅓ octave band should contain at least as many or more modes than the one below it. In large spaces, the modal density increases monotonically and smoothly with frequency.

Both Bolt's and Bonello's approaches suffer the same problem in that they do not take into account the geometrical properties of the eigenfunctions and the source and transmitter locations. Two different criteria are VSA, the degree of flatness of the spatial average of the receiver locations in the room, and MSV, a measure of how consistent the amplitude response is between receiver locations, as further discussed in Chapter 10 [8]. The curves in Figure 9.8 show the calculated VSA and MSV distributions at receiver points spaced 0.15 m, about 1.2 m above the floor in rooms of different l_x and l_y dimensions. The rooms were assumed to have a 2.7 m ceiling height.

In an experiment, a figure of merit based on the frequency response deviation from a linear frequency response was compared to subjectively determined preference ratings using a sound field simulation technique called auralization [20]. In this technique, binaural head related transfer functions are calculated or measured. One can then listen to the influence of the room *filter* on the transmission path from source to receiver by convolving speech and music with the transfer functions. One can either use complex transfer functions calculated, for example, by the

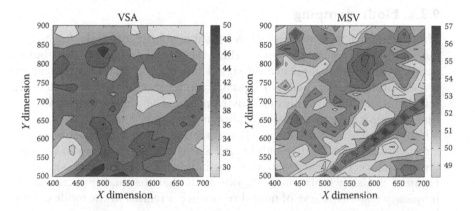

Figure 9.8 Welti's VSA and MSV quality criteria as functions of room dimensions for rectangular rooms with source in the front left corner (ceiling height = 2.7 m). Multiple receivers at 15 cm spacing. It can be seen that the optimum dimensions (small values) agree reasonably well with the lower left corner area of Bolt's criterion in Figure 9.7a. (After Welti, T. and Devantier, A., Low frequency optimization using multiple subwoofers, *J. Audio Eng. Soc.*, 54, 347, 2006.)

FEM or, since the transfer function of a filter is the Fourier transform of its impulse response, use calculated impulse responses more common with geometrical acoustics software.

In an experiment, the transfer function for the frequency range 20–250 Hz was used for auralization. The results showed that test persons prefer "a smooth characteristics that it brings, such as short decay times and punchy bass" [11]. An interesting finding was that in these tests, the musical samples were unimportant although these were fairly limited in scope. Regrettably, the electroacoustic part of the experiment is not specified in any detail.

Based on Reference 12, the recommendations by the European Broadcasting Union give a wide range of acceptable room aspect ratios as shown by the following mathematical formulas:

$$\frac{1.1L_y}{L_z} \le \frac{L_y}{L_z} \le \frac{4.5L_y}{L_z} - 4$$

$$L_x < 3L_z \qquad\qquad (9.2)$$

$$L_y < 3L_z$$

The British Standards Institute and the International Electrotechnical Commission give similar criteria and recommend a standard room of $7 \times 5.3 \times 2.7$ m^3 [12].

9.3.6 Mode damping

Mode damping is a much more reasonable subject of study for achieving good low-frequency response than the modal frequency distribution detail. For the least influence on signal modulation and thus *punch*, the reverberation time of the low-frequency modes should be shorter than the reverberation times of musical instruments. The shortest reverberation times of musical instruments were shown in Chapter 8 to be in the range of 50–300 ms. The quality factor (Q) is sometimes used as a measurement of a resonant system's relative bandwidth and is defined as the ratio between the resonance frequency and the one-half energy bandwidth shown in Figure 2.9. Using Equation 2.44, one finds that the quality factor of the frequency response curve of modal resonance is related to the mode's reverberation time T_{60} as

$$T_{60} = \frac{3\ln(10)}{\delta} = \frac{6\ln(10)Q}{\omega} \approx \frac{2.2Q}{f} \qquad\qquad (9.3)$$

The reverberation time in typical living rooms is about 0.4 s over a wide frequency range as mentioned in previous chapters. Subjective testing can be used to determine the subjectively acceptable Q-values and reverberation times of modes [13,14]. One such study indicates that Q-values lower than 16 corresponding to $T_{60} = 1.1$ s at 32 Hz and $T_{60} = 0.28$ Hz at 125 Hz could not be differentiated [14]. This corresponds well to the reverberation times of acoustically dry control rooms that are recommended to be lower than 0.3 s and to the reverberation times of musical instruments.

9.3.7 Modulation transfer function

The idea of the MTF was mentioned in Chapter 8. The MTF shows how much of the modulation depth in a signal that is retained as the signal is transferred over a path, for example, from loudspeaker input signal to microphone output signal [15]. Since both loudspeaker and room are resonant, they must both be included. The more transient the signal the more significant will the modulation reduction due to the resonances be. A value of MTF = 1 is ideal, that is, there is no modulation reduction. An illustration of the modulation reduction by resonances at midfrequencies can be heard in Reference 16. Figure 9.9 shows some measured results for the MTF of a small room in the 52–103 Hz frequency range for dipole and monopole low-frequency loudspeakers. The dipole loudspeaker measured here allows better retainment of the modulation than the monopole loudspeaker [17].

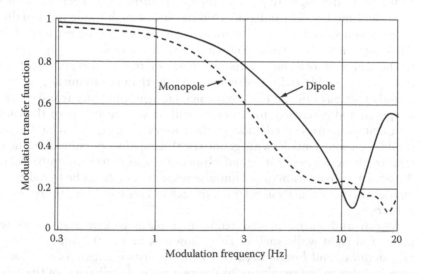

Figure 9.9 An example of the MTF of monopole and dipole woofers in a rectangular living room over a 52–103 Hz frequency bandwidth. (From Linkwitz, S., Investigation of sound quality differences between monopolar and dipolar woofers in small rooms, in *Proceedings of the 105th Audio Engineering Society Convention*, San Francisco, CA, Paper 4786, 1998.)

9.3.8 Source or sources

The frequency response at a number of subwoofer locations in a small room was calculated for various listener locations in an extensive numerical experiment as discussed in Chapter 10 [18]. It was also shown that four subwoofers surrounding the immediate listening area gave very good results indicating that listening locations that are in the near field of the subwoofers are a good choice [10].

A more advanced approach is to use compensating subwoofer loudspeakers at the rear of the room. These loudspeakers are fed the same signals as the front loudspeakers but with inverted polarity as further explained in Chapter 10 [19]. If there were no delay, the rear loudspeakers would simply increase the excitation of odd axial modes. With the delay, the forward propagating mode does not see an impedance discontinuity since the pressure at the rear wall will now be that expected for a plane wave (which has wave fronts parallel to the wall) in free space. The effect is that of a plane propagating wave as would be had in an infinitely long tube.

9.4 MID- AND HIGH FREQUENCIES

9.4.1 Reflections: Single and multiple

In the mid- and high-frequency range, it is difficult to keep track of the individual modes. The individual properties of each mode are usually of little interest since many modes are active at the same time because of their damping that leads to wide bandwidths and large modal overlap.

The transient behavior of the acoustics of rooms is typically of more interest in the mid- and high-frequency range than the frequency response. Sounds that have major high-frequency content and start transients such as speech and percussive instruments will be very revealing of the room's transient properties. Prediction of the room's acoustic behavior for these sounds is usually made with geometrical acoustics, possibly modified to take frequency-dependent sound absorption and scattering into account. A useful approximation of the impulse response can often be found by geometrical acoustics analysis adapted to each octave band for absorption and scattering.

Speech and music played back in a plain rectangular room with plane and rigid walls such as that shown in Figure 9.2 sounds unnatural, *metallic*, and *brittle* because of the repetitive nature of the impulse response due to the regularity of the geometric distribution of the mirror image sources and wall smoothness. Mirroring the source as shown in Figure 3.5 results in an infinite number of image sources. Diffusion must be added by wall unevenness, even so typically the floor and ceiling must be treated first.

By introducing frequency-dependent sound-absorption coefficients to simulate the effects of wall sound absorption, by simulating wall unevenness that creates diffusion, by slight randomization of the mirror image sources to simulate building imperfection, and by sound absorption by air, the sound quality of the auralization can be made plausible.

With randomization, and some filtering to simulate the high-frequency sound absorption by air, the reverberation can be made to sound very pleasant and natural. With randomization, and some filtering to simulate the high-frequency sound absorption by air, the synthesized reverberation can be made to sound very pleasant and natural. Physically, this corresponds to the unavoidable diffusion caused by surface and object irregularities that make a real rectangular room less so. The limited wall impedance contributes to sound absorption at both low and high frequencies. Impedance variations across walls contribute to diffuse reflection of sound.

9.4.2 Reverberation

For large rooms, the reverberation time is considered the most important room acoustic property next to loudness and background noise. The reverberation time recommended for various types of large rooms is shown in Figure 9.10 [21]. These requirements are based on subjective experience of the room for the best rendering of music and speech intelligibility. The

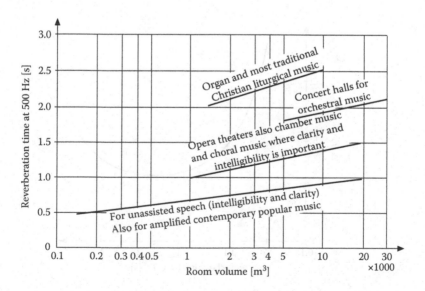

Figure 9.10 Typical recommended 0.5 kHz reverberation times for some common types of auditoria. (From Kleiner, M. et al., *Worship Space Acoustics*, J. Ross Publishing, Ft. Lauderdale, FL, 2011.)

reverberation time is accepted to be longer the larger the room as shown in the figure.

A characteristic of the reverberation in large rooms is that it starts to grow late: sound from the source must reach the walls to be reflected. A rectangular room that has a volume of about $12 \times 16 \times 25$ m³ can be estimated to have a mean-free-path length of about 10 m corresponding to a propagation time of about 28 ms. This implies that the reverberation would start to become appreciable after about 60 ms, since sound must have traveled at least two mean-free-path lengths for there to be an appreciable amount of diffuse reflections. This delay is one reason why we are very aware of the reverberation when listening in large rooms.

A second way of characterizing the start of the reverberation is to use the center-of-gravity time t_g introduced in Equation 8.3. For a perfectly exponential reverberation process, t_g is related to reverberation time T_{60} as [22]:

$$t_g = \frac{T_{60}}{13.8} \tag{9.4}$$

Finally, a third way is to analyze the impulse response of the room for when its energy integrated over about 30 ms has its maximum.

The ability of hearing to perceive changes in reverberation time is surprisingly good. As shown by the curve in Figure 9.11, a reverberation time change of less than 5% can be noticed (around 1–3 kHz) [23]. Usually, one

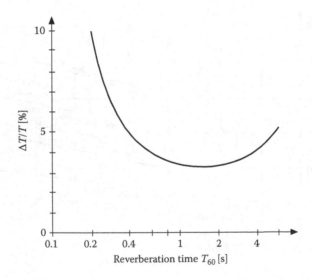

Figure 9.11 Just noticeable difference in percentage for various reverberation times. (After Cremer, L. et al., *Principles and Applications of Room Acoustics*, Vol. 1, Applied Science Publishers, New York, 1982.)

strives to achieve control of the reverberation time to an interval of ± 0.1 s at midrange frequencies in large auditoria.

In a small room, the mean-free-path length is short so the reverberation is also short compared to that of a large hall. Small listening rooms for loud-speaker reproduction of recorded voice and music (living rooms, control rooms, etc.) should generally have a reverberation shorter than that recorded.

They should also not have the reverberation time increase at low frequencies common in large rooms for *warmth* but have frequency-independent *flat* reverberation times in the audio range. Small listening rooms otherwise have a tendency to sound *boomy*.

Studies of the reverberation time measurements of domestic environments show a surprising agreement across continents for the reverberation time of domestic living room, with measured T_{60} in the range 0.4–0.5 s across the spectrum with a slight increase at very low frequencies in countries where masonry building is more common. Experience has shown that living rooms having reverberation times below 0.4 s are perceived as *dry*, whereas rooms that have reverberation times over 0.8 s as *live*. Figure 9.12 shows a recommended reverberation time range for control rooms [24].

The subjective preference for reverberation will be different depending on the type of playback used for the reproduced sound. The presence of listening room reverberation along with the recorded reverberation will be heard differently for mono, stereo, or multichannel surround sound playback. For the latter, the room reverberation contribution should be minimized. For mono and stereo playback, the room reverberation and early reflection characteristics should be fitted to the directivity of the loudspeakers.

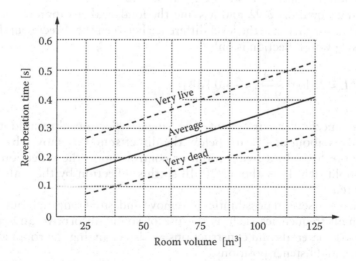

Figure 9.12 Suggested reverberation times for small rooms such as control rooms. (After Long, M., *Architectural Acoustics*, Academic Press, New York, 2013.)

9.4.3 Early reflections

An approximation from geometrical acoustics for a rectangular room that has mirroring walls and a volume of $V\,\mathrm{m}^3$ shows that the number of reflections ΔN within a time t after the radiation of the start of the sound is about [23]

$$\Delta N \approx \frac{4\pi c^3 t^3}{3V} \qquad (9.5)$$

Since classical music has an autocorrelation length of about 30–300 ms and because of the temporal integration properties of hearing as discussed in Chapter 7, it is reasonable to assume that the power of reflections is integrated by the ear over a time Δt of about 30 ms [25]. For a rectangular room having a volume of 50 m³, one then finds that the number of reflections over the first 30 ms is less than 100. Because of the power summation over this integration time, the sensitivity data for individual reflections as shown in Figure 8.10 should be reduced. The intensity of later reflections however is smaller both because of the extra traveled distance and the sound absorption (and scattering) by the surfaces.

The extra traveled distance by reflected sound components depends on room shape. Assume, for example, a small room of width × length × height 4 × 5 × 2.7 m³. With the listener on the center line and loudspeakers symmetrically placed on either side of the center line about 1 m from the front wall and about 0.7 m from the side walls and the listener about 2 m from the rear wall, the direct sound will arrive after about 6.9 ms and the first reflection (from the side wall) after about 9.8 ms.

Using Equation 2.92 and assume the loudspeaker omnidirectional, we find that for this case the level difference between the direct sound and the first side wall reflection is only

$$\Delta L \approx 20\log\left(\frac{6.9}{9.8}\right) \approx -3.0\ [\mathrm{dB}] \qquad (9.6)$$

In an anechoic environment, the single reflection threshold for click sounds is about at −50 dB below the direct sound as shown by the data in Figure 8.10. In a living room or control room, the reflection masked threshold (RMT) is about −20 dB so this reflection by the wall will still be noted.

There are several possibilities to remove undesired reflected sound at mid- and high frequencies such as the use of sound-absorptive and/or sound-diffusive materials and constructions as also changing the room geometry, source, and listener positions.

In laboratory environments such as anechoic chambers, the sound from the source will not be reflected by a floor, but in most practical listening situations, sound reflected by the floor will be a part of listening. It seems however that we are very skilled at not noticing this sound field component.

9.4.4 Sound absorption

The principles of sound absorption were discussed in Chapter 4. Data for sound absorption of porous materials are usually measured using either the impedance tube or the reverberation chamber methods. The tube method is roughly equivalent to measurement of the sound reflection for perpendicular sound incidence and reflection. The room method assumes that the incident wave field is diffuse, that is, all angles of incidence are equally probable and the intensity from all directions is the same, similar to that of the reverberant field in a room. In the listening room application of sound-absorbing materials, neither of these conditions apply to the early reflected sound by the surfaces. For a plane wave that is incident at an angle on a sound-absorbing material mounted on a hard surface, the sound absorption by theory follows the behavior shown in Figure 9.13.

Note that the sound absorption always has a maximum of $\alpha = 1$ at some angle and that a small wall impedance ratio $\zeta = Z_2/Z_0$ is favorable, that is, the absorber should be very porous and thick.

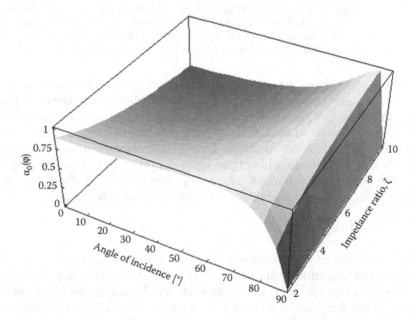

Figure 9.13 A surface showing the theoretical sound-absorption coefficient of a porous sound absorber as a function of the wall impedance ratio $\zeta = Z_2/Z_0$ and the angle of incidence of the plane sound wave.

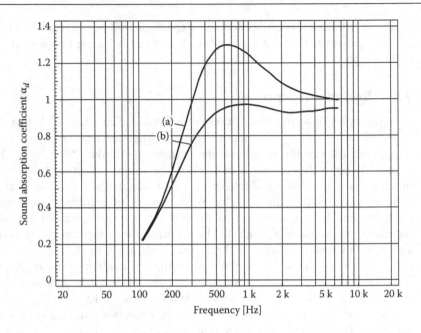

Figure 9.14 Calculated sound-absorption coefficients of a square patch of porous sound-absorbing material of different areas on a rigid surface, as a function of frequency. (After Kleiner, M., *Acoustics and Audio Technology*, 3rd edn., J. Ross Publishing, Ft. Lauderdale, FL, 2011.) The specific flow resistance was set to 10^4 kg/m^{3}s and the absorber thickness was set to 0.1 m. (Curves based on ⅓ octave-band values.) (a) 1 m^2, (b) 10 m^2.

Figure 9.14 shows the calculated sound-absorption coefficients of a square patch of a porous sound-absorbing material of different dimensions placed on a rigid surface, as a function of frequency [26]. Since the sound absorption can be made to peak for a certain wavelength/patch dimension ratio due to diffraction, subdividing a large absorbent surface to smaller, variously sized, widely distributed patches will optimize the sound absorption.

The sound level reduction ΔL of a ray due to a sound-absorbent specularly reflecting patch having a sound-absorption coefficient α is

$$\Delta L \approx 10 \log(1 - \alpha) \,[\text{dB}] \tag{9.7}$$

The intensity of late arriving reflections will depend strongly on reflection order (the number of reflections that they have been subjected to) and the sound absorption of the room surfaces. Typically, the room surfaces are assumed equally absorbing with a mean sound-absorption coefficient $\bar{\alpha}$ and a mean scattering coefficient $\bar{\delta}$. Both the sound absorption and the scattering by the reflecting surfaces are important for the strength of the sound in the room even for small values of $\bar{\alpha}$ and $\bar{\delta}$. In practice, $\bar{\alpha}$

is usually larger than 0.05 even for very rigid walls because of unavoidable porosity and heat transfer. A material that has a sound-absorption coefficient of 0.99 will reduce the level of the reflected sound by 20 dB and four reflections by 80 dB. This brings the level of the reflected sound components down to the characteristic background noise levels in test chambers, less than 15 dBA. The diffusion coefficient $\bar{\delta}$ is often in the range 0.2–0.8, resulting in an additional reduction of about –3 dB per reflection because of diffusion.

The geometrical sound attenuation in the example—due to the travel path for 30 ms (about 10 m)—is about 13 dB, which is quite negligible compared to the effect of the sound-absorptive or sound-diffusive treatment.

Some sound-absorption coefficients measured using the room method for materials commonly encountered in architectural acoustics are shown in Table 4.1.

9.4.5 Sound diffusion

Almost equally important to the sound absorption are the scattering properties of walls and objects in the room as discussed in Chapter 5. A prerequisite for the use of Eyring's and Sabine's formulas in Chapters 2 and 3 is that the sound field is diffuse, for example. Auralization of calculated room impulse responses clearly shows that adequate diffusion is very important for the subjectively experienced sound quality of a room [30].

The surface diffusivity and object scattering are difficult to estimate theoretically, and the acoustician will have to rely on experience of past use and the published measured characteristics of some few commercial products. A recommended alternative is to use physical scale modeling of diffuser properties, for example, in a 1:10 scale. Such diffusers can be machined quite easily or printed using a 3D printer.

It is important to note that the algorithms used for room acoustics prediction must take the actual diffuser response into account for best results which is difficult to achieve for practical reasons as discussed in Chapter 5. It is fairly easy to take surface diffusion into account approximately by using a Lambert diffusion characteristic for the wall. Surfaces such as those created by quadratic residue diffusers discussed in Chapter 5 have however very different characteristics that are best measured. The late reverberation is a result of many scattering processes and can be averaged, but the early reflections are best simulated with the response of the actual diffuser to be used.

9.4.6 Flutter echo

The comb filter and flutter echo effects of repetitive sequences in the impulse response were described in Chapter 8. Figure 8.16 shows the relationship between impulse response and its spectrum characteristics. The repetition is easiest heard with signals that have a wide spectrum (comb filter) and or are transient (flutter echo).

Typically, flutter echo is heard in small rooms where the sound bounces repeatedly between two opposite surfaces, for example, walls. For flutter echo to be heard, the repeated reflections must dominate the sound field because of insufficient scattering by the two opposing surfaces and excessive sound absorption by the remaining surfaces. Flutter echo is most easily heard for self-generated sound such as that from one's voice or musical instrument. If room conditions are prone to flutter echo, one can usually hear it by clapping one's hand. Flutter echo is most likely to be heard when the source of sound is omnidirectional, the sound is transient, and the listener is close to the line connecting the source and mirror images of the source.

Figure 9.15a shows a situation that is particularly prone to flutter echo. The situation shown in Figure 9.15b is less prone.

Industrially prefabricated room modules and corridors in which there are parallel surfaces that are undecorated and hard will often feature flutter echo. Splaying one wall by 7° or more as well as providing diffusion by random wall unevenness with an approximate depth of a quarter wavelength eliminates flutter echo. The wall unevenness can be achieved by using random roughness. Instead of splaying the wall, one can also treat it so that it has a somewhat curved shape so that the successive reflections are quickly dispersed.

Many small rooms have strong repeated sound reflections by floor and ceiling as shown in Figure 9.16. This may happen when the room is wide or when the walls are diffusive or absorptive. The situation is similar to that shown in Figure 9.15a but because the reflections are generated by surfaces above and below the listener, the reflections all arrive at about the same

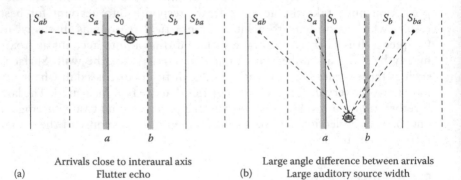

Arrivals close to interaural axis   Large angle difference between arrivals

(a)   Flutter echo   (b)   Large auditory source width

Figure 9.15 (a) A condition that is particularly prone to flutter echo. (b) A situation that is less prone to flutter echo.

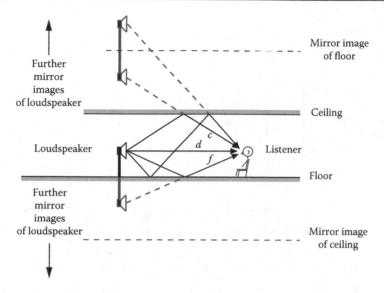

Figure 9.16 Floor and ceiling reflections close to the median plane are difficult for hearing to separate from direct sound by binaural means. They arrive in about the same cone-of-confusion.

angle of confusion. In most cases, the walls, floor, and ceiling of a room cause a multitude of reflections that arrive from many angles and remove repetitiveness.

The floor (*f*) and ceiling (*c*) reflections shown in the figure will arrive in approximately the same cone of confusion as the direct sound (*d*). In many homes, the ceiling height is about 2.4 m and adult listeners sitting in chairs typically have their ears at about 1.0–1.2 m above the floor. The ceiling and floor reflections then virtually coincide and the total level increase by about 6 dB at medium and low frequencies. The interference with the direct sound thus becomes very strong since the path length difference often only leads to an attenuation of a few dB in each path. The situation is somewhat ameliorated by the fact that the arrival angle relative the median plane of these reflections is reduced as a result of the larger distance to the image sources as the reflection order increases.

It seems that we subconsciously neglect this coloration when listening in situ. The reflections however are very noticeable on monophonic recorded sound and are then avoided by using microphones placed so close to the reflecting surface that there are no discernible frequency response dips; such microphones are sometimes called boundary microphones [38].

Figure 9.17 shows the 3D echogram of this case where repeated ceiling–floor reflections fall inside the 30 ms long time frame of analysis and will cause coloration. The simplest way to deal with the problem in a home,

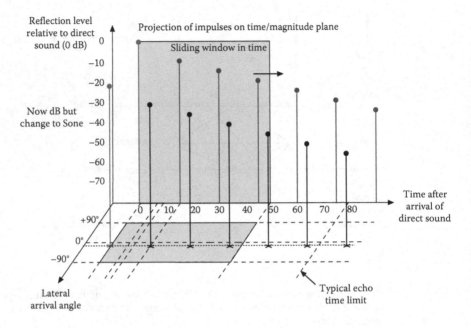

Figure 9.17 A 3D echogram that will result in rough and colored sound.

studio, control room, or other environment is to make the floor and ceiling sound absorptive or at least add a diffusing patch to the ceiling and a sound absorptive carpet to the floor. The use of vertical array loudspeakers that radiate less sound towards the floor and ceiling is another way to avoid the undesirable effects.

In many homes, the sofa or other listening location is often placed against a back wall resulting in a strong coherent reflection of sound from the back that cannot be separated binaurally from the frontal incident sound. Moving the sofa a meter or more away from the wall usually makes the reflection less noticeable as will of course applying a sound-absorptive or sound-diffusive patch at head height on the wall at the back.

In concert halls, there is little high-frequency sound reflected by the floor since the audience area usually has upholstered chairs. There is some scattering of high-frequency sound due to the heads of the audience and the top of the seat backs if these are not upholstered. The main difference at high frequencies between the small room case (with a sound-absorbing carpet) and the concert hall case is that the sound absorption by audience is much larger than that by a carpet and—more important—there is much larger time interval between the arrival of the direct sound and the center of gravity of the reflected sound, which gives the large room its auditory spaciousness. It is also common to design the ceiling so that it reflects sound to the side walls, instead of specularly down to the audience, by making the ceiling diffuse or by some other means.

At low frequencies, the sound from the basses and cellos in large concert halls is often subject to negative interference by reflected sound from the audience. The reflection results in considerable low-frequency loss on the orchestra floor in the 0.1–0.2 kHz range. At the front of the balcony levels, the sound of the orchestra is not subject to this interference. The difference between the two situations is usually very clearly audible illustrating the importance of the interference effects [27].

9.4.7 Timbre and coloration

Timbre depends on the spectral balance between various frequency ranges. In small rooms with short reverberation times, the timbre is mainly determined by the direct sound and the early reflections that are fused with it. Loudspeaker on-axis frequency response influences sound reproduction timbre, but it also depends on the directivity characteristics of the loudspeaker since most loudspeakers radiate sound that is reflected by the room. The sound absorption of walls and other surfaces is frequency dependent and so are reverberation level and reverberation time.

From the viewpoint of avoiding coloration, it is particularly important to treat ceiling and floor that cause reflections that cannot be separated binaurally, that is, occurring in the same cone-of-confusion as discussed earlier in this chapter. In control rooms, the mixing console will act as a barrier against the sound reflected by the floor. In living rooms, floor carpet and sofa tables can fill a similar function. The same treatment as previously discussed for lateral side wall reflections can be applied for ceiling reflections. It is important to treat all surfaces that carry first-order reflections arriving at the listening position since such reflections are less diffusely reflected than higher-order reflections.

Many commercial diffusers such as 2D quadratic residue and polycylindrical diffusers are only diffusing when sound is incident in a particular plane. If such diffusers are used, there is a choice to be made as to if the scattered sound is to be spread lengthwise or sideways. For ceiling, use of a 3D diffuser that redirects sound to the side walls and a scattering back wall is likely to be the better choice than a 2D diffuser since the acoustics will then sound more spacious.

Strong first-order sound reflections by the side walls affect the stereo stage accuracy (phantom source placement) as well as the auditory source width as discussed earlier. Sound-absorbing or sound-diffusing side and back walls are advantageous. The reflections by the rear wall should be made diffusive as described in Chapter 11. It is best to use a random distribution of spheres or half spheres (convex or concave) of varying sizes

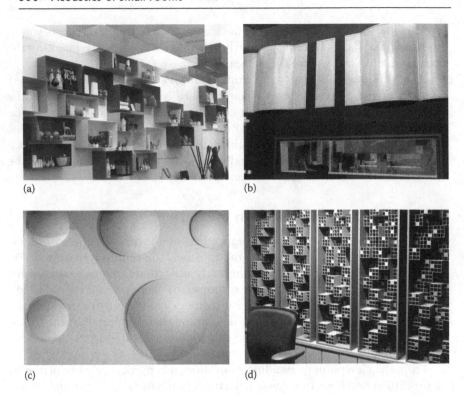

Figure 9.18 Some practical options for providing scattering: (a) disordered objects, (b) polycylindrical diffusers, (c) spherical caps, and (d) mathematically designed diffuser. (Photos by Mendel Kleiner.)

[28,30]. Balloons are useful for initial testing of the effectiveness of volume diffusers [29]. Figure 9.18 shows some options for providing scattering.

All diffusers are not created equal, however, the differences in acoustic properties between different surface irregularities can be heard. In an auralization study that used a 1:10 scale modeling five types of diffusers was compared using a listening test [30]. Linear arrays of 10 diffusers such as pyramids, polycylindrical, quadratic residue, wedges, and spheres were used. The arrays filled the same area. Sample drawings of single diffuser elements are as shown in Figure 9.19.

Wedges and pyramids affected the sound quality in a very negative way due their selective specular reflection of high-frequency sound, while a series of polycylinders resulted in coloration because of grating effects. Figure 9.20 shows partial results for the scale model frequency response of QRD diffusers versus a random arrangement of spheres used in the tests. The results showed clearly that the frequency response and its resulting timbre and coloration were decisive for the sound quality of the diffuse reflected sound. The spheres turned out to be the most pleasant sounding

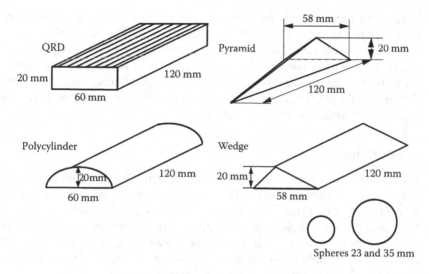

Figure 9.19 The five different diffuser types used in the 1:10 scale auralization study. All arrays used multiple units of each diffuser element on a large hard surface so that the arrays covered a rectangular area of 1.2 × 6 m² full scale. (From Kleiner, M., Svensson, P., and Dalenbäck, B.-I., Auralization of QRD and other diffusing surfaces using scale modelling, *Proceedings of the 93rd Audio Engineering Society Convention*, San Francisco, CA, preprint M-2, 1992.)

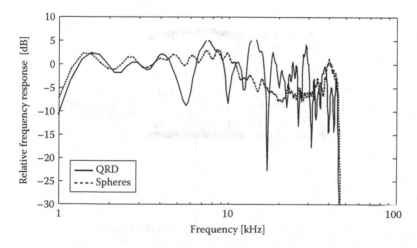

Figure 9.20 Frequency response characteristics of the scattered sound from QRD diffusers and spheres used in the scale model test. (From Kleiner, M., Svensson, P., and Dalenbäck, B.-I., Auralization of QRD and other diffusing surfaces using scale modelling, *Proceedings of the 93rd Audio Engineering Society Convention*, San Francisco, CA, preprint M-2, 1992.)

because of their smooth frequency response. Note that caps and inverted caps (cavities into the wall) will work similarly well for scattering.

In old-style control rooms, it was common to see polycylindrical diffusers, on walls and in corners. Particularly, arrays of polycylindrical diffusers on walls may cause unpleasant repetitive grating effects. They should have different radii of curvature and be placed at different angles or at random. Any orderly array should be avoided.

9.4.8 Echo

If the room has a hard plane floor and a focusing hard ceiling such as that shaped by a spherical shell with a large radius of curvature, there may be echo—even in a small room—because of repeated reflection. When the focal point lies well below the floor surface, multiple repeated reflections in a circular manner may result in echo as shown in Figure 9.21 [23]. Similar considerations apply to rooms with a reflective curved back wall sometimes found in control rooms and conference rooms.

Echoes due to such multiple reflections can be found in planetaria and other rooms that have a curved ceiling with a focal point at or below floor level but only if both ceiling and floor are hard and smooth. They are easily removed by the use of absorption or diffusion. The curved ceiling can be made of sound transparent material and absorptive treatment installed in the back of the ceiling. Rooms that have a cylindrical plan are also affected by focusing and repetitive echo unless walls are made absorptive or diffusive.

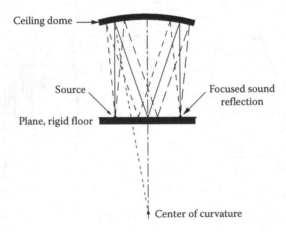

Figure 9.21 Echo caused by multiple *circulating* reflections occurring with a spherical shell ceiling because the center of the curvature is below the reflecting floor. (After Cremer, L., Müller, H.A., and Schultz, T.J., *Principles and Applications of Room Acoustics*, Vol. I. Applied Science Publishers, New York, 1982.)

9.5 LOUDSPEAKER AND ROOM

9.5.1 The room and its acoustical filter response

As mentioned earlier in this chapter the room can be considered as a spatially dependent multiresonant filter that acts on the sound radiated by the loudspeaker. The directivity characteristic of the loudspeaker is decisive for the influence of the room on the sound perceived by the listener.

Many mono or stereo sound reproduction systems need the reflected sound by the room to augment the direct sound arriving from the loudspeakers so that there is some feeling of envelopment. This means that the directivity of the loudspeaker is as important for the resulting sound quality as are the acoustics of the room. The more directional the radiation of the loudspeaker the smaller will the influence of the room be on the reproduced sound. It is well known that for the most neutral sound reproduction the loudspeakers should be highly directional so that the sound reflected by the surfaces of the room and objects in the room is minimized [31–37].

9.5.2 Loudspeaker characteristics at low frequencies

Many sound sources such as monopole (or rather single side driver) loudspeakers and many musical instruments generate diaphragm vibration that is not much affected by the acoustics of the environment. Such sources can be regarded as high internal impedance sources that are characterized by volume velocity that is not influenced by the radiation impedance of their environment. A typical example is closed-box (sealed enclosure) loudspeaker systems where the driver's diaphragm only radiates to the outside and the back side of the diaphragm faces the inside of the box, which is airtight. An alternative to closed-box designs are ported-box (vented enclosure) loudspeakers [37,38]. Their power output can also be regarded as approximately unaffected by the environment's radiation impedance at frequencies above the resonances of the driver and port [37,38].

From the viewpoint of low-frequency behavior, commercially available loudspeakers can be roughly subdivided into the following groups:

- *Small loudspeakers, free standing in room, well away from room surfaces*: Small loudspeakers will act as point sources at least at low frequencies roughly below 0.3 kHz. They will have high internal acoustic impedance and only add little damping to the room. Additionally, they will also not noticeably affect the modal structure of the room for frequencies where the dimensions of the loudspeaker are smaller than $\frac{1}{6}\lambda$. Mode frequencies and shapes can therefore be initially calculated using simple physics.

- *Small loudspeakers on wall or in bookshelf*: Loudspeakers placed among books in shelves will yield a flatter frequency response because of the sound absorption of the books, which results in more modal damping and the elimination of the mirror image caused by the front wall.
- *Large loudspeakers, free standing in room*: Large loudspeakers will because of their bulk change the effective room shape, which results in changed mode frequencies and patterns. Many such loudspeakers are ported-box loudspeaker systems and have driver and port resonances that often occur close to room resonance frequencies in the range of 30–40 Hz. They may thus add considerable damping to the lowest-frequency modes.

  Subwoofers are a special case of low-frequency free-standing loudspeakers in that they are often placed without regard for frequency response behavior above about 80 Hz. They can thus be placed to optimize the low-frequency spectrum for a flat frequency response. The use of multiple subwoofers allows better modal averaging and optimal excitation in this frequency region [39].
- *Large dipole-type loudspeakers, free standing in room*: Large dipole (*dual-sided*) loudspeakers with electrostatic and electrodynamic drivers will change the frequencies and patterns of the room modes since they act as major barriers. Full-range dipole loudspeakers, electrodynamic or electrostatic, often have resonance frequencies in the range of 30–40 Hz.
- *Horn loudspeakers firing into a room corner*: Although not commonly used, the mouth of low-frequency horn loudspeakers will effectively add a large sound-absorbing surface to the room. If a pair of horns are placed at the corners, they will be very effective in damping many low-frequency room modes.

9.5.3 Loudspeaker characteristics at mid- and high frequencies

At frequencies above about 0.3 kHz, most domestic (single-sided) loudspeakers will start to be directional, either because of the size of the driver or because of the size of the baffle or enclosure front. Small loudspeakers may have less directivity than large loudspeakers but this depends on the driver diaphragm sizes and mechanical mode behaviors [38]. The piston frequency range of conventional electrodynamic loudspeaker drivers using conical diaphragms is usually only 3–4 octaves wide. Since the audio range is about 9 octaves wide, clearly, multiple drivers, each designed to a part of the desired frequency range, need to be used unless wide range design is used such as a loudspeaker that is back loaded by a horn.

The directivity of the driver depends on the diaphragm geometry and its vibration pattern. Drivers for high-quality audio sound reproduction

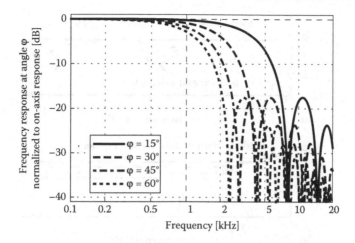

Figure 9.22 Frequency response at four off-axis angles φ for a flat circular rigid piston that has a diameter of 0.2 m and flat frequency response on-axis.

are generally designed to have diaphragms that function as rigid pistons [37,38]. Because of interference effects, the sound radiated to the sides will be attenuated as shown by Figure 9.22.

Most loudspeaker systems for high-quality audio are two or three driver designs. Each driver is then assigned a working frequency range with some small overlap and with an optimized enclosure type. The division of electric energy into different frequency ranges appropriate for each driver is done by electrical filters, usually called crossover filters or networks, which are also likely to influence the directivity [37,38]. Because of the different driver diaphragm sizes it is difficult to avoid jumps in directivity around the crossover frequencies. In the frequency range at and immediately below the crossover frequency, the combined frequency response will show directivity effects similar to those shown in Figure 9.23a. A high quality loudspeaker should have a directivity index which is approximately constant over the mid- and high-frequency ranges as indicated in Figure 9.23b [10].

Domestic loudspeakers often use dome-type mid- and high-frequency loudspeaker drivers that can be made to show even and wide frequency range on-axis measurement curves. If such loudspeakers are to be used in a critical listening environment, that room will need much room acoustic control such as room geometry adjustment to remove reflection paths and sound-absorptive and sound-diffusive surfaces.

One possibility in loudspeaker design is to strive for a multiway loudspeaker where all drivers radiate in omnidirectionally. Commercial attempts at making such loudspeakers have not been successful neither technically nor in the marketplace. Loudspeakers that are sold as being omnidirectional usually only have a wider radiation pattern than conventional

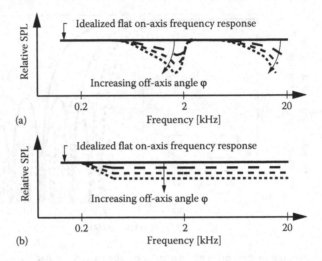

Figure 9.23 The frequency response of loudspeakers changes with direction. The dashed and dotted lines indicate how response changes for increasing off-axis angles. (a) A poor full frequency range loudspeaker using separate low- and high-frequency drivers shows directivity effect both at mid- and high frequencies. (b) A high quality full frequency range loudspeaker designed to have constant directivity index over the mid- and highfrequency ranges.

electrodynamic loudspeakers by having the sound bounce off the walls or some diffusing contraption before reaching the listener.

True wide frequency range omnidirectional loudspeakers are also likely to be unsatisfactory because of interference between the direct sound from the different drivers and between the direct sound and reflected sound. The interference generates peaks and dips in the frequency response both for on-axis and off-axis sound. The sound pressure contributions by the drivers add vectorially in or out of phase so there will always be interference between the drivers. The drivers cannot be at the same place physically and the crossover filters give only limited attenuation close to the crossover frequency.

In most cases, maximally flat on-axis frequency response is desired for the horizontal plane. Optimum directivity will depend on the sound-absorptive and sound-diffusive properties of the room surfaces and the reverberation time. Reviews of the effect of directivity and room acoustic properties show that the optimum often is to have constant directivity in the mid- and high-frequency ranges as indicated in Figure 9.23b [10].

Loudspeakers that use drivers on a vertical line on the front baffle make are good for many applications, but for near-field monitor applications in stereo listening, close, mirrored placement of the different drivers on the left- and right-side enclosures also gives acceptable results. If the drivers are not mounted on a vertical line or so that they are concentric but rather side by side, the directivity effects will be asymmetric. To further reduce diffraction effects, the front baffle can be covered by a sound-absorptive layer or have chamfered or rounded corners [31].

Since the time delay characteristic of the loudspeaker is important in listening, it should be noted that even if the loudspeaker is adjusted for frequency independent time delay characteristics on-axis, the placement of the acoustic centers of the drivers is important for sound radiation in off-axis directions. Remember though that most natural sound sources as discussed in Chapter 7 do not have acoustic centers in the sense used in loudspeaker design.

By mounting the loudspeaker enclosure recessed into a wall or into a bookshelf so that its front and driver are flushed against the wall or bookshelf front, the baffle effect will apply to the whole frequency range. This is useful in control rooms and domestic environments but care must be taken to have the enclosure vibration isolated so that sound is not radiated by vibration of the wall or bookshelf.

9.5.4 Arrays and large-diaphragm loudspeakers

Most large loudspeakers are very directional, for example, horns, loudspeaker arrays, and membrane loudspeakers that use weak force drivers such as ribbon, isodynamic, or electrostatic drivers. Directional loudspeakers limit the energy that can be returned as early reflections and reverberation.

An array can consist of closed-back designs with conventional drivers mounted in an enclosure so that the back of their diaphragms faces one or more closed volumes. The drivers may be mounted along a line (straight or curved) or on a baffle depending on the directivity pattern that is desired. Arrays that use individual drivers will often use four or more drivers along a line; such loudspeakers are common in sound reinforcement systems. The more drivers per wavelength the more controlled is the directivity pattern likely to be [38].

Another design alternative for array loudspeakers is to use open-back designs that allow the diaphragm to radiate from both front and back. Such arrays usually use isodynamic or electrostatic drivers mounted on a large baffle. Depending on the size of the driver and baffle as well as on driver placement, such arrays may work as dipoles at low frequencies and as dual front and back firing loudspeakers at mid- and high-frequencies.

The crossover between the two operational modes will depend on the baffle and the driver geometries and sizes. In any case, because of the size of the driver necessary for sound radiation, there will be directivity effects from the either side's radiation. For the most common design, the array is long, narrow, and mounted vertically. It has little radiation toward floor and ceiling but fairly wide horizontal plane directivity patterns for mid- and high frequencies similar to those of equally sized closed-box loudspeakers.

Note that if the diaphragm is many wavelengths long but only a fraction of a wavelength wide, then up to some distance away the wave front (on both sides) will be similar to that of a cylindrical wave. Such a wave has a distance attenuation of only –3 dB per distance doubling in contrast to the –6 dB per distance doubling characteristic of the direct sound of a small loudspeaker.

The mid- and high-frequency back-radiated sound will arrive delayed to the listener by way of reflection. To reduce the backside-radiated sound, the wall behind the loudspeaker can be treated with sound-absorptive or sound-diffusive material or an added sound-absorptive baffle, plane, or cylinder can be used. Such a cylinder or other diffusive device can also conveniently be used as a bass trap as shown in Figure 11.5.

Interaction between the dipole driver and the room modes will be different from that of conventional monopole loudspeakers in the low-frequency sparse mode region. Some early experimental data showing the two different type characteristics show that room reverberation results in different modulation transfer characteristics [17].

For best response at low frequencies, dipole-type loudspeakers must be placed away from corners and are best regarded as dipoles that drive the modes at the nodes [17,38]. Typically, the modes that have the lowest damping are the axial modes. In a room with modes, the coupling of the dipole loudspeaker to both axial and tangential modes can be adjusted by turning the dipole axis relative to the mode axes. The excitation of oblique modes is impractical to control in this way. For best power output in a nonmodal environment, the dipole should be placed so that it is at right angle with a hard side wall [38].

9.6 STEREO

9.6.1 Stereo sound reproduction

Sound reproduction of voice and music can take many forms depending on the recording venue, recording technique, playback venue, and playback system. In some cases, it is desired to hear the acoustical properties of the recording venue with as much detail as possible; in other cases, the desire is to fully control the spatial properties of the auditory event on playback.

In stereo playback systems, the listener and the two loudspeakers ideally form an equilateral triangle. Characteristics for all stereo and multichannel playback methods are the phantom sound sources that appear between the loudspeakers [41]. These are a result of the summing localization by hearing as explained in Chapter 7. Correct, or at least plausible, sound source localization is essential to high-quality sound reproduction and is a reason for the success of stereo sound reproduction. The *sound stage* between the loudspeakers in stereo sound reproduction is determined not only by physical but also by cognitive factors as mentioned in Chapter 8.

Recording techniques differ in their ability to generate proper phantom source location. For phantom sources to appear at distinct locations, the signals must be coherent, the interloudspeaker delays less than 1 ms, and the levels reasonably similar as shown in Chapter 8. Sounds that are long lasting may be located differently transient sounds in spite of the original location of the recorded sound source being the same. Phantom sources are best generated by transient sounds that represent well-known sound sources such as castanets and voice.

9.6.2 Sound recording for stereo

Many stereo recordings of classical music use a pair of microphones, possibly with some support microphones [42]. Three fundamentally different approaches illustrated in Figure 9.24 are the delay/level (*AB*) recording techniques (shown in Figure 9.24a) using two omnidirectional or wide cardioid microphones, level difference (*XY*) recording technique (shown in Figure 9.24b) using two cardioid or hypocardioid microphones, and the ORTF spaced cardioid technique in Figure 9.24c.

Since microphone types differ with respect to directivity functions, they will also differ in the relative amount of reverberant sound they pick up. *AB* recordings will sound much more reverberant than *XY* or ORTF. The latter typically use microphones that have cardioid or similar directivity that gives these microphones about 4 dB less sensitivity to reverberant sound than omnidirectional microphones. Note that the direct-to-reverberant sound ratio mentioned in Chapter 8 can be varied as a result of the microphone technique used for the recording and in that way the subjectively perceived reverberation time.

Figure 9.25 shows multimicrophone recording technique where the phantom sources are positioned using a mixing console. As pointed out in Chapter 7, phantom sources can be panned to any location on the baseline, and by digital signal processing, phantom sources can be placed also outside the triangle base by using cross-talk cancellation techniques [40].

Figure 9.26 repeats Figure 8.23 but with two added lines that roughly indicate how *AB* and *XY* microphone techniques differ in stereo rendering capability. *XY* intensity recording techniques use coincident cardioid or cardioid-like microphones that are angled suitably to give the desired level difference to generate stereo signals that result in a smooth distribution of phantom sources between the stereo loudspeakers. Line *XY* in Figure 9.26 shows the typical phantom sources distribution of *XY* recorded sound.

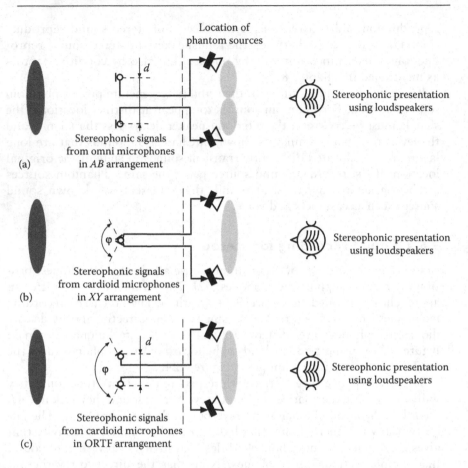

Figure 9.24 Some sound recording/playback systems: (a) *AB* spaced omni microphones, (b) *XY* microphones, (c) ORTF microphone arrangement technique. Many others are commonly used. Without reverberation in the recording and the listening room, the listener's auditory event will be placed between the loudspeakers.

AB recording on the other hand uses (in principle) two omnidirectional microphones. Thus, most of the difference between the microphone signals will be due to delays. A 3 m distance between the microphones results in about 9 ms delay. *A'-A-B-B'* shows rendering using poor *AB* microphone placement. In the latter case, often, most of the phantom sources will be located in either loudspeaker.

The ORTF technique is shown in Figure 9.24c where typically two cardioid or possibly hypocardioid microphones are used at a distance *d* of about 0.2 m. This recording technique can be thought of as a combination of *XY* and binaural recording. The linearity of sound stage pickup and

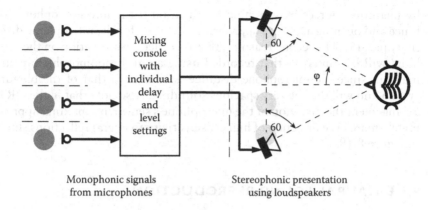

Monophonic signals
from microphones

Stereophonic presentation
using loudspeakers

Figure 9.25 Stereophonic playback to generate phantom sources between the loud-speakers at any desired angle φ.

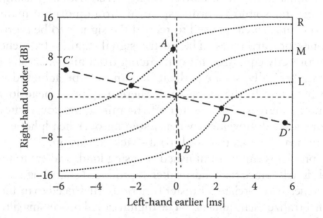

Figure 9.26 Lateral position of the phantom source in the case of mono sound source reproduced by two loudspeakers at φ = ±30° with simultaneous variation of delay and level. (After Franssen, N.V., Some considerations on the mechanism of directional hearing, Dissertation, Electrical Engineering, Mathematics and Computer Science, Technical University of Delft, Delft, the Netherlands, 1960.) Dashed lines explained in text. L, phantom source in left-hand loudspeaker; M, phantom source in the middle between loudspeakers; R, phantom source in right-hand loudspeaker.

phantom source rendering will depend on the distance, angle φ between the microphones, and the directivity characteristics of the microphones used [42,45].

ORTF recording of classical music in large venues often sound better than plain XY recordings particularly from the viewpoint of listener envelopment. The first reason is that the direct sound is appropriately rendered by phantom sources by summing localization if the intermicrophone distance and opening angle are appropriately chosen for the microphone directivity [42,45]. The second reason is that recorded reverberation on the other hand will be similar to that recorded using an anthropomorphic binaural manikin since the microphone distance is similar to that of the binaural system and will thus be very spacious sounding. Also note that in the ORTF arrangement the direction of the microphone's sensitivity maxima approximately match the direction of highest sensitivity to lateral reflections shown in Figure 8.19.

9.7 BINAURAL SOUND REPRODUCTION

The ability of binaural sound reproduction systems to render the spaciousness of large room sound fields is well documented but has the disadvantage of poor compatibility with conventional stereo listening using loudspeakers. In binaural sound recordings, an anthropomorphic manikin with a microphone in each ear is used to record the signals to be reproduced with headphones or earphones. The entire signal chain's frequency response must be correctly equalized for convincing front and back phantom source placement. Typically, such equalization can only be achieved using digital signal processing. In the worst case, there may be in-head localization of the phantom sources. The absence of the minute response cues created by the listener's head movement when listening over headphones contributes to both in-head localization and front–back confusion.

These problems can be diminished by using loudspeakers to reproduce the binaural signals. Since the loudspeaker-reproduced sound signals are affected by the listener's head-related transfer functions, they need to be filtered (*equalized*) to neutralize their influence. The digital signal processing filter also needs to include cross-talk cancellation to remove the left and right signals reaching the right and left ears, respectively, in loudspeaker listening to simulate headphone listening [43]. These modes of binaural sound reproduction are sometimes called *transaural* or *cross-talk compensated* sound reproduction.

With cross-talk cancellation, the listener is usually able to hear frontal spatial information correctly but sources to the back of the listener may be mirrored to appear in front. Since there is generally little sound from the direct back, this is of little practical importance. The principle of cross-talk cancellation for binaural systems is shown in Figure 9.27.

For cross-talk cancellation to function well, the listener must be in the sweet spot on the symmetry line between the identical loudspeakers facing directly forward. Loudspeakers that have very directional sound radiation characteristics are preferable to omnidirectional ones for this application. If small nondirectional loudspeakers are used, the room should preferably be hemi-anechoic although a reverberation time of less than 0.3 s is usually

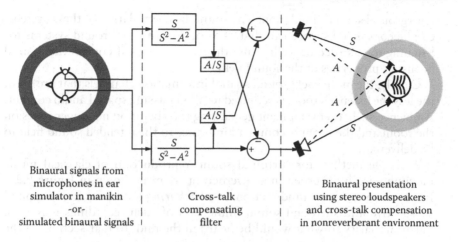

Binaural signals from
microphones in ear
simulator in manikin
-or-
simulated binaural signals

Cross-talk
compensating
filter

Binaural presentation
using stereo loudspeakers
and cross-talk compensation
in nonreverberant environment

Figure 9.27 A binaural system using cross-talk cancellation rather than headphone sound reproduction. A and S are the head-related transfer functions between the loudspeaker and listener's ears. Auditory events can (in principle) be placed anywhere in space around the listener. Unless the complete binaural chain carefully equalized there may be front/back confusion in reproduction. (The gray circle surrounding the listener roughly suggests from where envelopmental reverberant sound will appear to come.)

acceptable. The allowable head movement is typically ±0.1 m laterally and $\varphi < \pm 10°$ for acceptable sound reproduction of concert hall recorded music.

9.8 SURROUND SOUND SYSTEMS

In addition to stereo, modern sound reproduction systems can be in a multitude of formats offering multichannel sound with 5, 7, or more channels. Surround sound rendering is essential for film but also allows the music listener to have a better feeling of the envelopment of a large room. In surround sound recording of classical music, the ambition is usually to virtually move the listener to the *best seat in the house*. This requires not only a natural front sound stage reproduction but also the proper reproduction of the spatial properties of the reverberant sound field and its ambience.

It is important to observe that even mono or stereo sound can be convincingly processed by digital signal processing to render the acoustical qualities of a large room. Examples of such multichannel systems can be found in homes and in automobile sound systems [44]. While such systems do not render the exact acoustics of the actual recording location, they are able to plausibly recreate the acoustics of many types of rooms using only six loudspeakers. The acoustical contribution of the listening room can help fill in the gaps between side and rear loudspeakers and help provide smooth background spatial simulation. The sound field simulation of these systems can be considered superior to that of 5- or 7-channel surround systems for listening to recorded music because the listener can self-control the desired acoustics properties of the sound field.

Clearly, the more background spatial information channels that are used, the less the listener's room is required to fill in missing spatial audio content. Ambience such as envelopment generating reverberation no longer relies on the room and the room becomes a hindrance to the intended sound field to be delivered.

A simple method for surround sound reproduction of classical music monophonically recorded in a reverberant room is shown in Figure 9.28. In a live room using omnidirectional or *backfiring* loudspeakers, the sound perspective and surround sound immersion of such recordings may have much similarity to what would be heard in the mid- to rear section on the

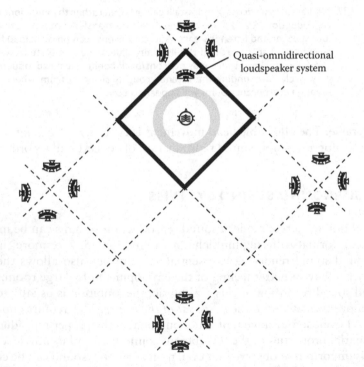

Figure 9.28 A monophonic recording may give desired envelopment effect in a small room if the loudspeaker is suitably quasi-omnidirectional. Adding a second stereo loudspeaker enhances the effect. The figure shows how the mirror images create a surround sound envelopment (shown as a gray circle surrounding the listener and roughly suggesting the angles from where reverberant sound will appear to come). Only some of the first- and second-order mirror images are shown here. Diffusive walls would further enhance the effect.

main floor of a good concert hall. A prerequisite is that the room is such that the first rear reflections arrive 30 ms or later after the direct sound. Of course, the stereo stage will be small but far from the stage in a concert hall, the auditory source width of the orchestra may appear small and the sound source fairly blurred in any case.

Figure 9.29 shows some methods for surround sound reproduction. The front loudspeakers are at $\varphi = 0°$ and $±30°$ so the angles are similar to the ones used for conventional stereo. The presence of the center channel improves dialog location in film and much improves the phantom source localization in music reproduction.

In most surround sound systems based on five loudspeakers as shown in Figure 9.29a, the angle between the rear loudspeakers is much larger than 60°. Typically the angle is 120° which leads to linearity and depth errors in rendering. By using properly adjusted angles and microphone distance, multimicrophone recordings can feature almost seamless surround sound rendering even with 5-channel surround sound setups [45].

Surround sound systems based on seven loudspeakers in the horizontal plane as in Figure 9.29b feature better rendering particularly in the rear of the listener. The rear loudspeakers are then at about $±140°$ and the side loudspeakers at about $±80°$.

Figure 9.29c shows how an array of loudspeakers, sometimes called a *sound bar*, can be used to direct *beams* of suitably processed sound to be reflected by the walls of a rectangular room so that diffusely reflected beams appear to come from loudspeakers to the side and the back of the listener. Because the array can only be suitably directional at high frequencies, the sound reproduction may be quite convincing since hearing mainly uses high-frequency cues for the placement of phantom sources as described in Chapter 8.

In multilayer surround sound systems that may use 10 channels or more, loudspeakers in two planes are used: six or more loudspeakers in a horizontal plane at about the height of the listener's ears and four or more loudspeakers in a top plane at a higher level so that they appear at about 30° over the listener. The loudspeakers in the horizontal plane carry the direct sound and the early reflected sound of the front sound stage, whereas the loudspeakers in the upper plane carry the reverberation and provide *height information*.

Ambisonics is a sound field recording system that uses four or more channels of coincident recorded sound, for example using a SoundField microphone a tetrahedral combination of cardioid microphones. These channels can be combined in various ways to fit the loudspeaker placement in the room [46,47]. Since such a microphone is based on the use of coincident cardioid microphones, it has more in common with coincident XY stereo microphone techniques than the simulation techniques that use signal processing to render simulated room ambience. The so-called

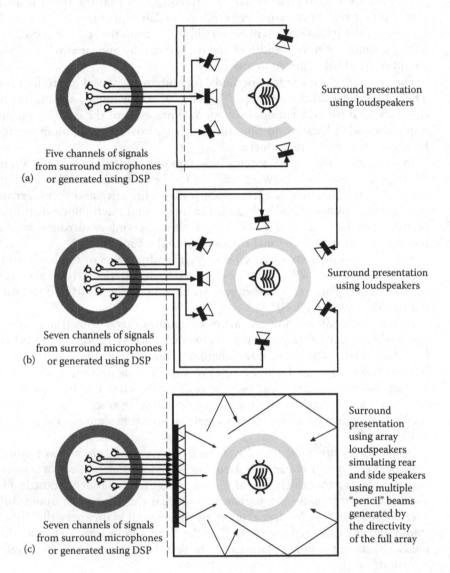

Figure 9.29 Three systems for recording/playback of surround sound: (a) using a 5-channel technique with spaced cardioid or hypocardioid microphones, (b) using a 7-channel technique with spaced cardioid or hypocardioid microphones, (c) using a 5-channel technique with an array loudspeaker simulating rear loudspeakers using reflection of sound beams by walls. (The gray circle surrounding the listener roughly suggests from where envelopmental reverberant sound will appear to come.)

higher-order ambisonics operates with more microphones sensing the sound field and more loudspeakers for reproduction and strives to achieve better directivity in reproduction.

The wave field synthesis (WFS) technique for surround sound uses large linear arrays of very closely mounted loudspeakers to effectively simulate the Huygens wave field recreation principle. Typically, 10 or more loud-speakers per meter are necessary to avoid severe interference lobes in the important audio frequency range. Each loudspeaker must be fed an appropriate impulse response filtered signal so that the wave front of the virtual sound sources can be created in the listening zone. An interesting property of WFS is that the sound field of sources inside the listening zone may be recreated [48].

REFERENCES

1. Kleiner, M. and Lahti, H. (1993) Computer prediction of low frequency SPL variations in rooms as a function of loudspeaker placement, in *Proceedings of the 94th Audio Engineering Society Convention*, Berlin, Germany.
2. Weisser, A. and Rindel, J.H. (2006) Evaluation of sound quality, boominess, and boxiness in small rooms. *J. Audio Eng. Soc.*, 54(6), 495–511.
3. Dueso Tejero, A. (2009) Audibility of frequency shifts in low mode density rooms, BSc thesis, Division of Applied Acoustics, Chalmers University of Technology, Gothenburg, Sweden.
4. Bolt, R.L. (1946) Note on the normal frequency statistics in rectangular rooms. *J. Acoust. Soc. Am.*, 18, 130–133.
5. Bonello, O.J. (1981) A new criterion for the distribution of normal room modes. *J. Audio Eng. Soc.*, 29(9), 597–606.
6. Louden, M.M. (1971) Dimension ratios of rectangular rooms with good distribution of eigentones. *Acustica*, 24, 101–104.
7. Walker, R. (1996) Optimum dimension ratios for small rooms, in *Proceedings of the 100th Convention of the Audio Engineering Society*, Copenhagen, Denmark, preprint 4191.
8. Welti, T. and Devantier, A. (2006) Low frequency optimization using multiple subwoofers. *J. Audio Eng. Soc.*, 54, 347–364.
9. Cox, T., D'Antonio, P., and Avis, M.R. (2004) Room sizing and optimization at low frequencies, *J. Audio Eng. Soc.*, 52(6), 640–651.
10. Toole, F. (2008) *Sound Reproduction: The Acoustics and Psychoacoustics of Loudspeakers and Rooms*. Focal Press, Oxford, U.K.
11. Wankling, M. and Fazenda, B.M. (2009) Studies in modal density—Its effect at low frequencies, in *Proceedings of the Institute of Acoustics. 25th Reproduced Sound Conference 2009*, Brighton, U.K., November 2009.
12. Walker, R. (1993) Optimum dimension ratios for studios, control rooms and listening rooms. BBC Research Department Report RD1993/8. Research Department, Engineering Division, The British Broadcasting Corporation, London, U.K.

13. Lokki, T. and Karjalainen, M. (2000) An auditorily motivated analysis method for room impulse responses, in *Proceedings of the Conference on Digital Audio Effects (DAFx-00)*, Verona, Italy, pp. 55–60.
14. Avis, M.R. et al. (2007) Thresholds of detection for changes to the Q factor of low-frequency modes in listening environments, *J. Audio Eng. Soc.*, 55(7/8), 611–622.
15. Schroeder, M.R. (1981) Modulation transfer functions: Definition and measurement. *Acustica*, 49, 179–182.
16. Lucier, A. (1990) *I Am Sitting in a Room*. Lovely Music, Ltd., New York.
17. Linkwitz, S. (1998) Investigation of sound quality differences between mono-polar and dipolar woofers in small rooms, in *Proceedings of the 105th Audio Engineering Society Convention*, San Francisco, CA, paper 4786.
18. Welti, T. (2002) How many subwoofers are enough, in *Proceedings of the 112th Audio Engineering Society Convention*, Munich, Germany, Paper 5602.
19. Celestinos, A. and Birkedal Nielsen, S. (2008) Controlled Acoustic Bass System (CABS) A method to achieve uniform soundfield distribution at low frequencies in rectangular rooms, *J. Audio Eng. Soc.*, 56(11), 915–931.
20. Kleiner, M., Dalenbäck, B.-I., and Svensson, P. (1993) Auralization—An overview. *J. Audio Eng. Soc.*, 41(11), 861–875.
21. Kleiner, M., Klepper, L.D., and Torres, R.R. (2011) *Worship Space Acoustics*. J. Ross Publishing, Ft. Lauderdale, FL.
22. Kirzenstein, J. (1984) An image source computer model for room acoustics analysis and electroacoustic simulation. *Appl. Acoust.*, 17, 275–290.
23. Cremer, L., Müller, H.A., and Schultz, T.J. (1982) *Principles and Applications of Room Acoustics*, Vol. 1. Applied Science Publishers, New York.
24. Long, M. (2013) *Architectural Acoustics*. Academic Press, New York.
25. Ando, Y. (2011) *Concert Hall Acoustics*. Springer, Berlin, Germany; Softcover reprint of the original 1st ed. 1985 edition.
26. Kleiner, M. (2011) *Acoustics and Audio Technology*, 3rd edn., J. Ross Publishing, Ft. Lauderdale, FL.
27. Schulz, T.J. and Watters, B.G. (1964) Propagation of sound across audience seating. *J. Acoust. Soc. Am.*, 36, 885–895.
28. Natsioupoulos, G. and Kleiner, M. (2006) Evaluation of a boss model and subtraction technique for predicting wideband scattering phenomena in room acoustics. *J. Acoust. Soc. Am.*, 119(5), 2798–2803.
29. Knudsen, V.O. and Delsasso, L.P. (1968) Diffusion of sound by helium filled balloons, in *Proceedings of the Sixth International Congress on Acoustics*, Tokyo, Japan, Paper E-5-5.
30. Kleiner, M., Svensson, P., and Dalenbäck, B.-I. (1992) Auralization of QRD and other diffusing surfaces using scale modelling, in *Proceedings of the 93rd Audio Engineering Society Convention*, San Francisco, CA, preprint M-2.
31. Olson, H.F. (1957) *Acoustical Engineering*. Van Nostrand Reinhold, New York.
32. Salmi, J. and Weckström, A. (1982) Listening room influence on loudspeaker sound quality and ways of minimizing it, in *Proceedings of the 71st Audio Engineering Society Convention*, Montreux, Switzerland, Preprint 1871.
33. Queen, D. (1979) The effect of loudspeaker radiation patterns on stereo imaging and clarity. *J. Audio Eng. Soc.*, 27(5), 368–379.

34. Kates, J.M. (1980) Optimum loudspeaker directional patterns. *J. Audio Eng. Soc.*, 28(11), 787–794.
35. Fryer, P.A. and Lee, R. (1980) Absolute listening tests-further progress, in *Proceedings of the 65th Audio Engineering Society Convention*, London, U.K., Paper 1567.
36. Russell, K.F. and Fryer, P.A. (1979) Loudspeakers: An approach to objective listening, in *Proceedings of the 63rd Audio Engineering Society Convention*, Los Angeles, CA, Paper 1495.
37. Colloms, M. (2005) *High Performance Loudspeakers*, 6th edn. Wiley, New York.
38. Kleiner, M. (2013) *Electroacoustics*. CRC Press, Ft. Lauderdale, FL.
39. Welti, T. and Devantier, A. (2006) Low frequency optimization using multiple subwoofers. *J. Audio Eng. Soc.*, 54, 347–364.
40. Aarts, R.M. (2000) Phantom sources applied to stereo-basewidening, *J. Audio Eng. Soc.*, 48(3), 181–189.
41. Franssen, N.V. (1960) Some considerations on the mechanism of directional hearing. Dissertation, Electrical Engineering, Mathematics and Computer Science, Technical University of Delft, Delft, the Netherlands.
42. Williams, M. (1991) Microphone arrays for natural multiphony, in *Proceedings of the 91st Audio Engineering Society Convention*, New York, Paper 3157.
43. Atal, B.S. and Schroeder, M. (1966) US pat. nr. 3236949 APPARENT SOUND SOURCE TRANSLATOR.
44. Zucker, I. (1989) Reproducing architectural acoustical effects using digital soundfield processing, in *Proceedings of the Audio Engineering Society Seventh International Conference: Audio in Digital Times*, Toronto, Ontario, Canada, Paper Number: 7-033.
45. Williams, M. (2004 and 2013) *Microphone Arrays for Stereo and Multichannel Sound Recording*, Volumes 1 and 2, Editrice Il Rostro, Milano, Italy.
46. Gerzon, M. (1971) Experimental tetrahedral recording, available at Michaelgerzonphotos.org.uk (sampled [May 2013] originally published in *Studio Sound*, Vol. 13, pp. 396–398 [August 1971], pp. 472, 473 and 475 [September 1971], pp. 510, 511, 513 and 515 [October 1971]).
47. Malham, D.G. (1998) *Spatial Hearing Mechanisms and Sound Reproduction*. University of York, York, England, http://www.york.ac.uk/inst/mustech/3d_audio/ambis2.htm (accessed May 2013).
48. Boone, M.M., Verheijen, E.N.G., and van Tol, P.F. (1995) Spatial sound-field reproduction by wave-field synthesis. *J. Audio Eng. Soc.*, 43(12), 1003–1012.
49. ISO 354:2003, Acoustics – Measurement of sound absorption in a reverberation room. ISO, International Organisation for Standardisation, Geneva, Switzerland.

Chapter 10

Low-frequency sound field optimization

10.1 BASIC STATISTICS OF STEADY-STATE SPATIAL AND FREQUENCY DISTRIBUTION OF SOUND PRESSURE

Research and publication in room acoustics is mainly concentrated with perfecting the acoustics of large rooms such as concert halls, large auditoria, assembly rooms, and other rooms with similar uses. But small rooms are different from large rooms because there are two major differences:

1. Due to the physical size of small rooms, the first reflections reach the listener much earlier than in large halls. This affects both live music played in the room and the playback of recorded music.
2. The modal density is much smaller at low frequencies in small than in large rooms, which has a substantial effect on the sound field for bass frequencies. Due to the low modal density in small rooms, individual modes are well separated in frequency. When the damping is low, the modal overlap is not sufficient to secure a reasonable uniformity of the spatial sound pressure distribution. Also, the sound field radiation and sound levels across the room depend on the source location, room shape, wall absorption, and objects inside the room (listeners, furniture). The creation of modes and their properties is described in Section 2.2. Figure 2.12 shows two important curves. Figure 2.12b shows the steady-state SPL as a function of position in the room, measured by moving the microphone along a straight path. The curve in Figure 2.12a was obtained by measuring the sound pressure at some observation point inside the room as a function of frequency. The response unevenness in both curves is the result of the sound waves interference between particular modes. The responses of the individual modes mix and, depending of the phase, the modes can add and create a maximum or subtract and create a minimum.

Although the curves in Figures 2.17a and b look similar, there is a substantial difference. The measurement of the spatial curve shown in Figure 2.17a was made at a high enough frequency so that both the modal density and overlap were high. The statistical average of the distance between the peaks in the curve is $\Delta x = 0.79\lambda$ [1]. This value, which applies to the 3D spatial field, is different from that of a 1D standing wave field. In such a field, the distance Δx between adjacent maxima or minima is only 0.5λ. The frequency irregularity at these high frequencies is a property of the interfering modes. The curve in 2.17b is usually called the frequency response curve. Both curves in Figure 2.12 depend on the modal density and damping. The damping of modes that can be changed by adding or removing absorption or by changing the geometry of the room.

The average frequency interval Δf between maxima and minima is $\Delta f = 4/T_{60}$, where T_{60} is the reverberation time. Obviously, the larger the damping the shorter the reverberation time and the larger the distance between peaks in the frequency response curve.

Another measurement of a frequency curve is shown on Figure 2.16 that shows an example of the frequency response of a room at low and middle frequencies. The right portion of the curve differs to some extent from the left portion. The modal density is the reason for the different appearance. Schroeder frequency can be used to define the two regions. At the Schroeder frequency f_S, the modal overlap is at least three modes [2]. The Schroeder frequency is calculated from

$$f_S = 2000\sqrt{\frac{T_{60}}{V}} \qquad (10.1)$$

where
T_{60} is the reverberation time
V is the volume of the room, all in metric units

In the upper region, above f_S, the shape of the frequency response curve is determined by the interference between a multitude of modes and their mixing according to statistical laws. In the low-frequency region on the left, the frequency response curve depends on the sparsely distributed room modes.

10.2 MODES IN SMALL ROOMS

A mode can be considered a resonance of the air in the room at a certain (modal) frequency. The mode is created when a wave reflects from room walls and other objects in the room and propagates in the room at such angles that the reflected waves overlap and the sound field is reinforced. In general the modes have a negative effect on the sound transmission from the source to the listening point. At the low end of the frequency spectrum the frequency response curves have peaks caused by the modes. At the lowest frequencies, we can usually identify the frequency peak of the curve with the resonance frequency of some mode. Because the modal density is proportional to the square of the frequency, the modes overlap with increasing frequency, and thus higher modal density, and the modal frequencies cannot be identified for higher frequencies. The irregularity of the room transfer functions may be heard as coloration of the sound. Therefore, it is important is to linearize the frequency function by lowering the maxima caused by the modes.

The strength of the sound radiation depends on the type of source (monopole, dipole, etc.), its location with respect of a mode, distance from the walls, and sound reflecting objects close to the source. The sound transmission to the listening point also depends on the modal amplitude at the listening location.

The axial modes carry more energy than the more common oblique and tangential modes at higher frequencies and are usually less damped, as discussed in Section 2.3. The axial modes have the largest effect on the sound transmission frequency curve at low frequencies. They carry more energy than the other modes and their damping by the wall absorption is smaller than for the other modes. The frequency response functions shown in Figure 2.11 illustrate the response for three different damping conditions. The effect of higher damping is clearly visible on the modal overlap.

Figure 2.4b shows the amplitude of one of the first low-frequency modes in the room, mathematically expressed by the cosine functions as shown in Equation 2.16. The figure shows that the different regions of the mode have different phase. This means that the radiation from a source on the left side can be canceled by a source located on the right side. Therefore, theoretically, the use of more sources could be used to eliminate the effects of such a mode on the frequency response curve.

In practical situations, it is desirable that the response from a source location has been sufficiently equalized not only at one point but also over a sufficiently large area, for instance, over a square patch of 4 m^2. The optimization process, which also includes practical aspects, has shown that the use of more sources can result in better sound transmission in the form of a smoother, less peaky frequency response. Therefore, the two approaches that have been followed are to increase the number of low-frequency sources (subwoofers) and to optimize their location.

10.3 SOUND FIELD OPTIMIZATION BY MULTIPLE SOURCES AT LISTENING AREA

As shown in the next section, the transmission curve can be electronically equalized for one listening location [4–8]. From a practical viewpoint, however, it is better to minimize the amplitude variations of the frequency curves at all listening points as much as possible. There are many variables that affect the frequency curves: room size, shape, wall absorption and furnishing, location and characteristics of sources, and listening location. It is practically impossible to analyze the effects of all these factors and variables for general conditions. However, extensive research has been done on the optimum loudspeaker number and location in a rectangular room $20 \times 24 \times 9$ ft^3 large with small wall absorption (0.05) with the goal of achieving minimum pressure variation over a grid of 16 listening points (2×2 m^2) assumed in the middle of the room [3]. In these studies, the frequency range studied covered frequencies where modal density is low and where the effect of modes could be seen in clear large single mode maxima and minima in the frequency response curve.

10.3.1 Sound pressure level metrics

The judgment on the frequency response curve variations can be done both objectively and physically. Physically, it is best done by statistical quantities applied on the SPLs measured or predicted at a grid of listening locations in the room center. The location of the sources has to be practical, close to the walls or at the walls. Although the number of sources using a computer program can be large, a real room could probably not have more than four sources.

The MSV in dB^2 defined as [3]

$$MSV = \text{mean}\{\text{var}_s[A(s,f)]\}$$

$$MOL = \text{mean}\{\text{mean}_s[A(s,f)]\}$$

and was selected as the principal criterion, as discussed already in Chapter 9. Here, A is the SPL of the transfer function from the source to the sth listening point and f stands for the desired low-frequency range of modal frequencies. The subscript s means that the quantity is calculated over s seats for each frequency. MOL is mean output level, calculated as the mean sound level over all seats in the frequency range of interest. The variance of the SPL and the mean is first calculated for each seat and frequency and then the values are averaged. The resulting data in the publications [3–6] show the results for 20 rooms with 16 listening locations and a maximum of 4 low-frequency speakers.

Figure 10.1 shows the calculated frequency response at 16 seating locations when a single loudspeaker is used in the room corner. The thin lines show the SPLs. Vertical lines were drawn at the modal frequencies.

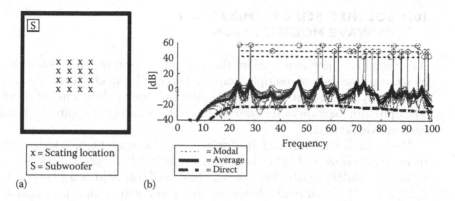

x = Seating location
S = Subwoofer

----- = Modal
—— = Average
—— ▪ = Direct

(a)   (b)

Figure 10.1 (a) Room and seating grid. (b) Amplitude responses calculated using closed-form solution for rectangular enclosure. Metrics are calculated from total acoustical responses at 16 seats. Axial, tangential, and oblique mode frequencies are also shown. Direct sound plot is for the center of the seating area. (From Welti, T. and Devantier, A., *J. Audio Eng. Soc.*, 54(5), 347364, 2006.)

The thick line is the SPL average. We can see that this line peaks at the modal frequencies for all three kinds of modes. The number of speakers and positions tested was 100,000 [3]. The placement of (a) four subwoofers either in the four corners or in the middle of the sidewalls or (b) two on the front and two on the rear wall provides the best values for MSV and MOL. The four loudspeaker locations can be either at room corners or at the middle of the four room walls. According to the investigation [7,8], the difference between one or four sources could not be heard for frequencies below 80 Hz. However, the size of the room and a small distance from one of the loudspeakers could have caused a nonuniform listening situation, and the conclusion could be due to the particular circumstances of the study.

Four loudspeakers might satisfy both technical and esthetic requirements and the solution is not costly. The discussed solution was calculated for steady-state sound, however, and so far there is no published psychoacoustic research for transient sound, time delays, and other time events. Nevertheless, application of the proposed solution should improve the listening conditions in small rooms. More details can be found in Reference 3 that also shows some additional configurations for seating arrangements further away from the primary source speakers. The conclusion that can be drawn from the paper is that the sound field sufficient uniformity can be achieved by source placement, although the optimization process can be quite elaborate.

10.4 SOUND FIELD OPTIMIZATION BY WAVE MODIFICATION

As shown in the previous section, the transmission curve can be smoothed by strategic location of multiple sources. This can be an efficient approach, particularly if the room has some irregularities due to objects in the room. The optimum location of the sources has to be found by measurements for each individual case.

The need for sound field uniformity can be very high, particularly in control rooms and other listening rooms where several persons are to listen simultaneously. Because the best uniform field is a plane wave field, a modification of the listening room to create such a propagating plane wave field is of interest and has been found. The basic idea is to place two sources on, or close to, the front wall of the room at the pressure minima of the first lateral cross mode. If the rear wall of the room is modified to be absorbing, then below some frequency, only plane waves will propagate along the y-axis in the room and the sound field will be uniform.

This idea has been developed and experimentally tested as reported in several publications [9]. Figure 10.2 shows the basic room configuration with two source loudspeakers in the front and two loudspeakers in the rear

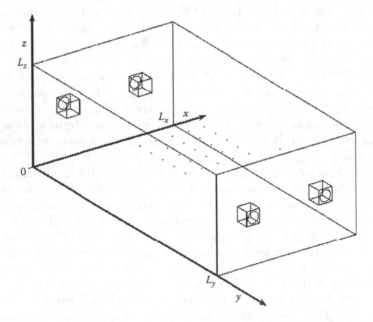

Figure 10.2 The 3D virtual room model and loudspeaker setup in room. (From Celestinos, A. and Nielsen, B., *J. Audio Eng. Soc.*, 56(11), 915, 2008.)

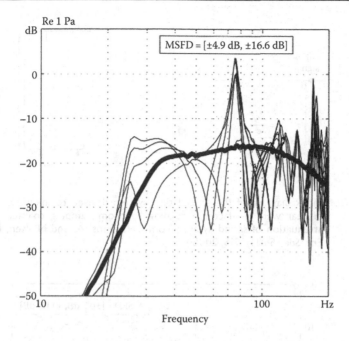

Figure 10.3 Traditional one-point equalization simulated for a virtual room. Thin curves, frequency responses at 25 virtual microphone positions throughout listening area after equalization; heavy curve, target microphone position equalized. (From Celestinos, A. and Nielsen, B., *J. Audio Eng. Soc.*, 56(11), 915, 2008.)

that have their phase controlled to absorb the incident wave, so that only a propagating wave can exist in the room. The signal to the rear speakers is adjusted so that their output is out of phase with the arriving sound from the front speakers by using a digital delay line.

Figure 10.3 shows the result of traditional single point equalization in the listening area, using a filter in series of the loudspeaker with a microphone at the listening point that in combination with the digital signal processing unit equalizes the transmission characteristics. The thin lines represent the SPLs measured at the twenty listening points in the vicinity of the controlled point. The thick line shows the response for the position of the equalized target microphone. The range of frequency response variation between the thin curves is ±16 dB and the results clearly show that single point optimization, while improving conditions at one location, affects the conditions at the other listening points negatively. Therefore, such equalization should be avoided.

Figure 10.4 shows the total system consisting of two signal loudspeakers and two delayed loudspeakers at the rear wall that absorb

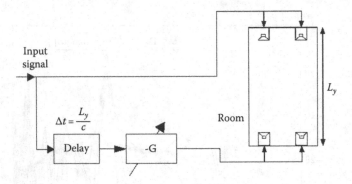

Figure 10.4 Block diagram of the cancellation system devised to minimize reflections of rear wall. *G*—factor according to room damping characteristics and attenuation of sound by air. (From Celestinos, A. and Nielsen, B., *J. Audio Eng. Soc.*, 56(11), 915, 2008.)

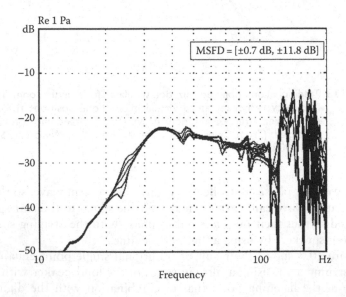

Figure 10.5 Simulation of the effect that the cancellation system had on the frequency response in the virtual room. Frequency responses of transfer function at 25 virtual microphone positions throughout listening area. (From Celestinos, A. and Nielsen, B., *J. Audio Eng. Soc.*, 56(11), 915, 2008.)

the incoming wave so that only a propagating wave travels through the listening points.

The system was tested on two small rooms with volumes 89 and 173 m³. The sound transmission curve simulation into 25 microphone locations is shown in Figure 10.5.

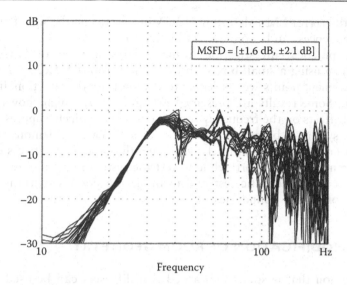

Figure 10.6 These measurement results from using the cancellation system in an IEC room show the frequency responses at 25 positions throughout listening area. (From Celestinos, A. and Nielsen, B., *J. Audio Eng. Soc.,* 56(11), 915, 2008.)

We can see that the sound field uniformity is now much better, compared to Figure 10.3. The actual measurements using the larger room are shown in Figure 10.6. There are some differences of the SPLs among the 25 listening locations, and the effect of the system is getting worse with increasing frequency. For high frequencies, the absorption of the waves by two loudspeakers is not sufficient and the system has to be expanded with more loudspeakers.

Studies on the effects of actual loudspeaker locations on the transfer function were also performed. Obviously, further research is needed to find out the effects of objects (such as furniture) on the sound transmission. Also, the effects of the damping and wall absorption irregularities have to be investigated.

Another approach to sound field equalization is based on the creation and control of plane waves to obtain a sound field that is spatially uniform at low frequencies [10]. This work was further expanded by reformulating the signal processing and documenting the system function by measurements in an experimental room [11].

In the approach, two sets of loudspeakers were located on planes, perpendicular to the direction of the plane wave propagation. A set of microphones was located in the zone of equalization. The strength of sources was determined from the cost function obtained from the microphones, using a multiple error LMS algorithm [12]. More details on the signal processing and references can be found in Reference 10.

The practical investigation, using a large number of loudspeakers, was done using a small box with inside dimensions $1.2 \times 2 \times 0.2$ m³. Four measurement points were selected to demonstrate the functionality of the system. Some results are presented in Figure 10.7, which shows how the modal effects on the frequency response were very nicely suppressed.

The system is based on several published ideas. As demonstrated, the sound field levels spatial levels were smooth out and the effects of modes were removed. It is worthwhile to test the system in a full-size room and conduct psychoacoustic test and examine the physics of transients, the effects of nonflat boundaries, etc.

10.5 MODIFICATION OF ROOM GEOMETRY

A question that is sometimes asked is if diffusers can be used similarly to absorbers to improve the low-frequency transmission properties of a room.

Analytical models for the transfer functions were formulated in equations in Chapter 2. The equations were, however, derived for rectangular rooms and idealized impedance conditions. More real-life cases can be modeled and analyzed using BEM or FEM. As an example we include the results of some calculations done by us using the FEM for four flat "2D" models of rooms.

In one of these models, shown in Figure 10.8, the height of a flat room was kept so low that there could only be 2D sound propagation. The ultimate purpose for this was to reduce the number of modes by eliminating the oblique modes and make the response and possible influence of modifications to the room geometry more clear. Figure 10.8a shows the simple rectangular flat room. To investigate the effect of a diffusing wall, one wall was then designed as a quadratic residue diffuser. By scaling the length of the diffuser wells, the active frequency range of the diffuser could be moved in frequency. The longest wells in Figure 10.8d had the same length as the shorter of the room walls. In Figure 10.8b andc, the wells were progressively shortened by a factor of 0.5 and 0.25, respectively. Finally, the damping of the modes was changed by setting the impedance of the *ceiling* to a finite value so the resonances of both the diffuser and room could be damped. Three impedance values were used: $Z_W = 300\rho_0 c$, $100\rho_0 c$, and $30\rho_0 c$.

It was decided to investigate the smoothness of the frequency response by inserting a source at the lower left-hand corner of the room and a receiving point at the lower right-hand corner, that is, both points were placed at the far end of the room away from the diffuser wall.

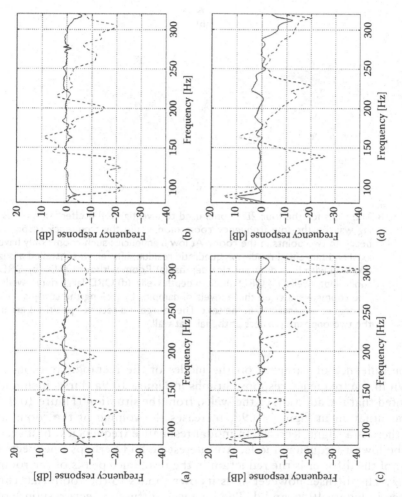

Figure 10.7 Frequency responses before (dashed line) and after (solid line) equalization at various points: (a) at the microphone used in the equalization, (b) a point within the zone of equalization, (c) a point just outside the zone of equalization, and (d) a point far from the zone of equalization. (From Sarris, J.C. et al., *J. Acoust. Soc. Am.*, 116(6), 3271, 2004.)

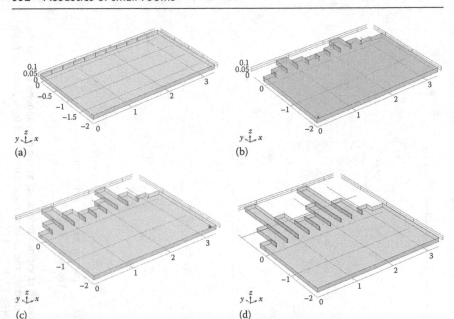

Figure 10.8 The plans for the four 2D rooms used to investigate the effect of a diffusing wall on the low frequency room modes and the frequency response
between two points in the room. At low frequencies such rooms only have
axial and tangential modes. A quadratic residue diffuser was installed along
one wall with four different slot depths. (a) Room without QRD, (b) QRD
short depth wells, (c) QRD medium depth wells, (d) QRD large depth wells.
The rooms were given three levels of damping by making the *ceiling* sound
absorptive as described in the text. The source and receiver positions are in
the two opposite corners at the far, flat wall.

The influence of damping on the modes of the rectangular room is
shown in the frequency response graphs in Figure 10.9a. First, it should
be noted that the addition of the wells, from the situation (Figure 10.9a)
to the situation in Figure 10.9d, increases the volume of the room so
that there is a slight decrease in modal resonance frequencies, best seen
for the lowest-frequency mode. An interesting and perhaps unexpected
effect of the diffuser is the reduction in the resonance peaks of the room
modes. The diffuser however adds its own modes to the room that can
be seen in Figure 10.9c and d. The long well diffuser affects response of
the room considerably particularly for the case of high wall impedance,
in which case the diffuser mainly reduces the room's mode frequencies by
about one-half octave. Simply adding volume to the room could however
have effected a similar change. Clearly, adding damping rather than diffusion is a better approach to control the low-frequency properties of a
room.

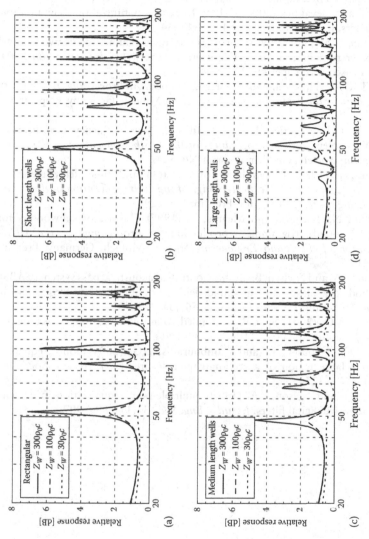

Figure 10.9 The frequency responses between the source and receiver positions for the four 2D rooms shown in Figure 10.8 calculated using the FEM. The source and receiver positions were at the two opposite corners at the far, flat wall. Each room was assigned three levels of damping by making the ceiling sound absorptive as described in the text. (a) Room without QRD, (b) QRD short depth wells, (c) QRD medium depth wells, (d) QRD large depth wells. The rooms were given three levels of damping by making the ceiling sound absorptive as described in the text.

REFERENCES

1. Kuttruff, H. (2000) *Room Acoustics*, 4th edn. Elsevier Science Publishers, London, U.K.
2. Schroeder, M.R. (1969) Effect of frequency and space averaging on transmission responses of multimode media. *J. Acoust. Soc. Am.*, 46, 278.
3. Welti, T. and Devantier, (2006) A. Low-frequency optimization using multiple subwoofers. *J. Audio Eng. Soc.,* 54(5), 347364.
4. Welti, T.S. (2004) Subjective comparison of single channel versus two channel subwoofer response, in *Audio Engineering Society, 117th Convention 2004*, San Francisco, CA, Paper #6322.
5. Welti, T. (2002) How many subwoofers are enough, in *Proceedings of the 112th Convention of the Audio Engineering Society*, Munich, Germany, Preprint No. 5602.
6. Welti, T. and Devantier, A. (2003) In-room low frequency optimization, New York, New York in *Proceedings of the 115th Convention of the Audio Engineering Society*, October, Preprint No. 5942.
7. Borenius, J. (1985) Perceptibility of direction and time delay errors in subwoofer reproduction, in *Proceedings of the 79th Convention of the Audio Engineering Society*, New York, Preprint No. 2290.
8. Zacharov, N., Bech, S., and Meares, D. (1997) The use of subwoofers for surround sound programme reproduction, in *Proceedings of the 102nd Convention of the Audio Engineering Society*, Munich, Germany, Preprint No. 4411.
9. Celestinos, A. and Nielsen, B. (2008) Controlled acoustical bass system (CABS), a method to achieve uniform sound field distribution at low frequencies in rectangular rooms. *J. Audio Eng. Soc.*, 56(11), 915–931.
10. Santillan, A.O. (2001) Spatially extended sound equalization in rectangular rooms. *J. Acoust. Soc. Am.*, 110(4), 1989–1996.
11. Sarris, J.C., Jacobsen, F., and Cambourakis, E.G. (2004) Sound equalization in a large region of a rectangular enclosure (L). *J. Acoust. Soc. Am.*, 116(6), 3271–3274.
12. Elliott, S.J. and Nelson, P.A. (1989) Multiple-point equalization in a room using adaptive digital filters. *J. Audio Eng. Soc.*, 37(11), 899–907.

Chapter 11

Rooms for sound reproduction

11.1 INTRODUCTION

The equipment and room acoustic conditions for optimal sound reproduction have been studied for more than 100 years. Subjective preference and the difficulty in describing what is heard have made work difficult. Additionally, any sound to be reproduced must also be recorded, so the sound reproduction is subject to the audio equipment used for recording and the acoustics of the recording venue. In control rooms, it is important for the engineer to hear every detail of the recorded sound without room reflections affecting the sound at the listening position. For music appreciation in homes, on the other hand, the reflected sound by the room may be necessary for audio quality.

Control rooms are necessary to provide a suitable listening environment to ensure quality and suitability of the audio signals for the end user that is likely to be in a very different environment. To hear all the properties of the recorded sound, it is necessary to have access to quiet rooms with well-defined room acoustic properties, but this is usually difficult to achieve since control rooms need considerable amounts of equipment that will be moved, changed, etc. The ±180° panorama photo in Figure 11.1 illustrates the problem.

The desired control room acoustics have changed as the audio distribution formats have developed. The number of channels used for the sound recording and playback has a dramatic influence on the desirable acoustics of the listening environment. While most audio is likely to continue being distributed for stereo listening—because of its suitability for both headphone and simple loudspeaker listening—much film sound is distributed for some form of surround sound listening. For film sound, the effect channel with its low-frequency sound content is particularly important and subwoofers are essential for adequate sound quality.

Figure 11.1 Ideal conditions are strived for but reality prevails as shown by this panorama ±180° photo of an LEDE control room. (Photo by Marco Verdi.)

11.2 ROOM STANDARDIZATION

There are many attributes of the reproduced sound that are due to the interaction between loudspeaker's sound radiation and the acoustics of the listening room. The standard IEC 268-13 defines three properties of the spatial quality of the reproduced (stereo) sound field: image localization, image stability, and width homogeneity [1]. Listening to sound reproduced in an anechoic or similar very sound-absorptive environment, one notices immediately that sound is only arriving from the loudspeakers, that there is little reflected sound, and that some reflected sound is beneficial for overall sound quality listening to this distribution format.

Audio equipment evaluation rooms are usually proprietary designs [27]. However, for loudspeaker testing, sound signal evaluation, and general listening, a *standard* listening room is a possible alternative. Guidelines can be found in the ITU-R BS.1116 recommendation that is summarized in Table 11.1 and in the commonly used IEC 268-13 standard [1,2]. Figure 11.2 shows a room that is an implementation of this standard. Table 11.1 shows how the ITU-R BS.1116 type of listening room is defined broadly by requirements for low noise, room proportions and reverberation time, and relative absence of early specular wall and ceiling sound reflections.

11.3 CONTROL ROOMS FOR STEREO

11.3.1 Symmetry

Early reflected sound will confuse hearing and make the stereo stage and its phantom sources appear incorrectly located or even blurred. As explained in Chapter 8 the listener's placement of the phantom sources is dependent particularly on the transient nature of the sound that comes from the loudspeakers so it will be affected by the early reflected sound from the room

Table 11.1 Excerpt from Recommendation ITU-R BS.1116-1

Parameter	Specifications
Reference loudspeaker monitors amplitude vs. frequency response	40 Hz–16 kHz: ±2 dB (noise in ⅓ octave bands, free-field) ±10° frontal axis ±3 dB re 0° ±30° frontal axis ±4 dB re 0°
Reference loudspeaker monitors directivity index	6 dB ≤ DI ≤ 12 dB 0.5–10 kHz
Reference loudspeaker monitors nonlinear distortion at 90 dB SPL	< −30 dB (3%) for f < 250 Hz < −40 dB (1%) for $f ≥$ 250 Hz
Reference monitors time delay difference	<100 μs between channels
Height and orientation of loudspeakers	1.10 m above floor reference axis at listener's ears, directed towards the listener
Loudspeaker configuration	Distance between loudspeakers 2–3 m Angle to loudspeakers 0°, ±30°, ±110° Distance from walls > 1 m
Room dimensions and proportions	20–60 m² area for mono/stereophonic reproduction 30–70 m² for multichannel reproduction 1.1 $w/h ≤ l/h ≤$ (4.5 w/h −4) l/h < 3 and w/h < 3 where: l is length, w is width, h is height
Room reverberation time	$T = 0.25 (V/100)^{1/3}$ for 200 Hz $≤ f ≤$ 4 kHz where: V = volume of the room The following limits apply: +0.3 s @ 63 Hz to +0.05 s @ 200 Hz ±0.05 s @ 200 Hz–4 kHz ±0.1 s @ 4 kHz–8 kHz
Room early reflections	< −10 dB for $t ≤$ 15 ms
Operational room response	≤ +3 dB, −7 dB @ 50 Hz to ≤ ±3 dB @ 250 Hz ≤ ±3 dB, @ 250 Hz–2 kHz ≤ ±3 dB @ 2 kHz to ≤ +3 dB, −7.5 dB @ 16 kHz
Background noise (equipment and HVAC on)	≤ NR15 in the listening area ≤ NR10 in the listening area recommended

Source: ITU-R BS.1116, *Methods for the Subject Assessment of Small Impairments in Audio Systems, Including Multichannel Sound Systems*, International Telecommunication Union, Geneva, Switzerland, 1998.

surfaces. The early reflected sound will also affect the global auditory source width for an orchestra for example and may make it extend considerably beyond the baseline between the loudspeakers.

In asymmetric rooms where the walls on the left and right of the listener have different acoustic properties, the stereo stage may become biased towards the wall that reflects the most. The curve in Figure 8.23 shows the dependency more clearly for different levels of unbalance as applied to the

Figure 11.2 The IEC listening room at the Aalborg University Centre's Acoustics Laboratory
is a good example of a listening room possibly similar to what may be achieved
in a domestic environment. Note the use of diagonal absorbers along upper
side walls to reduce low frequency reverberation time. (Photo courtesy of
Aalborg University Center [AUC], Aalborg, Denmark.)

center phantom source in a stereo loudspeaker system. The intensity will
then be higher at that ear and the sound stage distorted. This distortion is
usually compensated by changing the balance in amplification between the
stereo channels.

At low frequencies in the modal region, symmetry may not be desirable
since someone sitting in the middle of the room may be on or close to modal
node lines. One way of avoiding such node lines is to make the room asymmetric in the low-frequency region.

This can be achieved by having an asymmetric rigid shell surrounding the
inner room which is symmetric for mid- and high frequencies by suitably
reflective side walls, ceiling, and floor. The inner room must be open acoustically to the outer shell at low frequencies, for example through ventilation
vents, and similar large openings, for example at corners. In this way, one
can have the desired listening position sound field symmetry for mid- and
high frequencies while at the same time have asymmetric conditions in the
modal frequency range. Bass traps to control the damping—and thus the
reverberation times—of these modes can be placed between the outer and
inner shell. It is important to remember though that noise transmission to
the surrounding spaces will then be dependent on the sound isolation of the
outer shell that must be physically substantial.

11.3.2 Loudspeakers

Most control rooms use monitor loudspeakers flush mounted into the front wall, on either side of the control room window. The monitor's front baffle is then in the plane of one of the critically angled surfaces. This type of mounting reduces the risk of annoying early reflected sound that can affect the placement and properties of the phantom image created sound stage. Wall mounting allows the use of large loudspeakers, removes the influence of box diffraction, and often helps make the loudspeaker response more flat in the low and midfrequency regions. Such specialized flush-mounted loudspeakers can additionally be made to have controlled, reduced high-frequency dispersion. Less sound reflected by the walls contributes to a more direct sound-oriented approach that gives higher clarity.

A disadvantage of wall-mounted loudspeakers is that the loudspeaker may induce structure-borne vibration on the wall and other room surfaces. This structure-borne vibration travels faster than sound in air and can result in the *first arriving sound* coming from room surfaces rather than from the loudspeaker diaphragms. It is important that wall-mounted loud-speakers are vibrationally well decoupled from the wall.

Near-field monitor loudspeakers on the top of the mixing console are also often used in control rooms. An advantage of this approach is the possibility to use personally preferred loudspeakers. In this case, the room-reflected sound can be minimized relative to the direct sound at least for midrange and high frequencies since the monitor loudspeaker is close. In most cases, console top loudspeakers are used in small control rooms.

11.3.3 Control room types

One approach to stereo control room conditions would be to try to simulate some imagined domestic room. Obviously, this is not possible because of the large variations in the size, geometry, and acoustic conditions of domestic rooms. Idealized standard control room acoustics and engineer experience ensure that the audio that is distributed has the desired properties for enjoyment of stereo sound. Some well-known types of control room acoustics are [3,4]

- Nonenvironment
- Live end–dead end (LEDE)
- Reflection-Free Zone (RFZ)
- Controlled Image Design (CID)

11.3.4 Nonenvironment

A classical design of control rooms is to treat all walls and the ceiling with a wideband sound absorber to minimize reflected sound [5]. A drawback of such rooms, similar to anechoic chambers in a sense, is that while they allow very precise listening they are unpleasant to communicate with others in.

An anechoic chamber has all surfaces covered by sound-absorbing material and would thus be ideal as a control room for very precise listening. In practice, a hard floor is necessary for walking on so a control room along such lines would rather be similar to a hemi-anechoic chamber.

The sound absorption in classical sound-absorptive control rooms is usually achieved by using porous sound-absorbing materials such as sheets of glass wool or plastic foams. Conventional sheets of glass or mineral wool are usually sold in thicknesses of 0.02–0.1 m, some materials are available with cloth or other surface treatment to improve durability and minimize fiber shedding. Plastic foams tend to be sold in thinner sheets because of cost but are available in a multitude of shapes. If thin sheets would be mounted directly on a hard wall, they would give sound absorption only at mid- and high frequencies as described in Chapter 4. The absorptive sheets need either to be thick (and have low density) or to be mounted with an airspace backing to obtain reasonable sound absorption in the low frequency range. Sheets that have proper density to function optimally at low frequencies are generally not manufactured and must be replaced by the wedge construction used in anechoic chambers.

The wall surfaces of anechoic and hemi-anechoic chambers are thus usually covered by wedges cut from commercially available density glass wool, such as those shown in Figure 14.31. The wedges give a gradual change of impedance from that of free air to the full density of the glass wool so that the sound absorption will be high for a wide range of frequencies. Using an impedance tube, a sound absorption coefficient of about 0.99 is typically measured for wedges used for this application. The upper limiting frequency of an anechoic chamber is, in practice, determined by reflecting objects due to hardware needed for mounting of microphones and other equipment, rods, cables, etc., while the length of the wedges and their flow resistance determine the lower cutoff frequency. The function of porous sound absorbers is generally poor for frequencies where the distance to the hard back wall is shorter than one-quarter wavelength: for a 0.7 m long wedge, it will be about 70 Hz. The lower cutoff frequency can be slightly reduced by mounting the wedges with an airspace behind them, typically 0.2 m deep.

Because the type of wide frequency range sound-absorbing construction used for anechoic chambers requires considerable space, many have tried to achieve similar sound absorption in other ways. The design of the nonenvironment control room tries to address this problem by using a thick, somewhat resonant construction shown schematically in the drawing in Figure 11.3 [4,6–8]. The figure shows a plan and section through the room. The sound-absorbing construction is a *slot* design with thin hard sheets covered

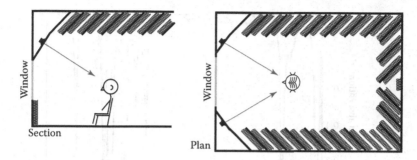

Figure 11.3 A *nonenvironment* room is an approximation to a hemi-anechoic chamber. The drawing does not show the user's mixing console that interrupts the first floor reflection.

with sound-absorptive material on the slot walls. For the incident sound wave, the situation can be thought of as a case of oblique sound incidence on a locally reacting medium.

The nonenvironment room is thus—as shown in the figure—in its broad outline similar to a hemi-anechoic chamber. Most control rooms need floors and usually also a wall that has a window for communication with the studio. Two monitor loudspeakers for stereo sound reproduction flank the window and are either dedicated designs or commercial designs for freestanding use but here flush mounted on the wall to minimize sound reflections other than those of the mixing console (not shown) and the floor.

The slots are at an angle to the wall as shown in the figure. One can therefore say that this is an approach at having a *longer* absorber than with the wedge system, resulting in a cutoff frequency that is about one-half octave lower than a wedge-style construction with the same depth. The sound absorption at very low frequencies is obtained because of the ¼λ matching action of the slot ducts. Some additional sound absorption is achieved by mounting the absorbing sheets resonantly.

Obviously, if this was a viable way of achieving low-frequency absorption, it would be used commercially in the design of anechoic chambers. Measurement results that allow direct comparison to traditional anechoic room design are not available. Figure 11.4 shows measurement results for one nonenvironment room [8]. The very low cutoff frequency is achieved at the price of less sound absorption overall.

It is clear from the cited measurements that the design, at least as implemented in the case reported, does not function well in the frequency range where the length of a slot is about one-half wavelength long. This is a result of the slots ending at a hard wall.

In summary, it can be said that the nonenvironment control room design strives to resemble a hemispheric anechoic environment using layers of a *thick* sound absorber. In its way, it features much higher sound absorption

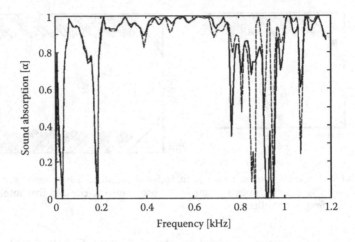

Figure 11.4 Example of measurement results for the sound-absorption coefficient of the sound absorber construction in a *nonenvironment* room. (From Torres-Guijarro, S. et al., *Acta Acust. Acust.*, 98, 411, 2012.) Dashed lines without and solid lines with sound-absorptive filling added to openings between sheets.

at low frequencies than that which can be achieved using even plane layers of .1 m thick sound-absorbing mineral or glass wool sheets. It misses the target though for frequencies where the length of the slots becomes comparable to about one-half wavelength. The sound absorption in this frequency range might however be improved by a more elaborate slot design.

11.3.5 Live end–dead end

Inspired by German research on the precedence effect, the large discussion topic in room acoustics in the 1960s was the relative importance of the time delay between direct sound and the first reflection, the so-called the initial time delay (ITD) gap in concert hall design [9,10]. This discussion also had effects on the thinking regarding control room design [11].

Many engineers prefer control rooms not to be as sound absorptive as the almost anechoic rooms discussed in the previous section. Reverberation times of about 0.4 s are often mentioned as a goal to have a pleasant spaciousness in the control room [12]. This means that there must be some room surface reflections, and the question then is, how can one combine spaciousness with clarity? It is important to remember, as was pointed out in Chapter 8, that the desired spatial effect is that of envelopment (background spatial impression) as indicated in Figure 8.27.

For the desired combination of reverberant envelopment with direct sound clarity, it is necessary to have one part of the room reverberant (*live*) and the other part nonreverberant (*dead*). Figure 11.5 shows the basic idea for how

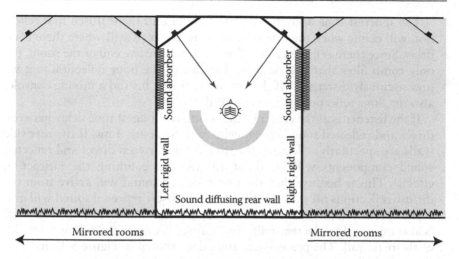

Figure 11.5 Plan of an LEDE control room (white center) with mirror images. The LEDE room has the rear half walls specularly reflecting and the rear wall covered by diffusers. Side wall mirrored versions of the LEDE room are shown in gray. The direct paths from the mirrored loudspeakers are interrupted by the side wall sound absorbers.

this could be achieved by a combination of anechoic and reverberant room halves. In any practical environment, it is necessary to have a floor so the nonreverberant end fulfills less of its ideal than the reverberant end.

Rooms built along these ideas are called live end–dead end (LEDE) rooms in the literature [11,13–15]. Figure 11.1 shows such a room. According to the originator, the main idea behind the use of the LEDE idea is to allow the listener to hear the influence of the first reflection of the studio rather than the control room. Because of the temporal masking, the first control room reflection is supposed to arrive after at least 20 ms, that is, the travel path must be about 6 m longer than the direct path. Such a delay is easily accomplished in a reasonably sized control room. Additionally, if such delayed sound arrives spread out in time such as that of a diffusely reflected sound rather than a sharp specular single reflection, its center of gravity so to speak will be even later, corresponding to an on the average longer travel time.

The control room sketched in Figure 11.5 is about 4 m wide and 6 m long. Assume the walls of the rear half of the room to be specularly reflecting but the rear wall covered by diffusers. The ceiling is absorptive in the dead front end but may be reflecting in the rear, live end of the room. If the loudspeakers are integrated in the front wall at the dead end, the first reflected sound will arrive from the start of the diffusing wall treatment (not counting the floor reflection, assuming sound to be diverted by the mixing console). Since the diffuser can scatter in many different ways, its effect can only be outlined as in the figure.

The shortest time delay between the direct and the diffused first reflection will occur when the listener is close to the rear wall where there is no delay. Since there is no discrete reflection from the live end of the room, the only comb filter that will occur is that due to the floor reflection that we unconsciously disregard. (If the listener is sitting behind a mixing console, also the floor reflection will be removed.)

If the listener is at the center of the room, the longest time delay between direct and reflected sound components will be about 3 ms. If the rear side walls are specularly reflecting, the time delay between direct and reflected sound components will be about 15 ms (not counting the diffraction effects). This is because then the first reflected sound will arrive from the diffuse reflections off the rear wall. Such diffusely reflected sound will give a pleasing envelopmental effect since the scattered sound by the rear wall is also mirrored by the rear reflecting walls, effectively extending the width of the rear wall. The precedence effect data shown in Figure 8.10 indicate that such reflections are likely to be masked and that the direct sound will be perceptually dominant and determine the direction of the sound sources although timbre and envelopment may be changed.

The rear parts of the side walls are supposed to be plane and rigid. This means that there is a risk for flutter echo in that part of the room. Since diffusion or absorption are not acceptable alternatives for these wall—they would reduce the reverberation time and remove the conditions to achieve the desired envelopmental effect—the rear parts of these walls can instead be splayed by an angle of 7°–10° or more. (Typically, the ceiling surfaces are splayed as well in control rooms.) The design shown in Figure 11.6b is common in many control room designs, LEDE, RFZ (see next section), and others. Sometimes the midsection of the diffuser on the back wall is angled downwards or replaced by some other arrangement to even further reduce sound arriving straight from the back. This can also be achieved by using the Galois sequence instead of QRD sequence-based diffusers since diffusers based on the Galois sequences suppress specular reflections [16].

An interesting requirement for a control room in order for it to qualify as an LEDE room is that there must be an inner symmetric shell that is open for low-frequency sound to reach an outer shell that is rigid. The outer shell must be asymmetric so that low-frequency resonance node lines through the listening position are avoided [15]. The intermediate volume between the inner and outer shell contains bass traps and other absorption to control reverberation time of strong low frequency modes.

11.3.6 Reflection-free zone

The Reflection-Free Zone (RFZ) room is an approach to reduce the first-order side wall-reflected sound even further than that possible with sound

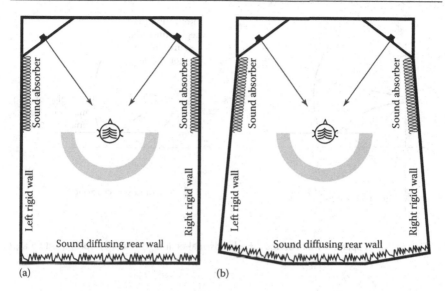

Figure 11.6 (a) In this LEDE room design the hard rear part of the side walls provides diffuse sound reflection by mirroring the sound from the rear wall diffusing surface. (b) In this fan-shaped LEDE room design the hard rear part of the side walls is splayed by an angle of 7°–10° or more to reduce the possibility of flutter echo which is always present when walls are rigid, smooth, and parallel. This design is commonly found in control rooms when diffusive or absorptive treatments of these surfaces are not acceptable treatment alternatives from the viewpoint of achieving the desired reverberation time.

absorption in the LEDE room [17]. The RFZ room is sometimes referred to as an RFZ/RPG room since it originally was designed with commercial diffusers made by the RPG company. In this type of control room, the stereo loudspeakers are placed close to the front corners of the room. Additionally, the side walls are splayed to change the geometry of the room and remove possible reflection paths that would otherwise direct first order reflections to the listener.

If an image source study is made instead of ray tracing, one can intuitively see how the reflected sound can be reduced. It is important though to remember that either type of analysis only applies when the assumptions made for geometrical acoustics are fulfilled. Consider a loudspeaker installed in baffle close to a corner as shown in Figure 11.7. The loudspeaker itself will not be mirrored in the baffle since it is approximately in its plane.

By using the idea of suitably splayed wall and ceiling patches, it is possible to further reduce mid- and high-frequency sound reflection (above about 2 kHz). A section and plan of such an RFZ design is shown in Figure 11.8. As in the case of the LEDE room, the surfaces in front of the listener are

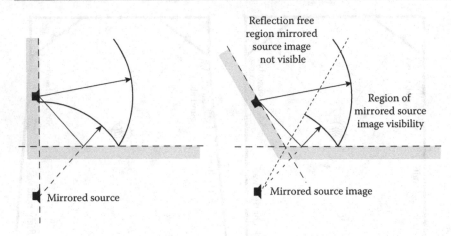

Figure 11.7 A splayed wall reduces the loudspeaker image visibility compared to a perpendicular wall.

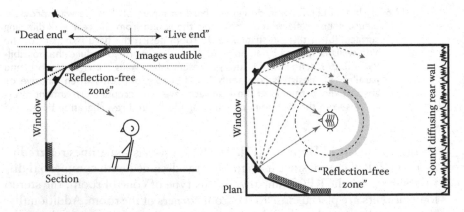

Figure 11.8 RFZ at high frequencies created by using splayed reflecting surfaces. Reverberation time controlled by sound-absorptive material distributed over room surfaces in the *dead end* of the room. Direct and some first-order reflection paths shown. Floor reflection path not shown since it will be removed by mixing console also not shown. Section shows how image visibility can be used to determine extent of RFZ. Plan shows how ray tracing can be used for the same purpose. Console (mixer) not shown.

treated to be sound absorptive (except the floor and of course the studio window). Typical reverberation times are about 0.4 s.

As do many other control rooms, the RFZ rooms often have an inner symmetric shell that is open for bass frequencies to an outer asymmetric shell that is rigid, and the intermediate volume contains bass traps and other absorption as for LEDE rooms.

11.3.7 Controlled image design

The Controlled Image Design (CID) idea is yet another variation on the use of reflecting patches angled in such a way as to avoid early high-frequency sound reflections at the listening position [18,19]. A difference between this idea and the earlier discussed LEDE and RFZ designs is that this control room design also considers the use of small, freestanding monitor loudspeakers as an alternative to flush-mounted loudspeakers. Some of the design requirements listed in Table 11.1 are from References 18,19. An important requirement is that discrete early reflections in the high frequency range should arrive more than 20 ms later than the direct sound and have levels –20 dB or less than the direct sound. Figure 11.9 shows the basic principle for the CID in the case of flush wall-mounted loudspeakers, whereas Figure 11.10 shows the basic design of CID rooms for freestanding loudspeakers. Because the reflecting patches are quite small, the design is limited to functioning as shown in the figures at medium and high frequencies only.

11.4 CONTROL ROOMS FOR SURROUND SOUND

Some control rooms are being developed with emphasis on surround sound monitoring for film and music [19,20,31,33]. A review of such surround sound distribution systems is found in Chapter 9. The dominant

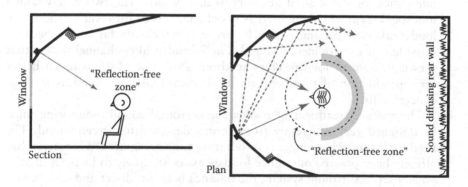

Figure 11.9 The plan and section of a CID-type control room for wall flush-mounted loudspeakers. The ray tracing shows how high-frequency reflections can be redirected away from the listening region. Console (mixer) not shown. (From Walker, R., Controlled image design: The management of stereophonic image quality. BBC Research and Development, Department Report RD1995/4, The British Broadcasting Corporation, London, U.K. 1995; Walker, R., A controlled-reflection listening room for multi-channel sound, *Proceedings of the 104th Convention of the Audio Engineering Society*, Amsterdam, the Netherlands, Preprint 4645, 1998.)

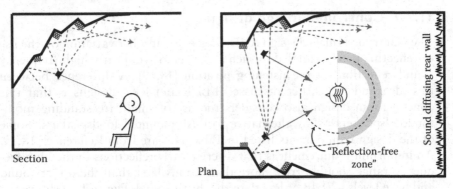

Section

Plan

"Reflection-free zone"

Sound diffusing rear wall

Figure 11.10 The plan and section of a CID-type control room for freestanding loudspeakers. The ray tracing shows how high-frequency reflections can be
redirected away from the listening region. Console (mixer) not shown.
(From Walker, R., Controlled image design: The management of stereophonic image quality. BBC Research and Development, Department
Report RD1995/4, The British Broadcasting Corporation, London, U.K.
1995; Walker, R., A controlled-reflection listening room for multi-channel
sound, in *Proceedings of the 104th Convention of the Audio Engineering Society*,
Amsterdam, the Netherlands, Preprint 4645, 1998.)

surround sound system is the 5-channel system with 3 front loudspeakers
to provide the phantom sources necessary to correctly achieve the desired
acoustic events in the front of the listener and to give the early spatial
impression of the desired auditory source width. The two rear channel
loudspeakers at ±110° are used as effect channels or for reverberation. The
loudspeakers at an angle of 110° are not sufficiently far to the back to
provide rear envelopmental sound. The 7-loudspeaker channel system that
uses envelopment loudspeakers at about ±80° and ±140° is much better
for reproduction of reverberation in classical music to sound as in large
concert halls.

The surface treatment of rooms for surround sound monitoring must
be designed very differently from rooms designed for stereo sound. The
need for a somewhat longer reverberation time being needed for stereo has
already been pointed out. Since loudspeakers are likely to be both smaller
and closer in surround system, the balance between direct and reverberant
sound intensity may be about the same as for stereo.

Figure 11.11a shows a 5-channel surround sound monitoring environment modeled using geometrical acoustics software. The room is LEDE
inspired and symmetrical along the long center axis. The impulse response
patterns are shown for the front (b), left front (c), and left back (d) loudspeakers. We see from figure (b) that the room as modeled would severely
color the sound from the front speaker because of the repetitive reflections
at 0°. For the front left loudspeaker, the tendency is the same although the

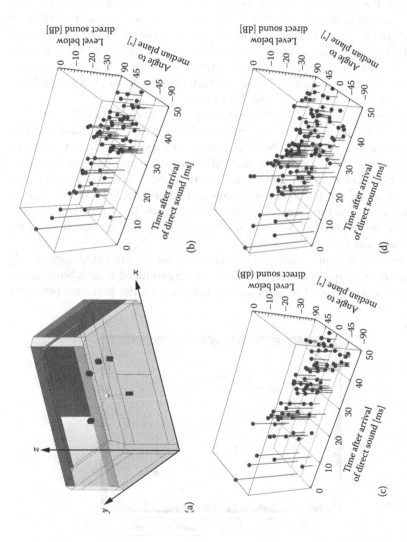

Figure 11.11 An LEDE-inspired listening room for 5-channel surround sound. (Model and data courtesy of CATT, Sweden.) (a) Drawing of room seen from above right (room sidewise symmetrical), 3D echogram response pattern for front (b), left front (c), and left back (d) loudspeakers.

first ceiling and floor reflections have angles to the median plane that vary slightly. The left rear loudspeaker has a more benign reflection pattern.

Three basic designs have been developed. The first is trivial in its use of a fully sound-absorptive treatment and similar to that of such stereo control rooms. Good sound absorption, $\alpha \geq 0.8$, can be achieved using thick glass or mineral wool absorptive sheets placed at some distance from the back surface as explained in Chapter 4. The sheets must be mounted adjacent to one another so that there are no air gaps between them to ensure good sound absorption also at low frequencies. Below the frequency at which the thickness of the construction is one-quarter wavelength, the sound absorption quickly drops to small values. To function down to about 0.25 kHz, the construction must be at least 0.3 m thick.

An absorber design along these lines that gives good sound absorption also at low frequencies and that can complement that of sound-absorptive sheets is the diagonal absorber that consists of mechanically and acoustically resonant sheets of compressed glass wool that are mounted as membrane absorbers along the corners of the room. Such absorbers can be seen in the photo of the IEC listening room in Figure 11.2 [21]. Depending on size, acoustic, and mechanical properties and fastening, high sound absorption may be achieved down even in the frequency range where modes are sparse and axial modes dominate.

As pointed out previously, hemi-anechoic rooms are often subjectively rated as overdamped except for listening to reproduced sound. Some reverberation is preferred. Listeners are accustomed to the reverberation time

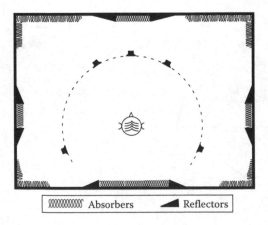

Figure 11.12 The plan of a CID-type surround sound control room for freestanding loudspeakers. Console (mixer) not shown. (From Walker, R., A controlled-reflection listening room for multi-channel sound, *Proceedings of the 104th Convention of the Audio Engineering Society*, Amsterdam, the Netherlands, Preprint 4645, 1998.)

of their living rooms which is often $T_{60} \approx 0.4$ s. The second type of design combines some reverberation with the removal of very-high-frequency reflection cues [19]. This design is shown in Figure 11.12. Patches of sound-absorptive treatment are combined with splayed wall sections that reflect high-frequency sound away from the listener so that there are no image sources visible from the listener's position.

The third approach is to use extensive sound-diffusive treatment of all surfaces except the floor. Such a control room is shown on the cover of this book, the Blackbird Studio C room that is described in Reference 43. Since diffusers need to be about ¼λ long to be effective, a room that uses diffusers to function down to about 0.25 kHz needs diffusers that are at least 0.3 m deep.

11.5 HOMES

The goal of the video and sound recording industry is generally to distribute recorded sound that will be listened to over loudspeakers in the home, the automobile, or over headphones. Home listening rooms have much larger variety than control rooms since they need to cover many needs and there are many ideas of decoration. In homes, the possibilities of remodeling to improve sound system requirements are often limited. At the same time, there will also be constraints by the presence of doors and windows. Floor plans may be L-shaped or have focusing niches. Since the listeners are not limited to a fixed seating by a mixing console, optimum loudspeaker placement is difficult to agree on.

The experience gained from control room design is useful in the design of domestic listening rooms. Figures 11.13 and 11.14 show the plan and section through a rectangular room, similar to the LEDE design used for control rooms. Both home theater and surround sound music are applications common in homes. For the first application, low-frequency sound reproduction is of particular importance, and for the second application, envelopment is important. The size of living rooms differs and with that the ratio between direct and reverberant sound. In houses that have masonry, concrete block, or concrete construction, the walls are likely be more rigid and offer less sound absorption and scattering than walls in houses that have lightweight construction using drywalls. Post and frame construction offers little sound isolation but much sound absorption. In homes or apartments that have heavy construction, the reverberation time will be longer particularly at low frequencies. In these dwelling, the reverberation time at low frequencies is often determined by the properties of the windows.

Several investigations have been made on the acoustics of domestic rooms [22–25,32]. These show a surprising agreement for the reverberation time of living rooms considering the different countries: Sweden, Britain, Canada,

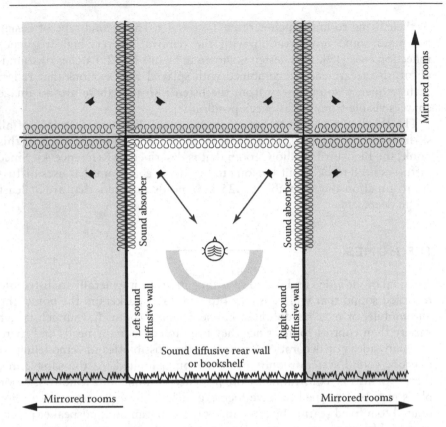

Figure 11.13 The plan of listening room for freestanding loudspeakers. Carpet on floor prevents sound reflected by floor. Sound absorbers on side walls may be replaced by (filled) book shelves or diffusers. Random diffusers are preferable.

and the United States. Accepting the possibility for bias in the selection of rooms, the results are nevertheless in surprising agreement and the typical living room was found to have $T_{60} \approx 0.4$ s with a standard deviation of 0.1 s. This is about the same reverberation time as that strived for in normal control rooms.

Also interesting in these references is that the reverberation time measured for the sparse-mode low-frequency region is about the same as in the midfrequency region with the exception of Sweden. There the reverberation time was generally slightly longer, particularly at low frequencies, probably because of the more common masonry building construction called for by sound insulation requirements for multifamily dwellings in that country.

In domestic environments, the reverberation time at high frequencies is typically determined by tapestries, carpets, drapes, and furniture. Carpets have little sound absorption and must be used in thick layers or with material

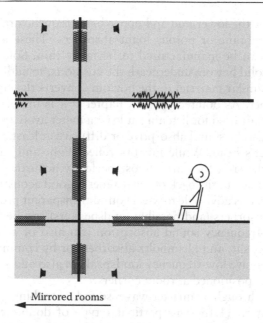

Mirrored rooms

Figure 11.14 Section of listening room for freestanding loudspeakers. Carpet on the floor prevents sound reflected by the floor. Sound diffuser in the ceiling may be replaced by sound absorbers or tilted reflector to direct sound towards rear, diffusive/absorptive wall.

that provides a porous layer between the carpet and floor to provide substantial sound absorption. The need for substantial flow resistance and air gap for textiles to provide sound absorption was pointed out in Chapter 4. Unless tightly woven, draperies and curtains usually have insufficient flow resistance unless used in deep folds or several layers. Fabrics made from natural fibers usually have higher flow resistance than smooth synthetic fibers. Mid- and high-frequency sound absorption for walls is also offered by designer acoustic panels that are covered by art print, for example. Ceiling sound absorption in this frequency range can also be provided by pressed and painted ceiling hanging panels similar to *acoustic clouds*. Shelves filled with books offer substantial sound absorption because of the porosity of books. For substantial sound absorption from 0.25 kHz and upwards, absorbers need to be thick. In the home environment, thick, tightly woven draperies or carpets hung ceiling to floor in front of a wall are good alternatives.

Sound absorption at low frequencies is as pointed out previously typically provided by leaks and drywalls in lightweight construction and by leaks and windows and doors in masonry construction. Upholstered furniture is also an important contributor to both low- and high-frequency sound absorption. Small furniture pieces may be more advantageous from the viewpoint of acoustics because of sound scattering. Additional sound absorption for the lower mid and upper bass frequency ranges can be provided by membrane or porous foam absorbers. These are commercially available or can be manufactured to be fairly thin. Sofas and armchairs often lack a solid bottom underneath the cushions, so adding a stiff sheet of plywood or similar material to the bottom converts the cushion to a membrane absorber. As pointed out in Chapter 9 it is important to bring the sofa or armchair used for listening at least a meter away from the back wall that should also be sound absorptive or diffusive at least around the height of the listener's head. While mirrors reflect light and can make a room seem twice the size, they are almost perfectly acoustically reflective and should be avoided to the back of the listener if good acoustics is desired. For home theater, it is advisable to use a sound transparent projection screen so that sound absorbers (and possibly loudspeakers) can be fitted behind the screen. Low-frequency sound absorption can also of course be provided by membrane, slit, and Helmholtz absorbers or by commercial bass traps, Additional passive low-frequency loudspeakers also add sound absorption, particularly if positioned at room corners.

A section through a quarter-wave sound absorbing resonator pair is shown in Figure 11.15. This particular type of double resonator will be quite useful in mode damping adjustment since it requires less exact tuning because it has two low-frequency resonances [26]. The design shown can also be implemented using a long rigid rectangular box with one open side. The length L determines the midfrequency of the two resonances and the ratio between a/b roughly determines the ratio between two resonance frequencies. Note that the dividing wall must be mounted so that it provides an air-tight seal between the two resonators. This can be achieved using foam rubber or draught proofing strips similar to those used for sash windows. Trimming of the flow resistance of Helmholtz and quarter-wave resonators can be done using gauze or similar textiles as indicated in the figure.

Slit absorbers are easiest *broadbanded* by filling their airspace with suitable glass fiber sheets as described in Chapter 4.

Extra sound diffusion on walls may be easier to achieve than sound absorption. If QR or similar diffusers that rely on resonant wells are used,

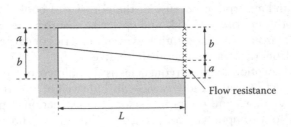

Figure 11.15 A double resonator pair. (From Kristiansen, U.R., *Appl. Acoust.*, 26, 175, 1989.)

it is necessary to consider their sound absorption, but also the diffuser shown on the book cover has substantial sound absorption [27,34]. A 20% coverage of the wall area by random decorative objects sized larger 0.1 m typically gives sufficient diffusion to remove flutter echo.

For surround sound reproduction, it is essential to avoid strong wall reflections. The best listening position from the viewpoint of mid- and high-frequency sound reproduction is at least about 1 m away from the rear wall in any case, similar to that suggested in Figure 11.12 or 11.13. This may result in the listening position to be at or close to bass frequency modal nodes. Most domestic environments are not very symmetric however because of the presence of large scattering objects, localized sound absorption by bookshelves and furniture, and asymmetrically placed building details such as doors and windows. If nodes are a problem, their location may be adjusted using resonators that alter the mode shapes.

One must also ensure good low-frequency sound reproduction by avoiding mode nodal lines at the positions of the loudspeaker, for home theater mainly the subwoofer. One approach is to use several subwoofers—operating in phase—close to the listeners. This approach places the users in the near field of the subwoofers and avoids the problem of the room modes. Another approach using several subwoofers is using four subwoofers placed along the walls of the room [28,29]. If only one subwoofer is used, the best approach is to place the subwoofer in a room corner to couple optimally to the room modes and then to adjust the modal pattern with Helmholtz resonators or other bass traps to avoid node nulls at the listeners. Correct use of Helmholtz resonators is requires the damping to be adjusted to that of the room mode to be controlled [30]. High damping is easiest achieved with Helmholtz resonators that have a comparably large volume flask and these may be difficult to hide. Quarter wave resonators made from tube stock may be easier to add inconspicuously than Helmholtz resonators.

REFERENCES

1. IEC Recommendation 268-13: Sound System Equipment, Part 13: Listening Tests on Loudspeakers. International Electrotechnical Commission, Geneva 20, Switzerland, Publication 268-13, 1985.
2. ITU-R BS.1116 (1998) *Methods for the Subject Assessment of Small Impairments in Audio Systems, Including Multichannel Sound Systems.* International Telecommunication Union, Geneva, Switzerland, 1998.
3. van Munster, B.J.P.M. (2003) Beyond control—Acoustics of sound recording control rooms—Past, present, and future. Thesis at Eindhoven University of Technology, Eindhoven, the Netherlands, FAGO Report nr: 03-23-G.
4. Newell, P. (2011) *Recording Studio Design*, 3rd edn. Focal Press, London, U.K.

5. Putnam, M.T. (1980) A thirty-five year history and evolution of the recording studio. *Proceedings of the 66th Audio Engineering Society Convention*, Los Angeles, CA, Preprint 1661.
6. Torres-Guijarro, S., Pena, A., Rodríguez-Molares, A., and Degara-Quintela, N. (2009) Sound field characterisation and absorption measurement of wideband absorbers. *126th Audio Engineering Society Convention*, Munich, Germany, Paper Preprint 7696.
7. Torres-Guijarro, S., Pena, A., Rodríguez-Molares, A., and Degara-Quintela, N. (2011) A study of wideband absorbers in a non-environment control room: Characterisation of the sound field by means of p-p probe measurements. *Acta Acust. Acust.*, 97, 82–92.
8. Torres-Guijarro, S., Pena, A., Rodríguez-Molares, A., and Degara-Quintela, N. (2012) A study of wideband absorbers in a non-environment control room: Normal absorption coefficient measurement and analysis. *Acta Acust. Acust.*, 98, 411–417.
9. Haas, H. (1951) Über den Einfluß eines Einfachechos auf die Hörsamkeit von Sprache. *Acustica*, I, 49. (Translation appears as Haas, H. (March 1972) The influence of a single echo on the audibility of speech. *J. Audio Eng. Soc.*, 20(2), 146–159).
10. Beranek, L.L. (1962) *Music, Acoustics and Architecture*. John Wiley & Sons, Inc., New York.
11. Davis, D. (1979) The role of the initial time delay gap in the acoustic design of control rooms for recording and reinforcement systems, *Proceedings of the 64th Audio Engineering Society Convention*, New York, Preprint 1547.
12. Long, M. (2005) *Architectural Acoustics*. Academic Press, New York.
13. Davis, D. and Davis, C. (1980) The LEDE™ concept for the control of acoustic and psychoacoustic parameters in recording control rooms. *J. Audio Eng. Soc.*, 28(9), 585–595.
14. Davis, C. and Meeks, G.E. (1982) History and development of the LEDE control room concept, *Proceedings of the 72nd Audio Engineering Society Convention*, Anaheim, CA, Preprint 1954.
15. Davis, D. (1980) Engineering an LEDE-control room for a broadcasting facility, *Proceedings of the 67th Audio Engineering Society Convention*, New York, Preprint 1688.
16. Schroeder, M.R. (2009) *Number Theory in Science and Communication: With Applications in Cryptography, Physics, Digital Information, Computing, and Self-Similarity*. Springer, New York.
17. D'Antonio, P. and Konnert, J.H. (1984) The RFZ/RPG approach to control room monitoring, *Proceedings of the 76th Audio Engineering Society Convention*, New York, Preprint 2157.
18. Walker, R. (1995) Controlled image design: The management of stereophonic image quality. Department Report RD1995/4. BBC Research and Development Department, Engineering Division, The British Broadcasting Corporation, London, U.K.
19. Walker, R. (1998) A controlled-reflection listening room for multi-channel sound, *Proceedings of the 104th Convention of the Audio Engineering Society*, Amsterdam, the Netherlands, Preprint 4645.

20. Holman, T. and Green, R. (2010) First results from a large-scale measurement program for home theaters, *Proceedings of the 129 Convention of the Audio Engineering Society*, San Francisco, CA, Paper 8310.
21. Hellström, P.A. (1982) Arrangement for damping and absorption of sound in rooms, United States Patent 4362222.
22. Burgess, M.A. and Uteley, W.A. (1985) Reverberation time in British living rooms. *Appl. Acoust.*, 18, 369–380.
23. Bradley, J.S. (1986) *Acoustical Measurements in Some Canadian Homes*. Institute for Research in Construction, National Research Council, Ottawa, Ontario, Canada.
24. Celander, F. (2008) Room acoustics for audio (in Swedish: Rumsklang för hifi-lyssning), Report, Kalmar College, Kalmar, Sweden.
25. Toole, F. (2008) *Sound Reproduction: The Acoustics and Psychoacoustics of Loudspeakers and Rooms*. Focal Press, Oxford, U.K.
26. Kristiansen, U.R. (1989) A different type of resonator for acoustics. *Appl. Acoust.*, 26, 175–179.
27. D'Antonio, P. and Massenburg, G. From mono to surround: A review of critical listening room design and a new immersive surround design proposal. Presentation by RPG Inc., Available from http://www.rpginc.com/docs%5CTechnology%5CPresentations%5CStudio%20Design%20From%20Mono2Surround.pdf (accessed June 2013).
28. Welti, T. and Devantier, A. (2006) Low frequency optimization using multiple subwoofers. *J. Audio Eng. Soc.*, 54, 347–364.
29. Welti, T. (2002) How many subwoofers are enough, *Proceedings of the 112 Audio Engineering Society Convention*, Munich, Germany, Paper 5602.
30. van Leeuwen, F.J. (1960) The damping of eigentones in small rooms by Helmholtz resonators, *E.B.U. Review, Part A*, No. 62, August 1960.
31. Williams, M. (2004 & 2013) *Microphone Arrays for Stereo and Multichannel Sound Recording*, Volumes 1 &, Editrice Il Rostro, Milano, Italy.
32. Jackson, G.M. and Leventhall, H.G. (1972) The acoustics of domestic rooms. *Appl. Acoust.*, 5, 265–277.
33. Storyk, J. and Noy, D. (1999) Acoustical design criteria for surround sound control rooms. *Proceedings of the 106 Audio Engineering Society Convention*, Munich, Germany, Preprint 4939.
34. Mr. Bonzai (2008) George Massenburg builds a blackbird room, *Digizine*, Digidesign. Available at http://www2.digidesign.com/digizine/dz_main.cfm?edition_id=101&navid=907 (Accessed May 2013). Also http://www.blackbirdstudio.com/#/home/ (sampled May 2013).
35. Olive, S.E., Castro, B., and Toole, F.E. (1998) A new laboratory for evaluating multichannel, audio components and systems, Report from R&D Group, Harman International Industries Inc., Northridge, CA. (Same title document also available as paper 4842 of the Audio Engineering Society 105th conference, San Francisco, CA [1998]).

Chapter 12

Small rooms for voice and music practice

12.1 INTRODUCTION

Many of the small rooms in our lives are work environments in which good acoustics may enhance motivation, efficiency, and comfort. The small office or cubicle is common to many of us as a workplace and because of their small volume voice communication is seldom a problem unless the office is plagued by noise. Solutions to such problems are outside the scope of this book.

For music students and musicians, small rooms such as rehearsal rooms, studios, and practice rooms are common work environments that need acoustic comfort: both quiet and harmonious acoustics suitable for the musicians' instrument [1,2].

12.2 SOUND POWER

Because of the high sound power emitted by some musical instruments such as drums, and the close use of instruments such as the violin, the sound levels at the musician's ears must be high. Additionally it must be remembered that many musical instruments are very directional and the sound intensity can be very high immediately in front of the instrument as in the case of brass instruments such as the trumpet and trombone [43].

A common rule of thumb is that a symphony orchestra playing forte radiate about 1 W of acoustic power (although peaks may be 10–20 dB higher). This means that on the average, we might use a sound power output of 1–10 mW per instrument as we calculate the sound level of musical instruments. Some instruments such as drums have very high output power; bass drums have been measured to emit a peak acoustic power of 20 W [3]. The large sound power in combination with small room volume results in loud sound, as can be calculated using Equation 2.92.

The voice has an acoustic power of about 10 μW in normal conversation and about 100 times stronger for a trained voice at maximum effort [4]. Some voice formants however have about the same power as musical instruments [5].

12.3 HEARING IMPAIRMENT AND DAMAGE

It is common for musicians to suffer hearing impairment because of the high sound pressure levels that they are exposed to while playing and rehearsing their instruments [6,7].

Sound levels of about 105 dBA have been measured at drummers' ears, for example. Sound levels at the ear of the musician's own instrument have been measured to be above 85 dBA in the case of violinists and above 95 dBA for the ear at the instrument side [8]. (Note that peak levels may be up 20 dB higher than rms levels.) In addition, there is all the sound from the rest of the orchestra. The sound levels due to one's own instrument cannot be much reduced using sound absorption as shown by Figure 2.36 since many instruments are within the reverberation radius and the direct sound dominates over the reverberant sound. To some extent, the musicians can play softly but not all instruments have a large dynamic range and the desired sound quality may not be possible to achieve [9].

Sound-absorptive treatment, along with cost, also leads to changed timbre, reduced support, and shorter reverberation times. For these reasons, many musicians use some form of flat frequency response hearing protection. Hearing protection however is likely to make it more difficult to hear details in the music and offset onstage communication.

On stage, the musician's sound level is lower than in the practice room because of the larger room volume. Loudness of one's own instrument on stage correlates with musicians' appreciation of stage acoustics since it helps the musician feel like the sound is carried to the audience [8,10]. By having good acoustic conditions in rehearsal and practice spaces, the often found tendency among musicians and others to play and sing louder when conditions are bad is mitigated. Good room acoustics also lessens the need to rely on the cocktail party effect for binaural sound source separation and thus results in less fatigue.

The precedence and cocktail party effects, the Haas time window, and temporal masking were discussed in Chapter 8. Because of the integrating time window, one can say that the small room allows us to hear the *full* output of our voice and of musical instruments. Small reverberant rooms are used for sound power measurement because of this effect. Especially for musical instruments that have a very directional sound radiation, such as the metal instruments, the sound that is returned to the musician on stage in a large room will be as reverberation and early reflections by the room. In a reverberant and small space, the loudness of such instruments will be much higher than on stage.

The typical recommended floor area of rehearsal rooms is about 1.5 m^2 per seated vocalist and about 3 m^2 per instrumentalist, with a ceiling height of 5 m or higher [11]. Small rehearsal rooms have a floor area of about 20–45 m^2. Experience shows that such rooms tend to be too loud for large ensembles [1]. The use of a diffusive and sound-absorptive ceiling is a way to adapt to low ceiling height, but rooms that have large volume are often though to sound better than those with low ceilings.

12.4 QUIET

When music students are asked for their thoughts about important issues in rooms for music education, the need for both visual and acoustical isolation is ranked highly. Many musicians are particularly annoyed by sound leaking in to their practice environment from others practicing the same type of instrument. Teachers and instructors need quiet to hear all nuances of a student's playing.

Background levels in practice rooms and studios should be below 20 dBA and in rehearsal rooms below 25 dBA. To achieve such low noise levels, functions such as heating, ventilation, and air conditioning must be designed for quiet or at least to have suitable masking properties. Table 14.1 shows a listing of suggested maximum background levels for various spaces [44].

The requirements for sound isolation of course depend on the sound source and the adjacent room size. Walls, doors, and windows should ideally be designed for sound isolation with sound reduction index R_w (sound transmission class, STC) larger than 65 dB. To achieve such high values, it is necessary to use heavy masonry for full-range high sound isolation [44]. Rehearsal rooms for drummers may need to be located in basements with stone or concrete walls for sound isolation as shown in Figure 12.1. Figure 12.2 shows a school environment with a number of small practice rooms next to a large rehearsal space. Such combinations require acoustically well designed doors and door seals to achieve sufficiently high sound isolation.

Multiple sheet drywall construction must use separate studs but can never achieve high sound isolation at low frequencies because of sheet-airspace-sheet resonances. When sound isolation requirements are very high, minute cracks in walls or missing wall seals cannot be allowed since they will set an upper cap on the sound isolation that can be achieved.

Figure 12.1 A drum studio. (Photo by Ioana Pieleanu, Acentech, Studio A.)

Also important are doors and windows with high sound isolation, as well as good door and window seals between the room and adjoining corridors or neighboring rooms. If suspended ceilings are used, walls must be full height and care taken to prevent flanking sound transmission through ducts and other conduits [44]. Reference 12 is a useful guide to the general planning of school music facilities.

12.5 STAGE VERSUS REHEARSAL AND PRACTICE ROOMS

Should rooms for rehearsal and practice have the same acoustics as that found on stage? For many reasons, this is not practical: the small room will inevitably be louder for the same playing, and for this reason, musicians may not use the same effort and this will affect their playing and the sound of the instrument.

Figure 12.2 A music rehearsal environment requires both walls and doors to have good sound isolation properties. Door seals are critical. (Photo courtesy of Acentech, Studio A.)

The acoustics of the stage that is a part of a larger room will be characterized by relatively long reverberation and comparatively few and late early reflections in contrast to those of the small room. Many musicians prefer tall, narrow stages probably because of the added support by the walls [9,10]. Tests were made using sound field simulation in an anechoic chamber on the stage acoustics preferences of flute players. The test results showed a clear preference by the musicians for reflections from the back but in the median plane, showing that the acoustic support by back wall is important, as can be expected [13].

Singers have been found to express preference for rather late reflected sound from the hall and this can only be achieved in rehearsal rooms that have electronically assisted reverberation and sound field simulation as described later in this chapter. The simulated reflected sound should be about 0.1–0.15 s late in time relative to the own voice [13–17].

Stage managers in large concert halls often bring down the overhead stage reflector to give added support to less well-trained groups. However, many musicians tend to dislike the unclear sound that results from low ceilings. In a rehearsal room that has a low ceiling, acoustic conditions similar to those of having a high ceiling can be implemented by a mix of sound diffusion, absorption, and electronically assisted reverberation enhancement. The low ceiling can then be made as a suspended ceiling with sound-absorptive and sound-diffusive tiles with additional tiles that have built-in loudspeaker driver units.

12.6 REVERBERATION TIME

Many musicians and music students train at home. Reported reverberation times for domestic rooms are quite short, typically in the range of 0.4–0.5 s and slightly longer 0.6–0.7 s in regions where masonry construction is used for building as described previously [18–22]. The frequency variation of domestic reverberation times is also measured to be quite small.

Many of the small rooms in music teaching facilities have timber floors combined with masonry or plasterboard walls. The ceiling is typically plasterboard or concrete. Unless treated with sound-absorptive materials, such rooms will feature cavernous and unacceptable acoustics. The amount of sound-absorptive treatment can be approximately estimated using Sabine's or Fitzroy's equations. Fitzroy's equation gives more correct results than Sabine's equation for rectangular rooms that have surfaces that differ in sound-absorptive treatment. If available, the use of a geometrical acoustics software is recommended for room acoustic analysis.

Any calculation using geometrical acoustics is only useful for frequencies from about 0.25 kHz and upwards for the rehearsal rooms and about 0.5–1 kHz for practice rooms since these rooms are not sufficiently large for geometrical acoustics to be applied successfully at lower frequencies. The adjustment of low-frequency reverberation time can be done in an approximate way by initially treating room corners with diagonal absorbers, low-frequency absorptive slitted ceiling tiles, or bass traps. Windows will help reduce the low-frequency reverberation times but windows that allow viewing of the musician should be avoided for privacy. In most cases, low-frequency room acoustic conditions will need to be adjusted after building completion, reverberation time measurements have been made, and musicians' comments collected.

It has been found that the preference for practice room and teaching studio reverberation times ranges from about 0.5 to 0.9 s among music students [23]. Rooms that had the lower value were considered as *dry* and those with the upper value as *live*. Most preferred a reverberation time midway, about 0.7 s. Students that had greater application preferred slightly shorter reverberation times than those that did not put in the same work. Interviews showed that students would appreciate a possibility to adjust reverberation time according to individual preference. Students that studied woodwind, brass, strings, and voice had different preference.

Practice rooms tend to have very short reverberation times because of their small volume, and many students would want to have rooms with a floor area of about 15 m², but this would be very expensive for institutions that have many students and rooms. A design compromise for the reverberation times in small practice rooms is 0.4–0.5 s, with a small rise to about

0.6–0.7 s at 100 Hz. [2]. Reverberation time should not vary by more than ±5% over the range of 63 Hz–4 kHz when the room is in use.

Teaching studios have reverberation times in between rehearsal and practice rooms because of their size, and floor areas are typically in the range of 15–20 m². For these, a design compromise is about 0.6 s, with a small rise to about 0.8 s at 100 Hz. [2]. A violinist needs more reverberation than a pianist. The latter has a sustain pedal that can be used between notes. A violinist also needs reverberation that has good timbre and an even T_{60} over the frequency range [8]. In a study of piano practice rooms it was found that $T_{60} \approx 0.3$ s over a wide frequency range was preferred and that the walls in the front and back of the pianist should be made sound absorptive (which follows from the discussion on coloration in Chapter 9). It was also found that the sound absorption of the right-hand wall depended on if the piano top lid was open in which case the acoustic treatment of left-hand wall was found unimportant [45].

A range of reverberation times are recommended for different instrument groups such as percussion instruments 0.3–0.5 s, bowed string instruments (violin, cello) 0.6–0.9 s, and wind instruments (trumpet, flute) 0.4–0.7 s [24]. A Swedish source cites different reverberation times that are recommended for the instrument groups of the orchestra [25]:

0.2–0.4 s for drums, low-frequency brass
0.4–0.6 s for brass, woodwind
0.6 s for flute, guitar, trumpet, cello
0.7 s for piano
0.8 s for clarinet and violin
1.1 s for song

If such a range of variation is desired, and individual preference is added, clearly, the best results will be obtained with individually variable sound absorption. If devices for variable sound absorption (and diffusion) are to be used optimally, then both students and instructors must be informed about their correct use. Figure 12.3 shows details of a practice room where both sound diffusion and absorption can be adjusted by the user.

12.7 ROOM GEOMETRY AND DIFFUSION

The small rehearsal, studio, and practice rooms typically have volumes in the range of 5–50 m³ and most will have a fairly rectangular shape. Rooms that have curved walls or ceilings may feature annoying sound concentration and echoes. Such shapes should be avoided unless diffusive or sound-absorptive treatments are used to prevent focusing as described in

(a)   (b)

Figure 12.3 A practice room where (a) a sound-absorptive curtain can be hidden behind a hinged panel and (b) a 7° splayed partial wall/barrier (in front of existing structural wall) that can be covered by a curtain to change absorption and diffusion. The splayed wall is partially covered by a high frequency diffuser. Low-frequency sound absorption on right in (b). (Photos by Mendel Kleiner.)

Chapter 3. Both approaches lead to shorter reverberation times since diffusers help reduce the mean free path length between sound reflections, and many diffusers that use wells (such as QRD and other types) have high sound absorption as described in Chapter 4.

Some consideration should be given to room dimensions. The length to width ratio might be chosen by the criteria mentioned in Chapter 9, but the ratio is not very critical. It is useful to choose the length to width ratio using a nonrational number or as the ratio between two low primes to make the arrival time distribution of the early reflections in the impulse response pattern maximally random as discussed in Chapter 8. Repetitive delays in the impulse response pattern cause coloration and flutter echo as discussed in Chapter 8, so square or circular rooms should be avoided [26,27].

Flutter echo is a result of the repetitive reflections that are not masked by reverberation and occur in rooms where reflections due to two opposing walls dominate, for example, in rooms that have sound-absorptive carpet and a wall covering sound-absorptive curtain or in long corridors that have sound-absorptive ceiling tiles. For this reason, it is common in narrow and bare offices that have hard opposing plane walls where it results in a *robotic* and poor voice quality such as in the office shown in Figure 12.14.

Flutter echo and coloration are primarily audible when sound source and listener are close, for example, with one's own handclaps or voice.

Repeating impulse responses can be eliminated using absorption or diffusion or by splaying the room surfaces so that mirror image contributions quickly become invisible. Often it is not possible to splay walls in an existing building but Figure 12.3 shows how a splayed barrier may be installed in a rectangular room. While absorption may not always be used because of reverberation time requirements, random patches of sound-absorptive material (about 0.2×0.2 m^2) on one of two opposing walls are usually sufficient to remove flutter. The use of a number of helium- or air-filled children's rubber balloons is an inexpensive way of trying the efficiency of a sound-diffusive treatment to specularly reflective surface [28].

If the height of the room is less than its width or length, ceiling diffusers will be helpful to achieve good sound quality. Diffusion should not be over-emphasized though as a means of improving small rehearsal and practice room acoustics. Diffusive surface treatments should simply be used to the extent necessary to remove coloration and flutter echo, for example. In a study of practice rooms for trumpet players in a school of music in Sweden, a strong negative correlation was found between (1-IACC) values averaged over the 0.5–2 kHz octave bands and preference, that is, the students preferred nondiffuse rooms [29].

Again, it must be stressed that numerical sound diffusers using wells have quite high sound absorption ($\alpha \approx 0.2-0.3$), and a random surface such as that shown in Figure 12.4 is preferable since any appreciable increase

Figure 12.4 A blasted rock wall has a quite ideal diffusion and little sound absorption. (Photo by Mendel Kleiner.)

Figure 12.5 Polycylindrical sound diffusers in a small practice room. (Photo by Mendel Kleiner.) Because the diffusers are made from strips of wood with narrow open slit between strips, they combine diffusion with resonator panel low-frequency sound absorption. The airspace inside the diffusers contains porous sound absorber to adjust bandwidth of low-frequency sound absorption.

in sound absorption will only be a result of added sound reflections. Stone without cracks has $\alpha < 0.02$. Other attractive solutions (from an acoustic point of view) are commercial concrete block diffusers, etc. Compare also to the various diffusive wall treatments discussed in Chapters 5 and 9.

Polycylindrical sound diffusers such as those shown in Figure 12.5 can be advantageously designed to be both diffusive over a wide mid- and high frequency range and sound absorptive at low frequencies by means of slit absorber action as explained in Chapter 4. Different radii of curvature are recommended if several diffuser units are to be used.

Some instruments that radiate to the back of the player—such as French horns—gain by some kind of diffusive wall for the instrument to radiate into since a specular reflection would generate severe coloration and

Figure 12.6 Ceiling with concave elements combined with slit sound absorbers at flat ceiling section next to walls. (Photo by Mendel Kleiner.)

absorption dull the sound of the instrument. Otherwise, seating next to a sound-diffusive wall eliminates the mirror image support that musicians like to have both in practice and on stage.

The use of sound-diffusive treatments in ceilings was mentioned previously as a means of accepting lower ceiling heights. This type of ceiling treatment is commonly used in choir and band rooms. If a sound-diffusive ceiling is used, it might be useful not to let it run all the way to the walls. Leaving about 1 m of specular ceiling next to the walls enhances the useful supporting wall sound reflections. Figure 12.6 shows such a ceiling with concave elements (just as diffusive as convex) where the ceiling surface close to the wall is used with slit sound absorbers to reduce low-frequency reverberation time.

Note that in some studios equipped for video, there may be many volume diffusers in the ceiling that together with cable runs, ventilation ducts, etc., will form effective diffusers. One such television studio ceiling is shown in Figure 12.7.

Finally, Figure 12.8 shows a detail of the walls and ceiling at IRCAM's Éspace de Projection (no longer existing) where pyramidal wall elements with a triangular cross-section were used and that could be rotated to show the desired acoustic characteristic (diffusive, absorptive, specular). The height of the entire ceiling could be adjusted as well. Using a small personal computer, the musician could control the settings of the elements over an intuitive graphics desktop interface.

Figure 12.7 Lighting equipment functions also as volume diffusers in the ceiling of a television studio. (Photo by Mendel Kleiner.)

Figure 12.8 Detail of IRCAM's Éspace de Projection showing the rotatable wall elements. (Photo by Leif Rydén.)

12.8 SMALL ROOMS OUT OF LARGE ROOMS

Sometimes a small practice or studio environment needs to be implemented inside a larger space. Then, an acoustical shell can be constructed using moveable acoustical barriers such as the ones shown in Figure 12.9 that are an attractive alternative for making a temporary *cubicle*. The barriers shown are made of glossy painted plywood sheets that have one side treated by acoustic foam. The holes on the second side introduce some diffuse reflection although at the cost of some absorption. The sound level inside the barrier cubicle will be relatively low because of the sound absorption resulting from the lack of a close ceiling and can be adjusted by spacing the barriers.

12.9 LARGE ROOMS OUT OF SMALL ROOMS

A more difficult situation arises when little space is available but the acoustic environment is desired to be similar to that of a large space. Then, the only available alternative is to use sound field simulation. In this approach,

Figure 12.9 Moveable barriers can be used to achieve small room conditions in a large space. These barriers have one sound absorptive and the other sound reflective. Units that have a diffusive side would be good to have available as well. (Photo by Mendel Kleiner.)

the room is treated to be fully sound absorptive to the degree practicable. The sound of the musician's instrument is picked up by one or more microphones, amplified, delayed, reverberated, filtered, and fed back to the room over several loudspeakers. Appropriate sound pickup for string instruments may be difficult since their sound radiation patterns are complex and most instruments are shadowed by the musician's body [7,30].

Sound field simulation systems require careful design using appropriate microphones, loudspeakers, and digital signal processing algorithms [31]. The sound from the loudspeakers will reach the microphones be amplified and set up a feedback loop that may start to oscillate and result in howl. By using somewhat intentionally oscillating time delays or adaptive notch filters, runaway oscillation can be avoided but successful implantation of the idea requires careful engineering [32–34]. A simple sound field simulation system that simulates stage conditions is shown in Figure 12.10.

The loudspeakers may be placed in the room so that they correspond to sound that comes from the stage side walls and the auditorium. While the microphone system can be made suitably directional and placed to primarily pick up sound from the performer practicing, the loudspeakers are typically quite omnidirectional because of the small-sized loudspeakers necessary for a small room. An electronic instrument does not need the microphones. The signal components in the room are shown in Figure 12.11.

The unavoidable feedback path H_{LM} between loudspeakers and microphones could have been used for reverberation but such a solution would suffer the coloration problems. Instead, the unavoidable feedback between loudspeakers and microphones is removed by time-variant filtering as mentioned that is added to all channels. The time variation is extremely subtle and will not be heard. To provide high-quality simulated reverberation and

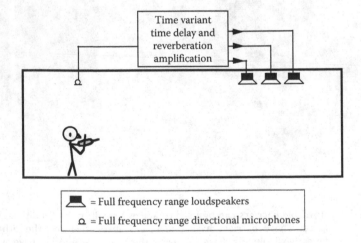

Figure 12.10 A sound field simulation system for a small rehearsal room.

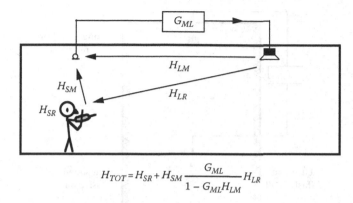

$$H_{TOT} = H_{SR} + H_{SM} \frac{G_{ML}}{1 - G_{ML}H_{LM}} H_{LR}$$

Figure 12.11 The transfer functions of the sound field simulation system (only a single channel shown). H_{SR}: instrument → musician (arrow not drawn in figure). H_{SM}: instrument → microphone. H_{LM}: loudspeaker → microphone. H_{LR}: loudspeaker → musician. G_{ML}: finite impulse response time-variant filter.

early reflections, the signal processing unit convolves the microphone signals with finite room impulse responses G_{ML} (which could, e.g., be sampled in a high-quality concert hall or be generated by the system designer). For large rehearsal rooms, several commercial systems exist that are essentially only expanded multichannel versions of the system sketched here. They make variable acoustics possible for groups of performers.

A practice room for a singer could, for example, have the plan shown in Figure 12.12. The room end close to the singer has a rigid back wall behind the singer to give the desired support, and the side walls of the close end are made diffusive to give some ambience. The far end simulates an auditorium in that it is made sound absorptive. The desired reverberation and supporting early reflections that singers need are instead provided over one or several loudspeakers. The signals for the loudspeakers are picked up by a directional microphone system that minimizes pickup from the loudspeakers. The electronic processing is that indicated in the figure and the transfer function G_{ML} contains early reflection delays, late reverberation, etc., implemented by finite impulse response filters.

Figure 12.13 shows a small commercial cabin for individual music practice that has adjustable acoustics by means of sound field simulation. An advantage of the approach is that the student's playing can be recorded and then reproduced as if listened to from various distances in a concert hall.

12.10 SPEECH

In hospital operating theaters, shops, process control rooms, offices, board rooms, car and truck compartments, and many other small rooms, the

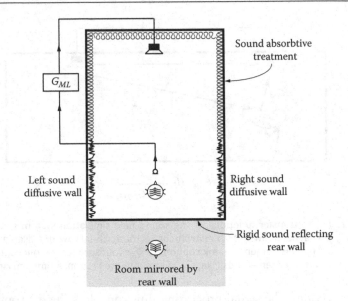

G_{ML}

Sound absorbtive
treatment

Left sound
diffusive wall

Right sound
diffusive wall

Rigid sound reflecting
rear wall

Room mirrored by
rear wall

Figure 12.12 A plan for a small rehearsal room suitable for a single voice or instrument simulating stage conditions.

acoustic support by the room acoustics is necessary for both speaking ease and decision-making confidence. The influence of room acoustics on speech intelligibility is well known and investigated. The detrimental effects of echo are also well known but the effect of flutter echo and coloration less so.

We hear our own voice over three feedback paths: the direct and reflected acoustic paths and the direct cranial vibratory path. The direct acoustic and cranial vibratory paths are about equally strong. The reflected path contains the early reflections and reverberation by the room and must be quite strong or displaced in time to overcome the masking by one's own voice over the direct paths.

When studying the conditions necessary for voice comfort in auditoria, several investigations have shown that diffusion is less important than strong support [14–17]. Research on the preference by talkers showed that most preferred high reflection levels, high interaural CCF, and low temporal diffusion in spite of the coloration that results from such conditions [35,36]. These findings correspond well to those for singers [14–17]. Singers have however been found to rely more on cranial vibration due to their own voice than on auditory feedback from the room [37]. For stage conditions, the average subject preferred a short distance to the back wall when that wall's reflection was the support. It must be noted however that these investigations using simulated sound fields were mostly concerned with single reflection sound fields as simple simulations of stage conditions and not with the repetitive flutter echo in offices.

Figure 12.13 A practice cabin by Wenger provides variable acoustics in a small space. (Photo by Mendel Kleiner.)

In an office or cubicle that has parallel, bare and smooth opposing walls and absorptive ceiling and floor, such as the one shown in Figure 12.14, the acoustic situation is similar to that shown in Figure 9.15a. As shown by Fitzroy's equation, two rigid and plane opposing walls cause long reverberation in spite of sound-absorptive ceilings and window curtains. The periodic linear pattern of image sources leads to an unpleasant voice quality and hard and brittle sound unless the walls are made diffusive at least for frequencies over about 0.5 kHz. Alternatively sound absorption can be used since the reverberation time is not critical in this application. Books and bookshelves add both diffusion and sound absorption.

For speech intelligibility under cocktail party conditions dialog, starting and finishing participants' speech is characterized by overlaps and simultaneous talking. Talkers listen to their own speech, discover errors and flaws, and may use hearing of their own speech to correct these [38]. In such situations, it is preferable that the reverberation time is short. In a conversation situation, there will be masking both due to other speakers and also due to the masking of one's own voice. The masking by one's own voice depends on physiological factors as well as room acoustic factors such

Figure 12.14 This photo shows a room that has an a sound absorptive ceiling and carpet along with mirroring plane and parallel side walls, all of which contributes to reflection patterns that result in colored sound and flutter echo. (Photo by Mendel Kleiner.)

as reverberation and adjacent noise sources. Such acoustic masking makes it difficult to control one's voice. The Lombard effect, the raising of one's own voice because of background noise, contributes to increased speech effort and tiredness [39–42].

REFERENCES

1. McCue, E. (1990) Rehearsal room acoustics, in *Acoustical Design of Music Education Facilities*, E. McCue and R.H. Talaske, eds. Acoustical Society of America, New York, pp. 36–41.
2. Lane, R.N. and Mikeska, E.E. (1955) Study of acoustical requirements of teaching studios and practice rooms in music school buildings. *J. Acoust. Soc. Am.*, 27, 1087–1091.

3. Rossing, T.D. (2000) *Science of Percussion Instruments*. World Scientific Pub Co Inc, New York.
4. Lin, E., Jayakody, D., and Looi, V. (2009) The singing power ratio and Timbre-related acoustic analysis of singing vowels and musical instruments. *Voice Foundation's 38th Annual Symposium: Care of the Professional Voice*, Philadelphia, PA, June 3–7, 2009.
5. Bloothooft, G. and Plomp, R. (1986) The sound level of the singer's formant in professional singing. *J. Acoust. Soc. Am.*, 79(6), 2028–2033.
6. Fearn, R.W. (1993) Hearing loss in musicians, *J. Sound Vib.*, 163(2), 372–378.
7. Royster, J.D., Royster, L.H., and Killion, M.C. (1991) Sound exposures and hearing thresholds of symphony orchestra musicians. *J. Acoust. Soc. Am.*, 89(6), 2793–2803.
8. Andersson, J. (2008) Investigation of stage acoustics for a symphony orchestra. Master's thesis 2008:40, Division of Applied Acoustics, Department of Civil and Environmental Engineering, Chalmers University of Technology, Gothenburg, Sweden.
9. Dammerud, J.J. and Barron, M. (2008) Concert hall stage acoustics from the perspective of the performers and physical reality. *Proc. Inst. Acoust.*, 30(3), 26–36.
10. Dammerud, J.J. and Barron, M. (2007) Early subjective and objective studies of concert hall stage conditions for orchestral performance, *Proceedings of the 19th International Congress on Acoustics*, Madrid, Spain.
11. McCue, E. and Talaske, R.H. (eds.) (1990) *Acoustical Design of Music Education Facilities*, Acoustical Society of America through the American Institute of Physics, New York, pp. 36–41.
12. Wenger (2013). *Planning Guide for School Music Facilities*. Wenger Corporation, Owatonna, MN. Available from www.wengercorp.com (V. 3.1 Accessed May 2013).
13. Nakayama, I. and Uehata, T. (1968) Preferred direction of a single reflection for a performer. *Acustica*, 65(4), 205–208.
14. Nakamura, S. and Shirasuna, S. (1989) On the acoustic feedback model for solo singers on concert hall stages, *Proceedings of the 13th International Congress on Acoustics*, Belgrade, Yugoslavia.
15. Nakamura, S. and Kan, S. (1988) Acoustic components supporting solo singers on concert hall stages, *Second Joint Meeting of the Acoustical Societies of America and Japan*, Honolulu, HI.
16. Nakayama, I. (1984) Preferred time delay of a single reflection for performers. *Acustica*, 34, 217–221.
17. Marshall, A.H. and Meyer, J. (1985) The directivity and auditory impressions of singers. *Acustica*, 58, 130–140.
18. Jackson, G.M. and Leventhall, H.G. (1972) The acoustics of domestic rooms. *Appl. Acoust.*, 5, 265–77.
19. Burgess, M.A. and Uteley, W.A. (1985) Reverberation time in British living rooms. *Appl. Acoust.*, 18, 369–380.
20. Bradley, J.S. (1986) *Acoustical Measurements in Some Canadian Homes*. Institute for Research in Construction, National Research Council, Ottawa, Ontario, Canada.

21. Burkhardt, C. (2004) Reverberation time in furnished and bare living rooms and its consequences for noise measurements (in German: Nachhallzeit in eingerichteten und leeren Wohnräumen und Konsequenzen für Geräuschmessungen), *Proceedings of DAGA 94*, Dresden, Germany.

22. Diaz, C. and Pedrero, A. (2005) The reverberation time of furnished rooms in dwellings. *Appl. Acoust.*, 66, 945–956.

23. Lamberty, D.C. (1980) Music practice rooms. *J. Sound Vib.*, 60(1), 149–155.

24. Osman, R. (2010) Designing small music practice rooms for sound quality, *Proceedings of the 20th International Congress on Acoustics*, August 23–27, 2010, Sydney, New South Wales, Australia.

25. Janson, F. (2006) Acoustic comfort in practice rooms for music (in Swedish: Akustisk komfort i musikövningsrum). Master's thesis Engineering Acoustics, Lund university, Lund, Sweden.

26. Halmrast, T. (2000) Orchestral Timbre. Combfilter-coloration from reflections. *J. Sound Vib.*, 232(1), 53–69.

27. Halmrast, T. (2013) Coloration due to reflections, further investigations, Report AKUTEK, Norwegian State, Directorate of Public Construction and Property, Oslo, Norway.

28. Knudsen, V.O. and Delsasso, L.P. (1968) Diffusion of sound by Helium Filled Balloons, *Proceedings of the Sixth International Congress on Acoustics*, Tokyo, Japan, paper E.5–5.

29. Olsson, O. and Söderström Wahrolén, D. (2010) Perceived sound qualities for trumpet players in practice rooms, Master's thesis 2010:127, Division of Applied Acoustics, Department of Civil and Environmental Engineering, Chalmers University of Technology, Gothenburg, Sweden.

30. Pätynen, J. (2007) Virtual acoustics in practice rooms. Master's thesis, Laboratory of Acoustics and Audio Signal Processing, Department of Electrical and Communications Engineering, Helsinki University of Technology, Helsinki, Finland.

31. Kuttruff, H. (2009) *Room Acoustics*, 5th edn. Taylor & Francis/CRC, London, U.K.

32. Kleiner, M. and Svensson, U.P. (1995) A review of active systems in room acoustics and electroacoustics, *Proceedings of Active 95, International Symposium on Active Control of Sound and Vibration*, Newport Beach, CA, July 6–8, pp. 39–54.

33. Svensson, U.P. (1998) Energy-time relations in an electroacoustic system in a room, *J. Acoust. Soc. Am.*, 104, 1483–1490.

34. Svensson, U.P. and Nielsen, J.L. (1996) The use of time-varying filters in control of acoustic feedback, *Proceedings of the Spring Meeting of the Acoustical Society of Japan*, Tokyo, Japan, March 26–28, pp. 499–500.

35. Berntson, A. (1986) Preferred distance to a wall behind talkers. *Nordic Acoustical Meeting*, Aalborg, Denmark, pp. 261–264.

36. Berntson, A. (1987) Preferred acoustical conditions for talkers: Judgements of simulated multiple early reflection patterns. Report F 87–02, Div. Applied Acoustics, Chalmers University, Gothenburg, Sweden.

37. Sundberg, J. (1974) Articulatory interpretation of the 'Singing Formant', *J. Acoust. Soc. Am.*, 55(4), 838.

38. Borg, E., Bergkvist, C., and Gustafsson, D. (2009) Self-masking: Listening during vocalization. Normal hearing, *J. Acoust. Soc. Am.*, 125(6), 3871–3881.
39. Junqua, J.C. (1993) The Lombard reflex and its role on human listeners and automatic speech recognizers. *J. Acoust. Soc. Am.*, 93, 510–524.
40. Lu, Y. and Cooke, M. (2008) Speech production modifications produced by competing talkers, babble, and stationary noise. *J. Acoust. Soc. Am.*, 124, 3261–3275.
41. Jung, O. (2012) On the Lombard effect induced by vehicle interior driving noises, regarding sound pressure level and long-term average speech spectrum. *Acta Acust. Acust.*, 98, 1–8.
42. Ryherd, E.E., West, J.E., Bush-Vishniac, I.J., and Persson-Waye, K. (2008) Evaluating the hospital soundscape, *Acoust. Today*, October 22–29.
43. Meyer, J. and Hansen, U. (2009) *Acoustics and the Performance of Music: Manual for Acousticians, Audio Engineers, Musicians, Architects and Musical Instrument Makers*. Springer, New York.
44. Kleiner, M., Klepper, D.L., and Torres, R.R. (2010) *Worship Space Acoustics*, J. Ross Publishing, Ft. Lauderdale, FL.
45. Fujita, T., and Yamaguchi, K. (1987) A Study on the Acoustical Characteristics of a Piano Practice Room, *Acustica*, 63, 211–221.

Chapter 13

Modeling of room acoustics

13.1 INTRODUCTION

The acoustic response of the room is described by its impulse response of rooms that can be both modeled and measured. Modeling is common in the design of large rooms. The room impulse response can be approximately measured using quasi-omnidirectional sources and receivers. Although modeling usually starts with omnidirectional sources and receivers, directivity can be included as well.

It is the mid- and high-frequency information that is of most interest because of the analysis features of human hearing and the spectra of many musical instruments.

In the design of small rooms, modeling is typically done to determine the frequency response between points and the frequencies and shape and damping of room modes. Reverberation time and the placement of absorbers, diffusers, and loudspeaker can be optimized using results from room models.

There are many methods available for acoustic modeling, each of which has application-specific properties. A review of the acoustic modeling methods used for large room can be found in Ref. [1]. Some metrics used in large room design may also be of interest in the study of small rooms.

In this chapter, we only outline those techniques that are applied to small rooms. The skilled acoustician–modeler can gain much information by comparing results from measurement to those from computer modeling of rooms. The modeler and the tool must be considered a combination.

Both physical and mathematical methods can be used to predict the room impulse response. The classical physical method is based on the use of ultrasound in physical scale models. Mathematical modeling methods are generally based on computer software, although hardware modeling using field programmable gate arrays is also possible.

Using the modeled binaural room impulse response we can listen to a simulation of the source–room combination rather than just study metrics. This method is called auralization. Auralization of an environment, for example, for speech and music can be considered analogous to visualization.

The ambition is to make it possible to listen to the sounds in the environment without being in the actual environment [2,3]. Auralization can be based on impulse responses acquired by both physical and mathematical methods.

13.2 LOW AND HIGH FREQUENCIES

Single-step prediction of the full frequency range characteristics of rooms is difficult. Different types of modeling must be used; geometrical, statistical, and wave equation-based methods are each insufficient for the modeling of small rooms over the full audio range. Typically, for small room modeling, the frequency range is divided into a low- and high-frequency part, each of which needs a frequency range of about four octaves. Ideally, the low- and high-frequency parts should fit seamlessly at the crossover frequency.

Superficially, it could be said that low-frequency modeling is simpler to do than high-frequency modeling. This is however only the case if just the transducers are considered; modeling the complex low-frequency behavior of drywall and chipboard walls, as well as floor and ceiling constructions, used in lightweight postframe buildings and in both residential homes and offices still requires careful modeling and much skill. Windows and doors have low-frequency properties that are likely to be as difficult to model.

There are many problems in modeling the mid- and high-frequency response. Comprehensive, full sphere measurements of midrange and high-frequency loudspeaker directivity are time-consuming and little used commercially, so there is little data available except for commercial sound reinforcement loudspeakers for large venues. The replication in model scale of midrange and tweeter loudspeaker directivity is difficult also because of the small physical size of models and the lack of available transducers. The directional properties of microphones are typically less difficult to model than those of loudspeakers.

Since none of the modeling approaches can be used by itself to model the full audio range, it is necessary to join their results to find the full frequency range result. This requires considerations similar to those made in the design of crossover filters for multidriver full frequency range loud-speaker systems [2].

13.3 PHYSICAL MODELING

13.3.1 Physical scale models

All physical scale modeling relies on measurement. The specialized measurement necessary for ultrasonic scale modeling is discussed here. Other acoustic measurement is discussed in Chapter 14.

A survey of physical modeling techniques can be found in References 4,5. The most common of these is ultrasonic scale modeling which allows simulation also of very irregularly shaped rooms and rooms that have unevenly distributed sound absorption. In addition, ultrasonic scale modeling inherently includes the effects of diffraction and scattering [5–18].

Any ultrasonic modeling needs a physical scale model, built to have the same boundary conditions as the full-scale room. If modeling is done at 1:N scale, the frequency scale is transposed by a factor of N:1 and the time scale by a factor of 1:N. Typical scale factors used are in the range 1:4–1:16. Sometimes, dollhouse miniatures made in 1:12 scale are useful for ultrasonic scale modeling of small rooms and can be used to provide objects such as tables, sofas, and chairs.

The upper frequency limit for modeling is usually determined by the electroacoustic transducers used and the sound absorption by air, whereas the lower limit is due to the difficulty of modeling the characteristics of walls and other room surfaces. Real walls such a drywalls display resonant behavior that is difficult to replicate in the model.

In ultrasonic scale modeling of large rooms a major problem is the correct scaling of the attenuation of sound by air [3]. The damping of sound as a function of frequency for different values of relative humidity is shown in Figure 13.1. This damping is typically not a problem in modeling small rooms for speech and music. In these the sound absorption by the surfaces of the room will dominate over that of air.

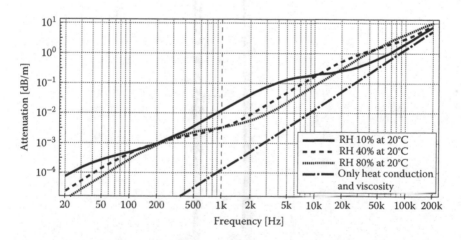

Figure 13.1 Attenuation in dB/m for sound propagation in air as a function of frequency at a temperature of 20°C and with relative humidity, RH in %, as parameter. (From Kuttruff, H., Ultrasonics: *Fundamentals and Applications*, Springer, 1991.) The graph can be used to calculate the attenuation coefficient m since the attenuation ΔL in dB over a distance x is $\Delta L \approx 4.3mx$.

Omnidirectional sound sources are used by default in much room acoustics research to avoid simulation of the directivity properties of real sources. Loudspeakers and microphones that can cover the frequency range of interest (typically 200 Hz to 200 kHz) with desired directivity and sensitivity are not available [8,11]. The influence of their directivity properties on measurement results is instead considered in the analysis of the measured room response.

13.3.2 Sources and microphones

The room impulse response can be measured directly using an acoustic impulse such as a sound from a spark source shown in Figure 13.2 or by sending sound from special electroacoustic high-frequency transducers. Since the spark can be made small, the source can be made virtually omnidirectional or—if the spark is placed next to a small sphere—it can be made to have an almost voice-like directivity.

The sound of the spark is generated by the heating of the ionized air that results from the electrical current and is determined by the electrode gap distance and the duration and strength of the current pulse. The necessary

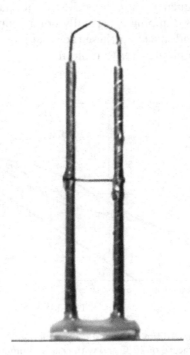

Figure 13.2 A spark gap (at the top of the figure) functions well as an omnidirectional sound source in ultrasonic scale modeling. Gap is about 1 mm wide. (Photo by Mendel Kleiner.)

electrical charge is stored in a high-voltage capacitor, and the current starts flowing when the isolating properties of the gap disappear when there is an excess electrical field strength at the gap tips that results in air ionization and vanishing electrical isolation. This can result from the application of a high-tension electrical pulse to the gap electrodes or to a special trigger electrode. Since the heating of air does not produce new air, the waveform of the sound pulse is not an ideally shaped Dirac delta function but averages to zero pressure over time [20,21]. The duration of the spark is determined by the available electrical charge in a capacitor and the air gap length. When the voltage of the capacitor drops as a function of the electrical discharge, the spark is quenched.

In practice, there will be scattering and reflections from various nearby surfaces, and the recorded waveform will also be affected by the measurement equipment. The measured trace will look similar to that shown in Figure 13.3 [16]. The ringing shown is a result of the microphone's low pass filter characteristics. Since the sound pressure is high close to the spark, care must be taken so that the sound pressure is not so high as to cause nonlinear behavior of the air at surfaces close to the spark gap.

The useful spark frequency range is determined by the waveform, primarily the duration of the positive pulse—unless there is substantial ringing. An example of a measured spark spectrum is shown in Figure 13.4.

The power output of the spark is small at low frequencies which results in poor signal to noise ratio in a single measurement. Since each spark travels a slightly different way between the electrode tips, it is also for

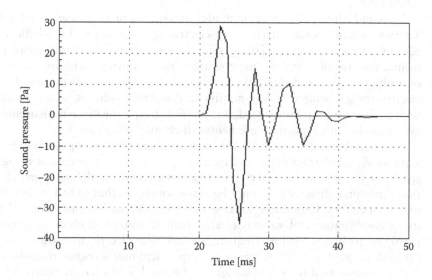

Figure 13.3 The measured acoustic waveform at about 15 cm distance from a spark, transposed to full-scale time. Note the high sound pressure.

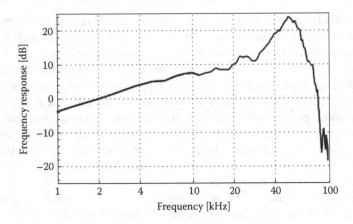

Figure 13.4 Example of the recorded spectrum of a spark. The high frequency limit is due to the microphone used for this measurement.

this reason necessary to synchronously time average a number of sparks, typically about 100–1000 sparks. The spark homogeneity can be improved by continuously blowing fresh air into the spark gap. Specialized software allows for automated averaging, and with the averaging, the low-frequency signal-to-noise ratio can be made sufficient for useful impulse response analysis [17]. Also, sparks have directivity, determined by the length of the—approximately cylindrical—ionized air path, which typically is about 1 mm long.

A second alternative in scale model measurement is to use an electro-acoustic sound source driven by an electric signal generated by dedicated equipment or a computer [10]. As in the case of spark source measurements, the use of electroacoustic sources also requires averaging because of ambient room and microphone electronic noise. An advantage of the electroacoustic source is less nonlinear distortion. Finally, since similar impulse responses are measured, it is possible to use the same measurement system as for full-scale measurements which may reduce cost.

The signal-to-noise ratio of the measurement can be optimized by a suitable signal characteristic and by averaging as for measurement using a spark source. A drawback of the electroacoustic scale modeling approach is that the source directivity cannot be made similar to that of full scale voice or loudspeakers. Omnidirectional sound sources can be approximated by using combinations of various electroacoustic sources such as electrodynamic transducers for low frequencies and piezoelectric film transducers for high frequencies [10]. Alternatively, spherical piezoceramic transducers such as those used in hydrophones can be used as shown in Figure 13.5. Generally though, the spark source is more omnidirectional than electroacoustic sources.

Figure 13.5 A scale model measurement. The microphone shown is equipped with a nose cone for reduced directivity. A hydrophone driven in reverse is used as an omnidirectional loudspeaker. Note the modeled audience in the background. (Photo by Kirkegaard Associates, Chicago, IL.)

The sound pressure in the scale model room is sensed using a miniature microphone as shown in the earlier figure. Typically, a ⅛″ (3 mm diameter) condenser microphone capsule is used. The condenser microphone shown in Figure 13.5 has a nose cone fitted to somewhat reduce directivity. It leaves a short cylindrical grill for the sound to enter through. Such an opening is more omnidirectional than a circular surface. Even such a small microphone has considerable directivity at the high frequencies used in ultrasonic modeling as shown by Figure 13.6.

For binaural measurement, a scaled down manikin with microphones at its ears can be used, such as the one shown in Figure 13.7. Because it is very difficult to correctly model the pinnae in small scales, the measured results must be used with caution.

13.3.3 Measurement systems

The room impulse response is often expressed as $h(t)$, the sound pressure referred to the sound pressure at 1 m distance from an omnidirectional source in an anechoic environment. This approach is used since it is difficult to determine the volume velocity of the source. Because the sound field, in the case of scale modeling, may be so strong close to the

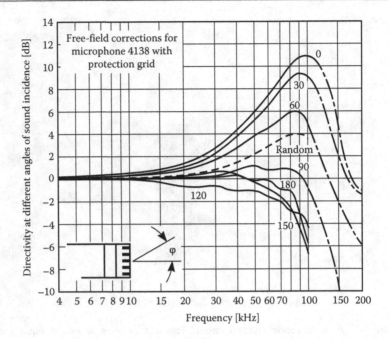

Figure 13.6 Curves showing the directivity of a ⅛″ diameter microphone on the measured SPL of a plane wave for some angles of incidence. (Data courtesy of Brüel & Kjær, Nærum, Denmark.)

spark that there are nonlinear effects, the reference measurement may be needed to be made at some larger distance. When electroacoustic sources are used, the strength of the sound is weak, so a quiet environment is necessary to determine the reference.

13.3.4 Scale model

The geometrical accuracy of the physical scale model is of course essential but so is also knowledge of the sound absorption and the acoustic impedances of the various surfaces and objects. Without knowledge of the properties of the real full-scale surfaces and objects, any acoustic simulation by scale modeling is at best an educated guess based on the modeler's assumptions and experience although sound absorption data are available for some scale model materials and constructions.

The postframe as well as the steel building techniques used for many offices, homes, and studios results in walls that are likely to have resonances in the 50–200 Hz frequency range that result in large sound absorption over the frequency range of the resonances. For small rooms, this absorption is a decided advantage because it helps in keeping the reverberation time short also at low frequencies, although it also contributes to poor

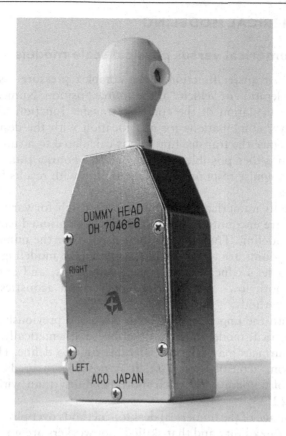

Figure 13.7 An *N* = 10 scale model manikin with microphones behind its pinnae. (Photo by Mendel Kleiner.)

sound isolation. Additionally, the resonant building parts affect the modal structure and mode frequencies of the rooms because of the relatively large modal coupling that results from the low impedance of the surfaces close to resonance. (Sometimes, up to four layers of gypsum, a total of 5 cm (2″), are necessary in large rooms that need long reverberation times in the low-frequency range.)

Rooms that have walls of heavy construction with high-density materials such as thick concrete, clay brick, and hollow concrete block walls are fairly easy to simulate. These constructions can be scaled using painted medium-density fiberboard that is reinforced to have sufficient rigidity and mass.

Note that scaling of Helmholtz resonators such as sound-absorbing structural masonry units and air bricks is difficult. Scaling the exact diffusive properties of walls may be more difficult than scaling their sound absorption.

13.4 NUMERICAL MODELING

13.4.1 Numerical versus physical scale models

The complex transfer function is the complex pressure response for the volume acceleration or velocity at the source position. Numerical modeling allows the calculation of the complex transfer function for the source–room–receiver combinations for any location with the desired frequency resolution. Once the transfer function is calculated to a suitable frequency resolution, it is then possible to use the inverse Fourier transform to obtain the desired impulse response for compatibility with results from geometrical acoustics.

Superficially numerical modeling seems straightforward, but in its use, difficulties are encountered similar to those mentioned earlier for physical scale modeling. The physical data needed for the numerical description of the rooms are generally not available, so modeling will be based on the modeler's educated guesses, assumptions, and experience. This applies to both low- and high-frequency room acoustics modeling by numerical methods.

Along with the impedance problem discussed previously, the geometry of any numerical model must be described mathematically. A model that has many surfaces may take considerable time to define. However, using suitable room acoustics software, it is possible to predict the acoustic properties of a room with much more flexibility than with analog scale modeling [22,23].

An advantage of the numerical desktop methods over physical scale modeling is their flexibility and that skilled shopworkers are not required. This makes them useful for the amateur and student.

13.4.2 Medium and high frequencies

For problems involving medium and high frequencies, geometrical acoustics using the ray tracing method (RTM) or mirror image method (MIM) is suitable. There are also hybrid geometrical acoustics methods based on the combination of features from the two basic approaches. Other methods such as cone tracing and pyramid tracing are derivatives of the RTM. Because of the wavelength criterion, any geometrical acoustics method can strictly only be used for calculation of the room impulse response (RIR) for high frequencies of rooms [24].

In principle, curved surfaces can also be used directly, but usually, curved surfaces are discretized to piecewise planar surfaces. Occasionally, curved surfaces are simulated by the methods of classical optics [24].

In calculation of the generic room impulse responses (GRIRs), wave phenomena such as scattering and diffraction are neglected. Commercial software based on geometrical acoustics may use various enhancements to simulate the presence of scattering and diffraction of sound in rooms.

13.4.3 Tracing rays and finding mirror images

Geometrical acoustics takes it starting point in the geometrical description of the room and thus is an optics-based modeling approach as was pointed out in Chapter 3. Numerical impulse response simulation will start by entering data for room geometry and objects into a database.

In using the RTM, a large number of rays (1,000–1,000,000) are assumed to be sent out from the imaginary sound source. The sound source directivity can be simulated by using different ray intensities or different ray densities. The software then follows each ray path and calculates how the ray is reflected by the surfaces and objects of the room. The equations used for ray tracing can be found in References 25–27.

The starting point of ray tracing is to define the planes of the model. Once the planes are defined, one can use the equations of reflecting planes and rays to find intersection points. New mirrored rays can then be drawn from the intersection point. The coordinates of a ray can be described parametrically by

$$x = x_0 + x_1 t$$
$$y = y_0 + y_1 t \tag{13.1}$$
$$z = z_0 + z_1 t$$

where t is a parameter. An infinite plane is determined by

$$Ax + By + Cz = D \tag{13.2}$$

and the value of the parameter t at the intersection is

$$t = -\frac{Ax_0 + By_0 + Cz_0 + D}{Ax_1 + By_1 + Cz_1 + D} \tag{13.3}$$

The intersection point on the plane must then be tested for validity. Rays will hit many surfaces, but it is only those patches that are within the room that are valid. This procedure must be repeated for all rays and planes.

There are many ways of testing if the intersection point is within the prescribed patch on the plane. A simple method is to sum the angles ϕ_{12}, ϕ_{23}, and ϕ_{31} shown in Figure 13.8. If the intersection point is within the triangular patch area, the sum of the angles will be identically 2π; if the point is outside, the sum will be smaller than 2π.

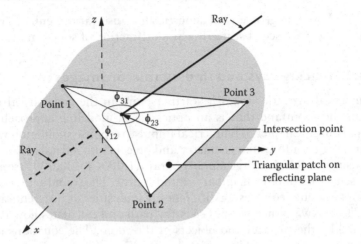

Figure 13.8 Simple principle for determining if ray intersects a triangular patch (out-lined in white) on an infinite reflecting plane (outlined by gray shaded area). If the intersection point is within the triangular patch area, the sum of the angles will be identically 2π; if the point is outside, the sum will be smaller than 2π.

A number of rays will eventually reach the vicinity of the receiver. The receiver cannot be modeled as a point since the chance of a ray passing through a point is almost nil. By studying the number of rays traveling through a test surface or test volume (such as a sphere) at the listening position, one can calculate the room impulse response at that position. The shape of the test area or volume affects the effective directivity of the imag-ined receiver. The receiver is usually modeled as a sphere (for omnidirec-tional sensitivity) or cube. A sphere is defined by

$$(x - x_2)^2 + (y - y_2)^2 + (z - z_2)^2 = R^2 \tag{13.4}$$

The effective intersection point is typically taken as the average of the two intersection points between the ray and the sphere. After all the intersec-tion points between rays and the sphere have been found, the impulse response is obtained by determining the distances traveled by the rays and their attenuation.

In practice, ray tracing is much more complicated if the speed of calcula-tion is to be optimized and errors are to be avoided. Since a sphere is used detection of arriving rays there may be multiple paths between the source and the receiver. Insufficient ray density on the other hand may lead to sur-faces being undiscovered in the tracing. Receivers and sources may also be shaded by obstructing planes.

Cone and pyramid tracing are improvements on the raw ray tracing that permit acceleration of the calculations but also introduce new errors. An advantage of these methods is that the receiver may be a point.

In many ray tracing models, the modeler adjusts the sound absorption after calculation of the room impulse response, for a better agreement between the model and reality. Scattering can be simulated by assigning various randomization rules to the angle of reflection after consideration of the scattering by surface irregularity and edges. A disadvantage of the MIM compared to the RTM is that it does not allow for easy inclusion of scattering and diffraction.

Modeling of sound propagation using the MIM assumes that sound reflected by a flat and rigid surface originates from a mirror source behind the surface. The principles of mirror image calculations are similar to those discussed for rays and the equations can be found in References 28–30. A normal to the plane is drawn from the source point to the image point. For a room having a complex geometry, it may be necessary to calculate many millions of mirror sources, as sound is reflected multiple times by the various surfaces. The mirror images that are created by the surfaces of the room are all constructed in the same way with higher-order mirror images following lower-order ones. As was described in Chapter 3, the valid mirror images must then be sorted out using rules for visibility. Once these mirror images have been found, the room impulse response is quickly calculated for any receiving point.

Using ray tracing, potential images are generated at a rate determined by the number of rays or beams, N, and the mean free path length, l_m. The ray tracing image rate is cN/l_m [images/s], whereas the mirror image theory predicts that the image rate is approximately determined by the room lattice as discussed in Chapter 3 and increases rapidly by time. For this reason, it is not immediately obvious how ray tracing and mirror image results can be joined.

Those methods that use a combination of the RTM and MIM are called hybrid methods. For example, ray tracing can be used initially to find the possible sound paths, and then mirror image modeling is applied to find the exact sound paths. As with ray tracing, a disadvantage is the need for modeling the propagation of a large number of rays.

The RTM and MIM are about equally computationally intensive. Neither method allows for exact room acoustic modeling since geometrical acoustics is used. Good modeling of absorption, diffraction, and scattering remains a problem for both methods, so accuracy is difficult to achieve without considerable experience on the part of the modeler. While the precision of the two methods is initially attractive, the presence of small building errors, surface irregularities, and uneven sound absorption makes the choice between them largely a practical one, where some forms of ray tracing typically win out. It is simpler and quicker to send out more rays than to calculate image sources and test for validity. A particular advantage of MIM over RTM is that back-propagation can be used instead of the forward-propagation described. This allows for instance that a large number of loudspeaker positions can easily be compared for a particular listener location.

The late part of the room impulse response corresponding to the late part of the reverberation is usually too computationally intensive to be handled by a direct approach in either method and various approximations are generally used. Some of these may be based on the use of statistical assumptions and random generated numbers.

Once the computer software has calculated the impulse response, it will typically generate a file containing a list of information about the paths; radiation and arrival properties of the reflected sound components are incident for the calculated listening positions:

- Reflection amplitudes relative to the direct sound
- Time delays relative to the direct sound
- Propagation path details
- Angles of sound radiation
- Angles of sound incidence

This information can then be used for further simulation of the binaural impulse response or the impulse response measured using omnidirectional or directional microphones. Figure 11.11b through d is based on the use of such output data.

13.4.4 Low frequencies

13.4.4.1 Mode superposition

The sound pressure in a room can be regarded as resulting from the contributions from a large number of modes that are all excited by the sound source as described by Equation 2.39. For low frequencies and small rooms, the simplest numerical prediction is done using mode superposition by which the approximate complex transfer function for the listening point pressure is a function of the volume velocity of the source. A summary of the most common numerical methods for solving the wave equation in small room acoustics (although as applied to loudspeaker enclosures) is found in Reference 31.

Using the simple method of just summing the pressure contributions over a reasonable number of modes, it is possible to obtain the approximate complex sound pressure under steady-state conditions for simple rectangular rooms with low losses and rigid walls.

The sound is usually assumed to be generated by a small-volume small, constant volume velocity source. In some cases, the acoustic impedance of the room may not be low compared to that of the source, and in those cases, it will be necessary to correct the volume velocity of the source by taking the acoustic load impedance of the room on the loudspeaker into account.

Theory shows that the mode superposition and mirror image contribution summation methods give equivalent results for a rectangular room that has rigid walls [32]. The number of mirror images necessary is high but such calculations are quite practical. Image contribution summation has given quite good results for rectangular chambers. The main problem with either method is to take the effects of deviations from ideal rectangular shape and diffusion into account.

13.4.4.2 Finite and boundary element methods

For general transfer function calculation, software based on the finite element method (FEM), the boundary element method (BEM) or finite difference time domain (FDTD) method can be used. These methods for solving partial differential equations are attractive because of their inherent accuracy and ability to handle arbitrary room shapes and surface properties.

Of these, only the FEM software has become sufficiently inexpensive and simplified in commercial software, for practical application to room acoustics problems. In room acoustics, the FEM can be used to extract the resonance frequencies of room modes, the frequency response functions for various loudspeaker and receiver positions, and the evaluation of the radiation impedance seen by the loudspeaker.

The FEM is based on the solution of the Helmholtz equation for room using a model where a number of volume elements are connected at nodes and prescribed conditions at the boundary, for example, wall impedance. The BEM is based on Green's theorem stating that pressure at a point can be calculated from knowledge of the acoustic conditions at the boundary. Both modeling approaches, however, require considerably more computational power than most models based on geometrical acoustics if the same frequency range is considered. This is due to FEM and BEM both being based on numerical solutions of large systems of equations. The upper frequency limit is set by the required accuracy, but an element subdivision smaller than one-sixth of a wavelength at the highest frequency is satisfactory.

The BEM is based on Green's theorem, according to which the pressure at a point can be calculated from knowledge of the acoustic conditions at the boundary. To the user, the main difference between FEM and BEM is in the requirement in FEM for modeling of a closed acoustic space and its boundary, whereas BEM can be used for an open space and only requires modeling and meshing of the boundary. With the FEM, results are obtained for all nodes, whereas the BEM only gives one point at a time. In both methods, all physical effects accounted for—such as surface scattering and edge diffraction—are included, provided the boundaries can be correctly described.

The collocation method is often used in BEM and only considers the pressure on the interior side of the boundary. Because the BEM only requires a mesh of the boundary, the size of the matrix equation is smaller than with the FEM. However, the matrices obtained with the BEM are full, whereas the matrices used in the FEM contain many zeros that result in a shorter processing time. The BEM is preferred when the shape of the model is very complicated, but the solution often recommended is to simplify the model to make the mesh creation easier and to then apply the FEM.

The numerical approximation in the FEM is achieved by searching a solution defined by a set of trial functions that fulfill certain criteria. Using such a selection of trial functions, it is easy to adjust for a complex geometry, and it is possible to vary material properties from element to element. It is also possible to include structural parts in the model, thus accounting for the detailed vibration of loudspeaker cones, membrane and Helmholtz absorbers, as well as sound transmission through the walls. The sound sources can be modeled both as acoustic model sources like point sources and as vibrating pistons simulating loudspeakers' membranes. The phase relations and the interaction between different sources can be taken into account. Multiphysics simulation can take electrical, magnetic, and acoustic properties all into account at the same time [33]. This allows for example the calculation of the loudspeaker's electrical input impedance for various positions of the loudspeaker in the room.

13.4.4.3 Meshing for FEM

The most tedious process in the use of the FEM is often the creation of the finite element mesh and specification of surface properties. The complex geometry and the detailed structure of the acoustic surface elements have long hampered the application of finite and boundary element modeling. The generation of a suitable FEM 3D mesh to describe the room by volume elements used to be a problem, but modern software has removed much of the effort needed to mesh small rooms such as a control room or studio [33,34]. Sometimes, it is useful to manually design a finite element mesh in 2D (such as over its floor surface) and then expand it (automatically) vertically at uniform intervals, to create a 3D mesh of the room volume.

The outline of a simple rectangular room that has some furniture and audiovisual equipment was shown in Figure 9.2. A free mesh of the room

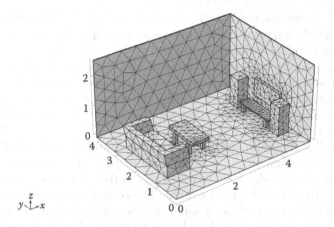

Figure 13.9 A software-generated free mesh shown for some of the surfaces of the room model shown in Figure 9.2.

was created by software with the condition that the distance between two nodes must not be larger than $\lambda/3$ at 200 Hz and is shown in Figure 13.9. The geometric multigrid solver used here allows larger element size than the commonly accepted maximum $\lambda/6$ size [33].

13.4.4.4 Acoustic properties of walls and other boundaries

Even with commercial meshing software, the modeler needs to critically asses which parts of the real object are necessary to include and which are not. In this way, the problem of modeling is similar to that using geometrical acoustics models. When the mesh has been created, the data for the physical properties of the boundaries (walls, ceiling, floor, etc.) and objects in the room are entered into the model.

Two approaches are possible. The first approach is to model the acoustic properties of the various boundary patches as correctly as one can. Parts of the wall surfaces can be ascribed for different properties; for example, patches can be given different wall impedances. The real part of the wall impedances will be responsible for most of the damping of the modes in the model. The wall impedances are important to know or estimate if the exact mode frequencies and shapes are necessary for the analysis. One can then use data obtained from impedance tube measurements (better) or (worse) approximated in some way from conventional measurement data of absorption coefficients. The latter approach is practical because at least some data are then already available from manufacturers. Surfaces that are vibrating mechanical systems will need to be modeled using coupling of the vibratory and the acoustic fields.

If only approximate frequency response and modal damping are needed, the sound absorptive surfaces can be assigned equivalent approximate real impedances calculated from sound absorption coefficients.

Control rooms and studios are usually treated with acoustic control elements—sound absorbers and diffusing elements—in order to adjust the reverberation time and to control the reflection patterns. It may be quite difficult to appropriately describe the impedance of such sound-absorbing surfaces and objects.

A small rectangular chamber with dimensions $1.5 \times 2.3 \times 1.8$ m³, similar to a car compartment, was modeled in an experiment [35]. The compartment was modeled as having rigid walls that were covered by an acoustically absorptive plastic foam. The complex impedance of the plastic foam was measured using Kundt's tube method and these values were used in the modeling. A 1:3 scale model of the chamber was built using 10 mm acrylic plastic, and the walls of the chamber model were covered by the foam.

Figure 13.10 shows the frequency response between two points found by the FEM versus those found by the scale model measurements. The curves show fairly good agreement but the scale model has higher damping than predicted by the FEM model using the measured impedance data. A probable reason for the added damping in the scale model in this case is probably the insufficient rigidity of its acrylic plastic walls that was

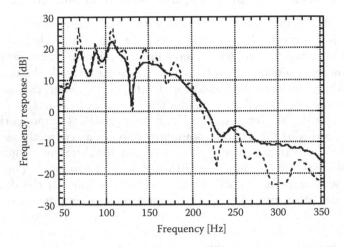

Figure 13.10 Frequency response results of the FEM model of the car compartment described in the text with wall surface complex impedance defined by Kundt's tube measurement. Solid line shows measured response; dashed line shows calculated response. Source located in lower part of front door, receiver point located at driver's head position. (From Granier, E. et al., J. Audio Eng. Soc., 44(10), 835, 1996.)

not modeled. Another reason could be that the acoustic impedance of the loudspeaker was not taken into account.

A second, more approximate, approach can be used if one is primarily interested in a particular transfer path in an already exiting room. This approach requires that one is reasonably far away from surfaces and objects or has a room with reasonably even distribution of wall imped-ance properties. Using the reverberation time data measured, one can then define a flow resistivity for the air volume of the room instead of a wall impedance. This way of handling the damping is attractive since one often knows the desired reverberation time of the finished room, and the exact ways of achieving this reverberation time will be determined empirically in any case.

13.4.4.5 Source and receiver

The sources can be defined as point sources with volume velocity as a known function of frequency or as vibrating membranes (or pistons) with known velocity. For low frequencies, the best solution is often to assume an omnidirectional point source that is usually acceptable. The source can of course also be introduced having an internal impedance.

The FEM allows the calculation of complex sound pressure at every mesh node, but when the room modeling is done with the aim of auraliza-tion of the source and room combination, it is necessary to include the head and torso of the listener among the boundary conditions and the two ear canal entrances as receiving points. Because of the way sound is reflected by the body, the sound pressures at the ears are going to depend on the angles of incidence and on the frequency as discussed in the earlier chapter.

Since both the MIM and RTM are basically geometrical optics models, they will readily (and intuitively) provide the angle and time delay informa-tion necessary for possible inclusion of head-/ear-related transfer functions provided the appropriate head-related transfer functions are taken into account. This is not the case however for FEM and BEM, in which it is pos-sible but impractical to model the head and torso in any great detail (even though this has been made by researchers). Possible simplified approaches for low frequencies include the use of sound pressure pickup points on a sphere/cube to obtain inter-ear phase and magnitude differences. It would also be reasonable to use two pickup points on both sides of a square or circular screen.

At low frequencies the pinnae however do not affect the pressure fields at the ears and it is sufficient to model the head and the ears by using two sound pressure pickup points at a distance of about 16 cm, which is the effective distance between the ears. Since the differences between the ear responses are primarily to be found in the phase difference between the ears, this is a reasonable simplification. The two-point pickup method will give some errors compared to what would be obtained by using a geometrical model of a real head, but these differences are thought to be negligible compared to other errors in room acoustic modeling.

13.4.4.6 Limits of FEM modeling

The highest frequency at which rooms can be modeled by the FEM is determined by the size of the computational task versus the computer available. The smallest element side length that can be used is determined by the shortest wavelength, that is, highest frequency, to be analyzed; typically six elements are used. Assuming the frequency range between 20 and 500 Hz to be of interest, the distance between the nodes in the FEM mesh should not exceed 0.1 m. This element size leads to more than 50,000 complex unknowns for the case of a volume that is the size of a control room of 250 m³ volume.

Figure 13.11 shows an elaborate mesh of a passenger car compartment model for which the manufacturer provided a *NASTRAN* mesh [36].

In this case, the mesh included extremely detailed geometries like the doors and dashboard with all their components, the enclosed air volume, the compartment interior, and the car structure, some of which had to be neglected for practical reasons. After editing the *NASTRAN* file, it was imported into *COMSOL* Multiphysics [33], and a new model was built and meshed. A constant volume velocity point source was located at the driver head position (front left) and three receiver points at the front right, back left, and

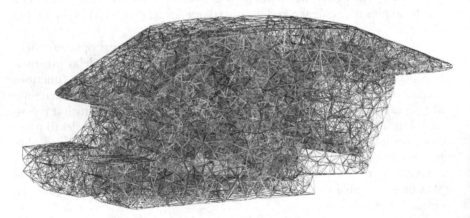

Figure 13.11 NASTRAN meshed interior of a passenger car. (From Guevara Flores, E., A FEM model for simulating reflection of sound by a car compartment's windscreen and lateral front windows, Report 2013:63, School of Building and Environmental Engineering, Chalmers, University of Technology Gothenburg, Sweden, 2013.)

back right head positions. To obtain reasonable accuracy, the new mesh had a resolution of ⅓λ. The FEM simulations were done for the frequency range from 50 Hz to 3.5 kHz, with a frequency step of 50 Hz. The maximum element mesh size differed between frequency ranges so that the most detailed mesh was only used for high frequencies to save calculation time.

The glass surfaces were modeled as fully reflective and therefore set as rigid boundaries; all the other surfaces were assumed somewhat absorptive and defined by a complex impedance obtained from a porous absorber model. In a second step, the model was changed so that all surfaces were considered fully absorptive to obtain the direct sound only. In this way, it was possible to separate the reflected sound from the direct sound by subtraction. The calculated real and imaginary parts of the acoustic transfer functions were analyzed for the frequency and impulse response of the reflection paths.

Figure 13.12 shows calculated impulse responses that were obtained after the Fourier transformation of the complex transfer function data into impulse response data after low pass filtering to avoid the Gibb's phenomenon. One can see that there are clear differences between passenger locations.

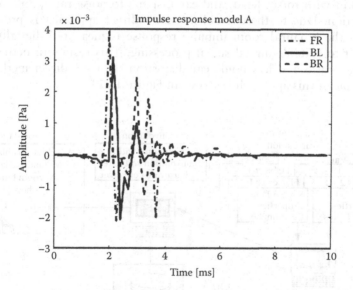

Figure 13.12 Calculated impulse responses from driver head to those of passengers for the car model shown in Figure 13.11 for one window condition. FR = front right, BL = back left, and BR = back right. (From Guevara Flores, E., A FEM model for simulating reflection of sound by a car compartment's windscreen and lateral front windows, Report 2013:63, School of Building and Environmental Engineering, Chalmers University of Technology, Gothenburg, Sweden, 2013.)

13.4.4.7 Auralization: Audible rendering of room acoustics

The idea behind auralization using MIM or RTM was mentioned in Chapter 3. Once the GRIRs for the positions of the source and receiver in the room have been found, these can be appropriately convolved with the head-related impulse functions and with the impulse responses of the sound-reflecting surfaces—the impulse responses of the various reflection coefficients described previously. This yields the binaural room impulse responses.

These can then be convolved with anechoically recorded speech and music so that one can listen to how the acoustic properties of the room would affect the sound radiated by the source. The convolution can be done using software in a general purpose computer or by a dedicated hardware convolver as shown in Figure 13.13. This is an attractive technique to study the individual importance of various parts of the impulse responses [2,16,37].

Auralization can also be done using a scale model approach. Direct modeling where the audio signals are played back as well as recorded in the model is quite problematic because of problems finding or building suitable electroacoustic transducers (both suitably scaled loudspeakers and listener manikin with torso, head, and ears), signal-to-noise ratio, and nonlinear distortion. Due to these problems, an indirect approach is preferred in which the binaural room impulse response is measured digitally and is subject to suitable digital signal processing. This results in better signal-to-noise ratio and less nonlinear distortion than the direct method. The principle of this approach is shown in Figure 13.14.

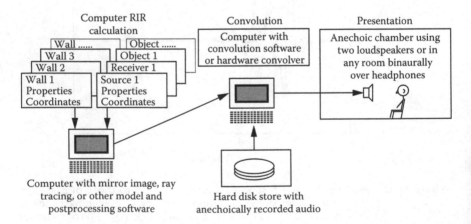

Figure 13.13 Functional diagram for fully computed auralization. (From Kleiner, M., Dalenbäck, B.-I., and Svensson, P., Auralization—An overview, *J. Audio Eng. Soc.*, 41(11), 861, 1993.)

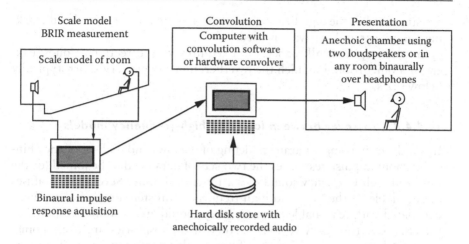

Figure 13.14 Functional diagram for auralization by indirect ultrasonic scale modeling. (From Kleiner, M., Dalenbäck, B.-I., and Svensson, P., Auralization—An overview, *J. Audio Eng. Soc.*, 41(11), 861, 1993.)

The main difficulties in the indirect method are again electroacoustic, but averaging of the impulse response can be made; about 1000 averages typically need to be made for sufficient signal-to-noise ratio if a spark source is used. This average is then convolved with audio so a much better signal quality results than with direct scaled audio playback in the model. An example of a scale model manikin is shown in Figure 13.7. Once the binaural impulse responses have been obtained, these are then convolved with the desired anechoic audio signal as in the case of fully computerized auralization described previously.

There are two main presentation methods used for auralization: binaural playback and playback by surround sound systems such as *Ambisonics*. Surround sound systems need at least four loudspeakers: four loudspeakers are the minimum used with ambisonics. Binaural playback can be done directly by using either headphones or loudspeakers using cross-talk cancelation, described in Chapter 9.

The idea behind crosstalk cancelation is to provide the listener or listeners with sound signals at the ears corresponding to those that would be obtained when listening directly using headphones. This requires that the leakage of sound between the ears due to the loudspeaker presentation is canceled using cancelation signals generated using knowledge about the transfer paths between the loudspeakers and the ears. Presentation using the cross-talk cancelation technique described in Chapter 9 is generally perceived as more natural than ordinary headphone presentation since the virtual sound sources are perceived to be outside the listener's head. This is seldom the case with binaural presentation unless the listener's own pinnae are simulated in the equalization or—better—simulation using head-tracking equipment and software is used.

In practice, crosstalk cancelation requires very directional loudspeakers or a quiet room in which the reverberation time is very short, typically below 0.2 s.

13.4.4.8 Crossover between low- and high-frequency models

In auralization using separate modeling of the low- and high-frequency binaural room impulse responses the two sets of data need to be joined (or the low- and high-frequency sound files after convolution). Several possibilities are available. If the low- and high-frequency transfer functions have been calculated with reasonable accuracy, they should give a similar response in the crossover frequency region. Further, if the responses are overlapping, a simple low-pass and high-pass filter combination can be used, or one can simplify the crossover by simply switching from the low- to the high-frequency response at a given frequency in the frequency domain. A third possibility is to go back to the time domain and simply add the impulse responses provided that the appropriate crossover filtering or limiting has been done. In any case, the responses must overlap by about ±2 octaves around the crossover frequency for the filters to be noncritical.

The choice of crossover frequency is not immediately obvious. Since the low- and high-frequency limiting frequencies are quite arbitrary depending on the the type of errors that one is prepared to accept there is no ideal crossover frequency. Depending on the application, one or the other of the frequency ranges may be of particular importance, and errors in the other may be tolerated.

Due to the small size of purely reflecting patches such as windows, doors, desks, and other objects in the small room, one would assume that geometrical acoustics could not be used below 1 kHz. Comparisons of measured and calculated reverberation for a car compartment-sized space indicate however that the geometrical acoustics model used in that study seems to be applicable even at lower frequencies and that approximately 350 Hz would be a lowest possible crossover frequency for a space this size [35]. The geometrical acoustics model is the more approximate in comparison to finite and boundary element models and in the study was only useful down to 1 kHz. For this reason it is attractive to try to put the crossover frequency high. The use of the FEM will result in long calculation times for high frequencies, and the problem of requiring detailed modeling of the head and torso. In any case, it is typically necessary to calculate the response to at least four times the crossover frequency since the frequency response of the filters cannot be allowed to have steep roll-off characteristics. This practice means that the calculation effort will be four times than that given by the crossover frequency.

If Linkwitz–Riley-type filters are used, the contribution by the response over four times the crossover frequency will be negligible.

REFERENCES

1. Kristianssen, U.R. and Viggen, E.M. (2010) *Computational Methods in Acoustics*. Department of Electronics and Telecommunications, Norwegian Technical University, Trondheim, Norway. Available at http://www.iet.ntnu. no/courses/ttt12/compendium.pdf (accessed May 2013).
2. Kleiner, M., Dalenbäck, B.-I., and Svensson, P. (1993) Auralization—An overview. *J. Audio Eng. Soc.*, 41(11), 861–875.
3. Grillon, V., Meynial, X., and Polack, J.-D. (1995) Auralization in small-scale models: Extending the frequency bandwidth, in *Proceedings of 98th Audio Engineering Society Convention*, Paris, France, Paper Number 3941.
4. Cremer, L., Müller, H.A., and Schultz, T.J. (1982) *Principles and Applications of Room Acoustics*, Vol. 1. Applied Science Publishers, New York.
5. Spandöck, F. (1934) Raumakustisch Modellversuche. *Ann. Phys.*, 20, 345.
6. Xiang, N. (1990) Ein Miniaturkunstkopf für binaurale Raumsimulation mittels eines verkleinerten raumakustischen Modells im Maßstab 1:10 (A miniature dummy head for binaural room simulation using room-acoustic tenth-scaled models), in *Proc. Fortschritte der Akustik, DAGA' 90*, pp. 831–834.
7. Hegvold, L.W. (1971) A 1:8 scale model auditor. *Appl. Acoust.*, 4, 237–256.
8. Satoh, F., Shimada, Y., Hidaka, Y., and Tachibana, H. (1997) Auralization using 1/10 scale dummy head and transaural system, in *Proceedings of the International Symposium on Simulation, Visualization and Auralization for Acoustic Research and Education ASVA 97*, Tokyo, Japan, pp. 337–345.
9. Tachibana, H. (1997) Scale modeling technique for room acoustic design, in *Proceedings of the International Symposium on Simulation, Visualization and Auralization for Acoustic Research and Education ASVA 97*, Tokyo, Japan.
10. Els, H. and Blauert, J. (1986) A measuring system for acoustic scale models, in *Proceedings of the Vancouver Symposium Acoustics and Theatre Planning for the Performing Arts*, Vancouver, British Columbia, Canada, p. 65.
11. Xiang, N. and Blauert, J. (1991) A miniature dummy head for binaural evaluation of tenth-scale acoustic models. *Appl. Acoust.*, 33, 123–140.
12. Hidaka, Y., Yano, H., and Tachibana, H. (1989) Scale model experiment on room acoustics by hybrid simulation technique. *J. Acoust. Soc. Jpn.* (E), 10(2), 111.
13. Day, B.F. (1968) A tenth scale model audience. *Appl. Acoust.*, 1, 121–135.
14. Brebeck, D., Bücklein, R., Krauth, E., and Spandöck, F. (1967) Akustisch ähnliche Modelle als Hilfsmittel fur die Raumakustik. *Acustica*, 18, 213–226.
15. Xiang, N. and Blauert, J. (1989) Binaural simulation technique for scale modelling, in *118th Acoustical Society of America Meeting*, St. Louis, MO.
16. Kleiner, M., Svensson, P., and Dalenbäck, B.-I. (1992) Auralization of QRD and other diffusing surfaces using scale modelling, in *Proceedings of 93rd Audio Engineering Society Convention*, San Francisco, CA, preprint M-2.

17. Polack, J.-D., Marshall, A.H., and Dodd, G. (1989) Digital evaluation of the acoustics of small models: The MIDAS package. *J. Acoust. Soc. Am.*, 85, 185–193.
18. Kleiner, M., Orlowski, R., and Kirszenstein, J. (1993) A comparison between a physical scale model and a computer mirror image model. *Appl. Acoust.*, 38, 245–265.
19. Kuttruff, H. (1991) *Ultrasonics: Fundamentals and Applications*, Springer.
20. Wyber, R.J. (February 1974) The application of digital signal processing to acoustic testing. *IEEE Trans. Acoust. Speech Signal Processing*, ASSP-22, 66–72.
21. Wyber, R.J. (1975) The design of a spark discharge acoustic impulse generator. *IEEE Trans. Acoust. Speech Signal Processing*, ASSP-23, #2, 157–162.
22. CATT Acoustic, www.catt.se. (Accessed May 2013).
23. Odeon, www.odeon.dk. (Accessed May 2013).
24. Cremer, L., Muller, H.A., and Schultz, T.J. (1982) *Principles and Applications of Room Acoustics*, Vol. 1. Applied Science Publishers, New York.
25. Krokstad, A., Strøm, S., and Sørsdal, S. (1968) Calculating the acoustical room response by the use of a ray tracing technique. *J. Sound Vib.*, 8, 118–125.
26. Kulowski, A. (1985) Algorithmic representation of the ray tracing technique. *Appl. Acoust.*, 18(6), 449–469.
27. Long, M. (2005) *Architectural Acoustics*, Academic Press, New York.
28. Lee, H. and Lee, B.-Y. (1988) An efficient algorithm for the image model technique. *Appl. Acoust.*, 24(2), 87–115.
29. Allen, J.B. and Berkley, D.A. (1979) Image method for efficiently simulating small-room acoustics. *J. Acoust. Soc. Am.*, 65(4), 943–950.
30. Kirszenstein, J. (1984) An image source computer model for room acoustics analysis and electroacoustic simulation. *Appl. Acoust.*, 17, 275–290.
31. Karjalainen, M. (2001) Comparison of numerical simulation models and measured low frequency behavior of loudspeaker enclosures. *J. Audio Eng. Soc.*, 49(12), 1148–1166.
32. Kuttruff, H. (2009) *Room Acoustics*, 5th edn. CRC Press, London, U.K.
33. COMSOL Multiphysics, http://www.comsol.com/products/multiphysics/ (accessed May 2013).
34. Pietrzyk, P. and Kleiner, M. (1997) The application of the finite-element method to the prediction of sound fields of small rooms at low frequencies, in *Proceedings of 102nd Audio Engineering Society Convention*, Munich, Germany, Paper 4423.
35. Granier, E., Kleiner, M., Dalenbäck, B.-I., and Svensson, P. (1996) Experimental auralization of car audio installations. *J. Audio Eng. Soc.*, 44(10), 835–849.
36. Guevara Flores, E. (2013) A FEM model for simulating reflection of sound by a car compartment's windscreen and lateral front windows (Report 2013:63). School of Building and Environmental Engineering, Chalmers, Gothenburg, Sweden.
37. Kleiner, M., Svensson, P., Dalenbäck, B.-I., and Linusson, P. (1994) Audibility of individual characteristics in reverberation tails of a concert hall, in *Proceedings of the W. C. Sabine Centennial Symposium*, Cambridge, MA, June 5–7, pp. 227–230.

Chapter 14

Measurement

14.1 INTRODUCTION

Measurement of the acoustic properties of rooms and the interaction between audio, audio equipment (such as loudspeakers), and room acoustics is an essential part of acoustic engineering, both for audio product development and for room acoustics design. It is necessary to prove that a proposed design functions the way it is promised, and measurement is thus important from the viewpoint of contractual issues and quality control. It is virtually only sound pressure measurements that can be done over a wide frequency and amplitude range. For this reason, most measurements of other quantities such as particle velocity, intensity, and power are done by the use of several pressure-sensing microphones and the other quantities calculated from these microphone signals. For audio, it is primarily the frequency response, background noise, and nonlinear distortion that are important.

For decisions on room acoustics, it is typically the measured room impulse response between various source and receiver points that contains the salient information. The room impulse response is however only one link in the measurement chain. Any room acoustic measurement is affected by the exciter (loudspeaker or other device) as well as by the microphone, stands, and cables, as well as of signal properties and signal analysis techniques.

It could be argued that measurement for audio and acoustics should take its start in imitating the properties of hearing. An approach can then be to use a binaural manikin instead of a single microphone when measuring room acoustic conditions. It is difficult, however, to specify and check the acoustic properties of binaural manikins. Their directional frequency response characteristics can be specified using HRTF measurements, but these require a well-equipped acoustic laboratory. For this reason, omnidirectional sound pressure measurements prevail.

Care is necessary in specifying measurements and using measurement results. Extensive documentation of the way the measurement was done, physical circumstances, standards used, photos, etc., are all useful. It is often good to redo someone measurement for education and validation. The standards described in this chapter are likely to change over time, so the reader should always consult the issuing organization for the most recent version. Many acoustics standards are described in the two books of Reference 1.

14.2 MICROPHONES AND LOUDSPEAKERS

14.2.1 Sound pressure

Sound pressure as a function of frequency is the key quantity to the determination of many sound field properties such as particle velocity, sound intensity, and sound power as well as transfer functions, spectra, and reverberation [29].

Microphones, specially made for measurement, are the only ones considered in this chapter. Such microphones are nominally pressure sensing so they will have omnidirectional characteristics, being equally sensitive to sound from any direction. The microphone converts the sound pressure time signal at its diaphragm into an information-equivalent electrical signal, for example, a voltage. Figure 14.1 shows a typical precision condenser microphone used for sound pressure measurement in the frequency range 20 Hz–15 kHz and sound pressure in the range 30–140 dB.

Figure 14.1 Typical 12 mm (½") diameter microphone used for sound pressure measurement. (Photo courtesy of G.R.A.S., Copenhagen, Denmark.)

Commercial microphones for sound recording are available with which pressure, pressure gradient (*velocity, bidirectional*), or combinations (e.g., *cardioid*) can be measured, but these are of little use for measurement because of uncertain stability under varying heat and humidity and may be sensitive to stray electric and magnetic fields. Such microphones should not be used for room acoustic measurements unless special requirements make them necessary. If directivity is needed, then arrays of pressure microphones are used, and the desired directivity of such virtual directional microphones is synthesized by the combination of their electrical output signals. Additionally, microphones for sound recording are difficult to calibrate except in special measurement chambers, such as anechoic and reverberant chambers.

Pressure microphones are based on the sensing of the motion of a membrane due to the force of the sound pressure on the membrane. Most laboratory measurement systems use condenser microphones for sound pressure measurement because of their small dimensions, high stability, low vibration sensitivity, large dynamic range, and wide frequency response. Condenser microphones use a microphone capsule that incorporates the acoustically sensitive capacitor and a microphone preamplifier that converts the capacitance time variation into an equivalent electrical signal. To convert the capacitance change due to sound pressure to voltage the microphone capsule must have constant electrical charge. Commercial microphones come both as high voltage externally polarized (charge supplied by a bias voltage supply) and as prepolarized (electret) in which the charge is permanently injected into a plastic film. Both types also need an amplifier/impedance converter to overcome the losses that would otherwise result due to the capacitive load of a conventional connector cable.

Many microphones also contain electronic signal amplification, signal enhancement, and analog-to-digital conversion circuitry. Such microphones must use special power supplies and cables to feed the microphone with the electrical power needed for their operation. Self-powered microphones, such as those based on piezoelectric or electrodynamic signal conversion, usually do not have sufficiently good electroacoustic properties to be used as measurement microphones.

Capacitive microphone capsules are very sensitive to shock because of the small distance between their thin diaphragm and the close fixed electrode. Dropping a microphone even from a small height onto a hard surface may result in permanent damage. Cleaning the microphone diaphragm should preferably be avoided since it can only be done using special solvents. Cleaning fluids that evaporate rapidly such as alcohol will cause water droplets to be formed on the back of the diaphragm because of the cooling of the air inside the capsule, effectively short-circuiting the capacitance.

Figure 14.2 A small low internal impedance calibrator for use with standard size cylindrical pressure microphones. Microphone is inserted in the hole at cylinder
end on the left. (Photo courtesy of G.R.A.S., Copenhagen, Denmark.)

The sensitivity of a microphone is specified as the ratio between its rms
open-circuit electric output voltage e to the rms sound pressure p in the
sound field:

$$e = K_m p \tag{14.1}$$

The sensitivity K_m is usually expressed in mV/Pa. Alternatively, the sensitivity
can be specified as the electrical output level L_e relative to 1 V output to an
SPL L_p = 94 dB. This is the SPL in a plane wave field with sound pressure 1 Pa.

Since the microphone will affect the sound field and because of the variability in sensitivity due to production spread, handling, and atmospheric
conditions, it is necessary to calibrate the microphone. The calibration can
be done in a number of ways, for example, by comparison to a normal, by
reciprocity, or by use of a pressure chamber. The most common and practical approach is to use a low internal impedance pressure chamber calibrator shown in Figure 14.2. Such calibration is usually done before and after
each measurement using the microphone.

Most condenser measurement microphones are shaped as cylinders with
the microphone capsule at one end as shown in Figure 14.1. Typically, condenser microphone capsules for laboratory measurement come in 1", ½" and
¼" standard diameters. Many electret microphones for amateur use are available in ½" and ¼" diameters. Generally, the larger the microphone, the higher
its sensitivity. The sensitivity of any microphone varies with the direction of
the incident sound field as shown in Figures 13.6 and 14.3. To have little
directivity, the microphone has to be much smaller than the wavelength of the
sound. Since the wavelength of sound at 20 kHz, the maximum frequency of
interest in audio, is 17 mm, ideally, the microphone should have a diameter
less than about 6 mm to ensure omnidirectional response to at least 10 kHz.

Selecting the right microphone for measurement depends on many factors
as outlined in Figure 14.4. Even within the group of pressure-sensitive

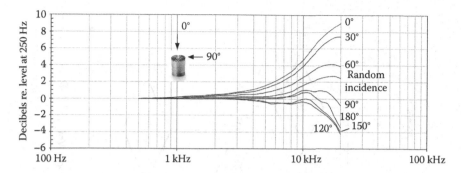

Figure 14.3 The influence of sound field incidence on the sensitivity of a ½" diameter microphone capsule. (Graph courtesy of G.R.A.S., Copenhagen, Denmark.)

microphones, there are different types depending on how the microphone has been equalized for flat frequency response:

- Free field: This type is usually equalized for the flattest frequency response with a plane wave field incident at 0° microphone pointed at source (on axis) or at 90° (off axis).
- Diffuse field: This type is usually equalized for the flattest frequency response in a random-incidence (diffuse) field. When used with a plane wave field, the wave should be incident at about 70°–80° relative to its axis.
- Pressure: This type is usually equalized for the flattest frequency response flush mounted in the wall of a cavity or tube. The microphone is not equalized for the influence of sound reflection by its body.

For more extensive documentation and advice on the choice and use of measurement microphones, the reader should consult the References 2–4.

Practical microphones will feature various forms of undesired nonlinear distortion at large sound levels and self-noise at low sound levels. The use of small microphones is advantageous since these microphones are usually designed to have taut diaphragms, resulting in relatively little nonlinear distortion even at very high sound levels. A ½" microphone typically will have less than 3% harmonic distortion at $L_p = 146$ dB. Generally, within a *microphone family*, use of smaller diaphragm microphones results in a reduced signal-to-noise ratio but less nonlinear distortion.

14.2.2 Particle velocity

Accurate and precise measurement of sound pressure by condenser microphones is relatively easy, but the measurement of the particle velocity is

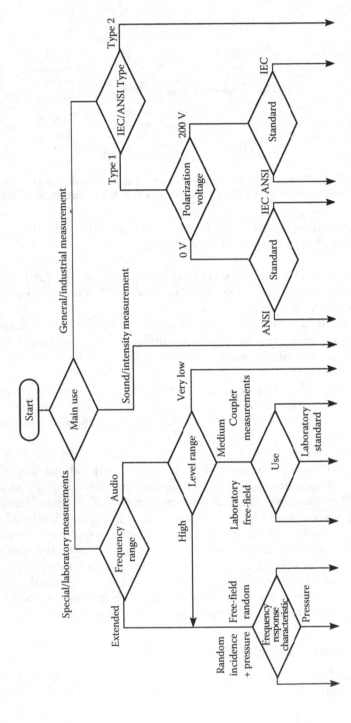

Figure 14.4 Flowchart for selecting an appropriate microphone. (From Brüel & Kjær, *Microphone Handbook*, Vol. I, BE1447-11, Nærum, Denmark, 1996.)

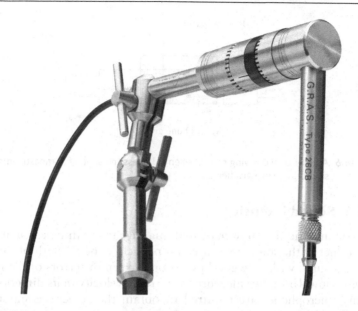

Figure 14.5 A two-microphone probe for measurement of particle velocity and sound intensity. (Photo courtesy of G.R.A.S., Copenhagen, Denmark.)

more difficult. Measurement of particle velocity is of interest to characterize the vibrations and sound radiation of structures interacting with air. It is also used in the determination of volume velocity, which is used to determine the excitation of sound fields, and in the determination of sound intensity, that is used to study energy flow and sound power emission. Such a velocity/intensity microphone probe is shown in Figure 14.5.

In a plane wave, the particle velocity and sound pressure are simply related over Equation 1.30. For a spherical wave, the particle velocity grows quickly as distance to the sound source is reduced as shown by Equation 1.38. The particle velocity is a vector quantity and can be measured indirectly by measuring the pressure gradient between two close microphones in a simple array such as the one shown in Figure 14.6 and using Equation 1.92 to calculate the particle velocity from the pressure differential measured.

Measurement of particle velocity generally involves measurement of the particle velocity in three perpendicular directions from which the resultant particle velocity vector can be determined. This requires the use of special probe assemblies using an array of at least four pressure sensing microphones.

There are also sensors for a direct measurement of particle velocity based on various principles such as doppler shift, hot-wire cooling, etc., but these were developed much later than those of pressure microphones and are not used as much.

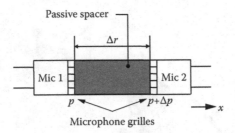

Figure 14.6 A functional drawing of a differential pressure probe for measurement of the sound pressure gradient.

14.2.3 Sound intensity

Acoustic intensity is a power-related quantity that can be calculated from the product of the sound pressure and particle velocity. Both of these have to be measured with very good precision to keep the errors of the intensity estimates low. Unless we measure the particle velocity in its direction, three velocity microphones are required to obtain the velocity components in three directions.

Most intensity measurement is done, however, using a single sound pressure gradient probe microphone such as the one shown in Figure 14.5. The pressure gradient microphone operates as shown in Figure 14.6 and gives a possibility to measure sound pressure and particle velocity simultaneously so that the sound intensity component in the direction of the microphone axis can be calculated.

Once the pressure time signals have been acquired, they will need to be converted to the frequency domain for simple signal processing. First, the particle velocity is calculated according to Equation 1.92. Then the particle velocity is multiplied, frequency by frequency, with the pressure average according to Equation 1.91 and so that medium impedance, potential and kinetic energy densities, as well as active and reactive intensity can be calculated as shown in Section 1.8.

Figure 14.7 shows a block diagram of a computer-based intensity meter. The signal acquisition part of the meter has to be constructed with extreme care, so that the sound pressure phase error between channels is smaller

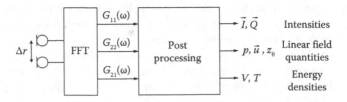

Figure 14.7 A block diagram of computer-based intensity meter.

than 0.05°. The description of the required measurement methods and equipment properties is defined by the ISO 9614-1 standard [30].

14.2.4 Loudspeakers

Special loudspeakers are often used for transfer function measurements of room characteristics. For the best signal-to-noise ratio, the low-frequency loudspeakers must be capable of generating large volume velocity, which is easily accomplished using drivers that have large diaphragms and consequently quickly become very directional for high frequencies. Traditionally, omnidirectional loudspeakers have been used in architectural acoustics to generate the most diffuse and unbiased sound field. This requires separate loudspeakers for low- and high-frequency measurement to avoid unacceptably long averaging, which would otherwise be necessary to have an acceptable S/N ratio. Examples of such loudspeakers are shown in Figure 14.8. An advantage of the cylindrical loudspeaker shown in Figure 14.8b is that it can be used as an omnidirectional volume velocity source because of its small opening (at top). Because of the small source opening this loudspeaker can be fitted with a particle velocity sensing microphone pair to provide volume velocity values for accurate impulse response measurement.

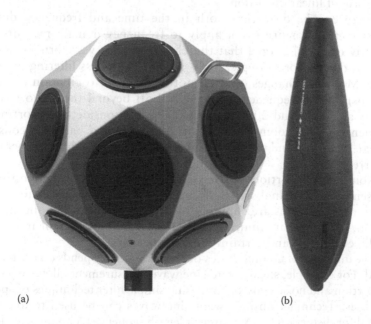

(a)   (b)

Figure 14.8 Two omnidirectional loudspeakers: (a) midfrequency range and (b) high-frequency range. (Photos courtesy of Brüel & Kjær Sound and Vibration A/S, Nærum, Denmark.)

For measurement in control rooms and domestic environments, transfer function measurement is better made with the loudspeakers that are to be used for monitoring/listening. There, the omnidirectional loudspeakers are best used for comparing transfer function measurement results with predicted transfer functions.

14.3 NOISE AND DISTORTION

14.3.1 Distortion

Two types of distortion, linear and nonlinear, are of particular interest in audio and acoustics. Linear distortion, such as changes in frequency response, does not create new frequency components. The frequency-dependent response of a loudspeaker or room and the frequency-dependent attenuation by an amplifier's tone control are examples of linear distortion. Commercially available microphones are usually designed to have linear frequency response, that is, frequency-independent electroacoustic conversion in the audio range. It may be necessary to add amplification or passive filtering to achieve a linear frequency response. That the microphone's frequency response varies with the angle of incidence of the sound is a form of linear distortion.

There can be distortion both in the time and frequency domains. Most metrics for distortion apply to frequency domain measurements. Strictly one could argue that the changes in the waveform of transient signals due to, for example, high-pass and low-pass filtering are distortion. Moderate changes (that do not incur major oscillation in the time signals) due to frequency response roll-off beyond the audio frequency limits at 20 Hz and 20 kHz are usually not considered distortion. The influence of the room filter action discussed in Chapter 2 constitutes linear frequency and time domain distortion, although it is generally not referred to as such.

Nonlinear distortion creates additional frequency components not present in the original signal. Some important nonlinear distortion generation mechanisms are clipping, hysteresis, frequency/amplitude modulation, and slew rate limiting. All of these may be present in electronic audio equipment and certainly in loudspeakers.

The influence of nonlinearities on measurement depends on the technique used. For example, steady-state sine wave measurement allows rejection of distortion and noise components using simple filter techniques of spectrum analysis. Techniques using swept sine waves can be used to analyze the signal for distortion [5]. Maximum-length sequence techniques for impulse response measurement are inherently sensitive to nonlinear distortion that may cause spurious impulses in the measured impulse response.

14.3.2 Different types of noise

Noise is often defined as an undesirable sound. Room noise is due to human activity, generation by heating, ventilation and air condition systems, as well as outside noise that arrives through the building elements. Room noise is often identifiable by hearing because it is not truly random. White noise is defined as a signal that has an autocorrelation function that is zero for all delays except for no delay. It is random in time and has the same spectral density at all frequencies. The terms pink and red noise are used for noise that has a spectral density that drops by $1/\sqrt{f}$ and $1/f$, respectively.

14.3.3 Spectrum and measurement uncertainty

Since hearing has frequency-dependent sensitivity, it is necessary to analyze noise for its frequency content. This is done by filtering the noise signal using electrical filters such as bandpass filters that allow a frequency band Δf wide of signals to pass through. Any real frequency band-limited system will have limited time resolution. For a physical filter the relationship between signal duration Δt and signal bandwidth Δf is approximately given by

$$\Delta t \cdot \Delta f = \frac{1}{2\pi} \tag{14.2}$$

We see that for frequency response analysis with good frequency resolution at low frequencies, the analysis window needs to be long. When analyzing stationary random or deterministic signals, there is added uncertainty because of the limited length of the time under which the signal is analyzed. It is advisable not to use random noise signals for the analysis of transfer functions.

A white noise signal $s(t)$ has an amplitude that varies at random over time. A signal's effective value or mean square value is denoted s_{rms}^2 or $\tilde{s}^2$:

$$\tilde{s}^2 = \frac{1}{T} \int_0^T s^2(t)\, dt \tag{14.3}$$

where T is the integration time over which the effective value is measured. The integration time T is usually long to improve measurement precision.

One can show that the estimate of the rms value $\tilde{s}$ of a bandpass-filtered white noise has a relative standard deviation that, for large values of BT, is

$$\xi_{rms} \approx \frac{1}{2\sqrt{BT}} \tag{14.4}$$

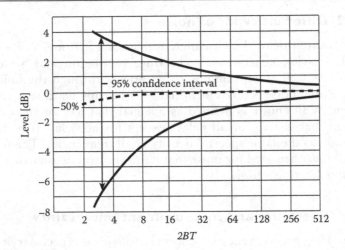

Figure 14.9 The 95% confidence interval for the level of band-limited white noise as a function of the 2BT-product. (After Broch, J.T., *Principles of Experimental Frequency Analysis*, Elsevier, London, U.K., 1990.)

where B is the bandwidth in Hz of the noise. Since the standard deviation determines the confidence interval of the estimated value, to make good noise measurements, one must use either wide band filters or long averaging times to obtain precise values. Figure 14.9 shows the 95% confidence intervals for a band-limited white noise signal as a function of the variable $2BT$.

14.3.4 Practical filters

Practical spectral analysis is often done using octave- or ⅓-octave-band wide bandpass filters since such filters give adequate resolution in the analysis of spectra that vary relatively little over the frequency range of interest. The bandwidth B of a bandpass filter having an upper cutoff frequency, f_u, and a lower cutoff frequency, f_l, is

$$B = \Delta f = f_u - f_l \qquad (14.5)$$

The geometrical mean frequency f_m, usually called the center frequency, is used to denote the filter:

$$f_m = \sqrt{f_u f_l} \qquad (14.6)$$

The ratio between the bandwidth and the center frequency is always constant for any particular type of octave- or ⅓-octave-band filter.

For octave-band filters, the following applies:

$$\frac{B}{f_m} = \frac{\Delta f}{f_m} = \frac{f_u - f_l}{f_m} = \sqrt{2} - \frac{1}{\sqrt{2}} \approx 0.71 \qquad (14.7)$$

Similarly, for ⅓-octave-band filters,

$$\frac{B}{f_m} = \frac{\Delta f}{f_m} = \frac{f_u - f_l}{f_m} = \sqrt[6]{2} - \frac{1}{\sqrt[6]{2}} \approx 0.23 \qquad (14.8)$$

The center frequencies f_m (in Hz) of ⅓-octave-band filters have been internationally standardized as

← ..160 200 <u>250</u> 315 400 <u>500</u> 630 800 <u>1000</u> 1250 1600.. →

The underlined numbers are the internationally agreed center frequencies of octave-band filters (see the IEC 651 and ANSI S1.16 standards [34]). Sometimes, ⅙-octave filters are used since they better resemble the characteristics of the ERB filters described in Chapter 7. Measurements to simulate the frequency response characteristics of hearing also use the so-called gammatone filters [7].

When analyzing sounds for information about tonal components, nonlinear distortion components, or narrowband resonances due to modes or resonators, for example, it is often practical to use filters having constant narrow bandwidth. Such filters for use in audio and room acoustics have bandwidths as small as 0.3 Hz.

Narrowband filters are usually implemented digitally. Such filters can be both finite and infinite impulse response filters. The bandwidth can be set as small as determined by the available measurement time as given by Equation 14.2. The measurement is done by sampling the signal to time values and then either using direct digital filtering or doing a digital Fourier transform of the time data.

Determination of the signal's Fourier transform has become a common part of most signal acquisition equipment. It is important though to consider the transform measured as a result of a narrowband filtering process. The result will depend on both the signal and filter properties. If measurement is done without period synchronization, there will be signals leaked into adjacent frequency bands. The leakage can be reduced using suitable *windowing* of the signal. To obtain octave or octave-band data, the Fourier transform's narrowband data can then, in principle, be combined to wider band filters such as octave or finer-resolution filters. The reader is directed to relevant literature in experimental digital signal analysis such as Reference 6.

14.3.5 Sound pressure level and sound level: dBA

The determination of the loudness level is difficult because it must take frequency domain masking into account. The principle is described in the ISO 532-1975 standard [31]. The loudness level of a sound can be measured approximately by using appropriate weighting filter. Common weighting filters are so-called A, B, and C filters. When used in measurement, these give the sound level reading expressed in dBA, dBB, or dBC, depending on the filter used. The frequency response curves of these filters are shown in Figure 14.10. SPL values measured without filtering are often expressed in dBlin to clearly mark out that no frequency weighting was used. The term sound level is usually reserved for measurement values in dBA.

The filter characteristics of the A, B, and C filters were chosen to some-what resemble the shape of the equal loudness contours in the intervals 20–55 Phon, 55–85 Phon, and 85–140 Phon, respectively. Because of the availability of data and considerable experience, the sound level expressed in dBA has become the most commonly used, particularly for the determination of the risk of noise-induced hearing loss. One also finds that often the dBA value correlates well with the Phon value for many common sounds.

14.3.6 Room noise

Most spaces for voice and music require quiet; the output power of these sources is small. Low background noise will be a requirement for most rooms intended for music performance, recording, or reproduction. Recording studios, control rooms, concert halls, operas, and theaters are examples of venues that are particularly sensitive to noise.

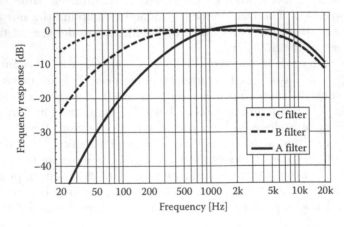

Figure 14.10 Attenuation characteristics for A, B, and C filters.

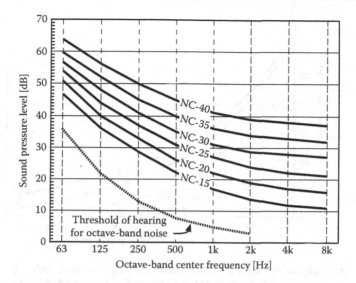

Figure 14.11 These NC curves apply to octave-band-filtered noise levels and are often used in room acoustics. (After Tocci, G.C., *Noise News Int.*, 8(3), 106, 2000.)

Significant experience and extensive subjective testing have provided a series of curves that show the allowable SPL in each octave band. The long established noise criterion (NC) curves, shown in Figure 14.11, are still recommended by the authors for most uses, while Beranek's balanced noise criterion (NCB) curves, shown in Figure 14.12, were developed for concert hall and opera house applications and feature extended low-frequency ranges, and the more recent room criterion (RC) curves, shown in Figure 14.13, are recommended for most other noise-critical spaces. Table 14.1 shows some typical NCs for various spaces.

The NC, NCB, and RC curves are often used in determining whether a noise spectrum is acceptable or not and in specifying acceptable or maximum noise conditions. Such curves, specifying maximum octave-band SPLs from 31 Hz to 16 kHz, are useful but do not take the spatial characteristic or the temporal and cognitive character of noise into account. Note that for wide band noise, the ⅓ octave-band values are about 5 dB lower than those of the octave-band.

Noise shaped along the NC curves has a tendency to sound both *rumbly* and *hissy*. The NC curves were not shaped to have the best spectrum shape but rather to permit satisfactory speech communication without the noise being annoying.

The RC curves on the other hand have been shaped to be perceptually neutral; they are straight lines with a slope of –5 dB/octave. The RC method involves the determination of both an RC rating and a spectrum quality descriptor that determines if the spectrum is *rumbly* or *hissy* [8].

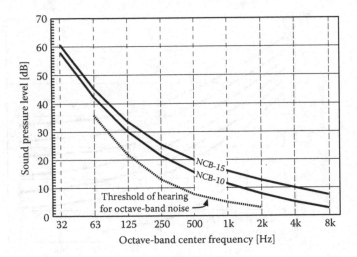

Figure 14.12 Beranek's NCB curves for octave-band-filtered noise. (After Beranek, L.L., *Concert Halls and Opera Houses: Music, Acoustics, and Architecture*, Acoustical Society of America, New York, 1996.)

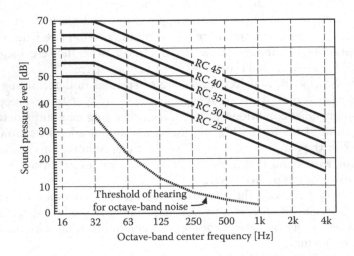

Figure 14.13 The RC curves also apply to octave-band-filtered noise levels. Note that there are several editions of these curves—the ones shown are the Mark II curves. (After Beranek, L.L., *Concert Halls and Opera Houses: Music, Acoustics, and Architecture*, Acoustical Society of America, New York, 1996.)

Table 14.1 Suggested NCs for various spaces

Worship spaces	NCB10–NCB15
Music and speech recording/radio/TV studios	NCB10–NCB15
Control rooms for music and speech recording	NCB10–NCB15
Music teaching classrooms and practice rooms	NCB15–NC20
Homes, listening rooms, and bedrooms	NC15–NC20
Board rooms, conference rooms	NCB15–NC25
Libraries and reading rooms	NC25–NC30
Private enclosed offices	NC25–NC30

Sources: Tocci, G.C., *Noise News Int.*, 8(3), 106, 2000; Beranek, L.L., *Concert Halls and Opera Houses: Music, Acoustics, and Architecture*, Acoustical Society of America, New York, 1996; Harris, C.M. (ed.), *Handbook of Acoustical Measurements and Noise Control*, McGraw-Hill, New York (reprinted by Acoustical Society of America).

In some cases, one has to be content with a measurement value for sound level in dBA, although in such cases, a dBC value may also be available to characterize the low-frequency noise. Spaces designed for quiet such as concert halls typically require background noise sound levels below 20 dBA, but preferably below 15 dBA.

The cognitive properties of noise are usually very important. It may sometimes be easier to tolerate a noise having a random character than a noise that is transient, which has a specific time pattern or which carries information content. The directional incidence of the noise is also important. A diffuse sounding noise is often less disturbing to some people than a noise that can be localized in space (although in the latter case, one can perhaps more often find it possible to eliminate the noise source).

One often uses various metrics such as tonality and sharpness to further describe the properties of noise. Some examples of terms sometimes used to describe noise are bassy, booming, buzzy, clanking, damped, deep, droning, grating, grinding, groaning, gurgling, harsh, hissing, humming, melodic, modulated, pinging, pounding, pulsing, purring, rapping, rattling, resonant, rhythmic, ringing, roaring, rumbling, sharp, shrill, smooth, soft, squeaking, swooshing, ticking, tinny, tonal, uneven, whirring, whistling, and muffled [11].

It is also important to avoid audible pure tones in the noise spectrum. For tones to be inaudible their level must be below the ⅓-octave band level of the noise at their frequency.

It should be kept in mind that most NC values have been set with persons having normal hearing in mind. Persons having hearing impairments or hearing loss generally have difficulty in understanding speech buried in noise and are often more annoyed by noise than persons having normal hearing. Besides causing annoyance, noise may also be tiring and may have physiological effects. For speech, the S/N ratio (assuming typical speech and heating, ventilation, and air-conditioning [HVAC] noise spectra) must be at least 15–25 dB for speech intelligibility to be unaffected by noise in the speech frequency range.

14.3.7 Microphone noise

All microphones are subject to internally generated noise. Many microphones, such as condenser microphones, also contain electronic signal amplification, signal enhancement, and analog-to-digital conversion circuits that will add to the nonlinearities and the background noise. Externally polarized condenser microphones also suffer from leakage currents between the diaphragm and the backplate that cause a special type of noise. Typically, the noise from the microphone is required to be lower than that in the environment where the microphone is being used so that sufficient S/N ratio can be obtained.

It is convenient to express the noisiness of a microphone and preamplifier system by an equivalent room noise level, for example, as an equivalent sound level in dBA. A 25 mm (1") diameter, externally polarized, condenser microphone typically has a background noise level L_{pA} below 15 dBA. A similar electret condenser microphones will have background noise level slightly higher at about 20–30 dBA equivalent room noise level.

It is important to note that room noise and microphone noise generally have different spectral properties, so even at the same dBA level number, they will feature very different noise quality. The typical noise spectra for room noise and the electrical noise from microphones are shown in Figure 14.14. The ambient noise in a room typically has a roll-off of about –5 dB/octave, whereas the microphone electrical self-noise at medium

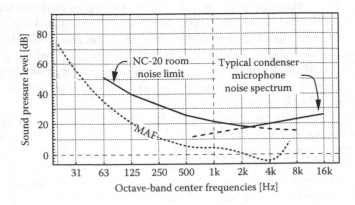

Figure 14.14 Examples of typical octave-band noise spectra for ambient and electronic noise, respectively. The curve marked MAF illustrates the minimum audible sound pressure level for pure tones at the respective frequencies.

and high frequencies typically increases by +3 dB per octave. The very different noise character means that microphone noise may still be quite audible even if the equivalent room noise sound level of the microphone is nominally the same in dBA as the noise sound level in the room in which the microphone is used.

Because any practical microphone is also a mechanical device, microphones will also be sensitive to mechanical vibration, and such vibration in the audio frequency range will result in an electrical output signal.

14.4 LINEAR SYSTEMS

14.4.1 Transfer function and impulse response

The transfer function and the impulse response were introduced in Chapter 1. They are used to characterize the properties of linear systems that affect signals, such as electric and acoustic filters. For a linear system, the transfer function and the impulse response are related over the Fourier transform by Equations 1.52 and 1.53, which is measured is a matter of convenience, hardware, and software availability. The measurement of time-varying systems should be left to be measured and analyzed by a specialist. Strictly, most rooms have time-varying acoustic properties (people movement, air movement, etc.), but these usually cause slow changes so that the system for practical can be regarded as stable during the time of the measurement.

14.4.2 Frequency response

It is usually more convenient to express transfer function and frequency response properties by using frequency f in Hz rather than as radial frequency ω in rad/s.

The transfer function $H(f)$ of a system (see Chapter 1) is the complex ratio of the spectrum of the output signal $S_o(f)$ divided by the spectrum of the input signal $S_i(f)$:

$$H(f) = \frac{S_o(f)}{S_i(f)} \tag{14.9}$$

The (power) level of the transfer function L_H can be written as

$$L_H(f) = 20\log(|H(f)|) \tag{14.10}$$

In audio engineering, the amplitude frequency response is usually taken relative to its level at 1 kHz. Some of the terms used in characterizing frequency response curves are shown in Figure 14.15.

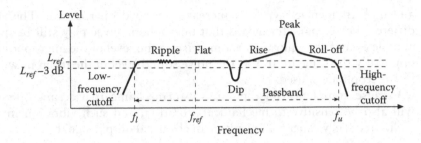

Figure 14.15 Various terms used in the characterization of the frequency response.

The amplitude frequency response function FR (usually just called the frequency response or simply the response) is

$$FR(f) = \frac{L_H(f)}{L_H(1000)} \qquad (14.11)$$

The band limits are usually taken as the frequencies where the response is –3 dB down relative to the level at some reference frequency, often 1 kHz. In most cases, one will require amplifiers, loudspeakers, microphones, etc., to have a frequency response that is flat, that is, having a minimum of ripple, dips, and peaks in the passband.

Strictly, the measurement of the frequency response even for a stable system should be done one frequency at a time, but a comprehensive measurement using this technique requires considerable time so two other techniques are usually used.

Using analog techniques, the frequency response can be measured using a sweep sine wave generator set for slow sweep speed and a level recorder. The sweep speed (frequency increment per time unit, Hz/s) usually is set to increase with the signal frequency. Typically, the time necessary for a sweep measurement depends on the accuracy by which one wants to measure peaks or dips in the response. Assuming the resonant system (or the filter used for the analysis) to have a quality factor Q and that the resonance frequency is estimated as f_0, one can write the maximum sweep speed as

$$S_{MAX} < 0.9\left(\frac{f_0}{Q}\right)^2 \text{ Hz/s} \qquad (14.12)$$

for linear sweeps (i.e., constant sweep rate in Hz per second) and

$$S_{MAX} < 1.3\frac{f_0}{Q^2}\text{Octaves/s} \qquad (14.13)$$

for logarithmic sweep rates [12].

It is common practice for electroacoustic equipment to be measured at excessive sweep rates so that the equipment seems to have a smoother frequency response than it really has.

For measurement using digital signal processing fast sweep sine wave signals called *chirp* signals are often used. Note that such signals are unsuitable for analog measurement for the previously mentioned reason but may be used successfully with digital fast Fourier transform (FFT)-based measurement. The FFT window length must be large enough for any transient to decay to some sufficiently small value within the analysis time. For analysis of room modes and their decay (reverberation time), the time window length must be at least about ⅔ of the reverberation time. The window function should be chosen appropriately for the signal type. In the case of room impulse response, a half cosine window is usually satisfactory. Further, the sampling frequency must be high enough for aliasing not to occur, and the number of samples within the time window must also be large enough so that peaks (or dips) in the frequency spectrum can be adequately resolved. This requires at least ten spectrum lines within one-half power bandwidth of a resonance frequency peak. Dips caused by interference may be very difficult to resolve adequately since they are often very sharp.

In any case, measuring the properties of high Q room or Helmholtz resonator, resonances require long measurement (and window) times as described in the discussion on experimental modal analysis in Section 14.5.3.

14.4.3 Phase response and group delay

Any acoustic system will be characterized by unavoidable time delays such as the delay of sound due to its travel from loudspeakers to microphones or listeners. Characteristic for the direct transfer path is that its time delay for point sources and receivers is frequency independent.

Most audio transducers have resonant characteristics at the spectrum extremes and peaks and dips in the frequency response. These will cause additional delays because of the time it takes for energy to be stored in the resonant systems, no matter whether these are electrical, mechanical, or acoustical. Many physical filters are characterized by a minimum phase response. For minimum phase filter systems, the phase response is linked to the frequency response via the Hilbert transform.

The rate of phase shift in the transfer function (at some frequency) as a function of frequency can be shown to be coupled to the group delay of the signal. The group delay was defined in Equation 6.2 but can also be written as

$$\tau_g = -\frac{1}{360}\frac{d\varphi}{df}$$

(14.14)

where, $d\varphi$ is given in degrees and df in Hz.

The group delay is a function of frequency and is of considerable interest in the design of loudspeaker systems. The more resonant the system, the larger will be its group delay deviations from the purely geometrically caused delay because of the travel time for sound. Loudspeaker system crossover filters may have large group delays.

A system having constant group delay has a linear phase, that is, the phase of the transfer function increases proportionally to frequency. This preserves the shape of the wave. Mathematically, it means that all frequency components are time-synchronized as they were in the original signal.

In practice, group delay is best measured using digital techniques, for example, using maximum-length sequence analyzers such as MLSSA [29] or by FFT analyzers using chirp signals. In these systems, the phase response is *unwrapped* and the phase versus frequency slope is calculated to find the group delay.

14.5 ROOM MEASUREMENT BELOW THE SCHROEDER FREQUENCY

14.5.1 Frequency response

Using a loudspeaker and a sine wave generator, one can study the damping and spatial properties of the room modes. Each mode in a rectangular room excited by a volume velocity sound source Q is characterized by a spatial distribution of sound pressure given by Equation 2.39. For a tonal sound source, that is, a sound source emitting sound having just one frequency, all modes will oscillate at its driving frequency. How *strongly* any particular mode will be excited depends on how close its resonance frequency is to the excitation frequency, how damped the mode is, and where the source is located in the room. Qualifying a control room or domestic environment for low-frequency noise requires detailed specification of where the microphones were positioned.

The summed sound pressure of several excited room modes is going to depend on how the sound pressure contributions of each mode add up in magnitude and phase. This, in turn, depends on the particular modes that contribute at a particular location, so the sound pressure varies from one position to the next. All eigenfunctions will have a large magnitude at the corners of the room in the simple case of a room that has high-impedance walls. The ratio between the sound pressure and the particle velocity is highest at the corners, that is, the sound field impedance is highest at the corners. If a source has high internal impedance its volume velocity is unaffected by the acoustic load impedance and is called a constant volume velocity-type source. Such a source will excite the room modes best when placed in the corner of a room. Most electrodynamic loudspeakers can be regarded constant volume velocity–type sources.

At high frequencies where many modes leak into one another, the SPL will vary considerably from one position to the next as shown by the experimental curve in Figure 2.10. The interference between the pressure contributions of many mode gives sharp minima and maxima in both frequency and spatial response. Even so, a single mode may be measured by using a large number of microphones positioned at the mode's maxima, for example, so that all the microphone signals for that particular mode are in phase. Summing the microphone time signals will enhance the desired mode and reject others. This of course only works for well known mode patterns, otherwise the experimental modal analysis techniques described in Reference 14 need to be used.

The frequency response fluctuations shown in Figure 2.13 illustrate the importance of frequency (and spatial) averaging when the SPL in the diffuse sound field of a room is desired. In practice, one does well to measure the SPL average of noise only in frequency bands that are wide enough to span over at least 10 modes. Since ⅓-octave bands are quite narrow, this leads to a lower-frequency measurement limit of only about 250 Hz for such noise measurements in a 100 m³ volume room.

It is difficult to avoid spurious sound reflection by room surfaces affecting the measurement of sound pressure. Even in anechoic chambers, there will always be some sound reflected by the room surfaces and objects in the room that results in random error in measurement as shown in Figure 14.16. Moving closer to the sound surface may help, but then other systematic errors may be introduced.

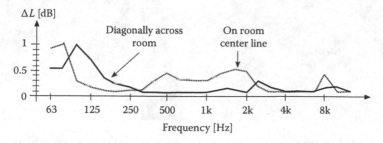

Figure 14.16 Measured deviation from the geometrical distance attenuation from a small sound source at 1 m distance in the anechoic chamber shown in Figure 14.30. If the room were truly anechoic, then ΔL should be 0 dB.

14.5.2 Reciprocity

An interesting property of Equation 2.39 is that the source and observation points r and r_0 can be exchanged without change to the equation. This property is an example of reciprocity.

The principle of reciprocity states that the sound pressure $p_{12}(\omega)$ produced at a receiver location 1 by a point source with a volume velocity $U(\omega)$ at a location 2 is the same in amplitude and phase as the pressure $p_{21}(\omega)$ produced at location 2 if the source would instead be radiating the same volume velocity $U(\omega)$ at location 1.

This allows the use of reciprocal measurements that are practical in finding suitable placement positions for low-frequency loudspeakers. It is easy to place a loudspeaker where the listener's head is supposed to be and then move the microphone around until a position with a suitable low-frequency response is found. Figure 14.17 shows an example of the use of reciprocity to find the best position for a woofer in a car door.

Figure 14.18 shows results of reciprocity measurements of frequency response in a laboratory test chamber. The two measurements were made switching microphone and loudspeaker between positions in the chamber.

The frequency responses are quite similar, mostly being within ±3 dB. The method can be used advantageously in determining *optimally flat* loudspeaker positions since moving the microphone instead of the loudspeaker allows much quicker measurement.

 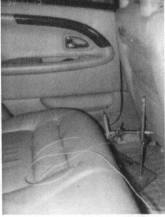

Figure 14.17 In reciprocity measurement, the loudspeaker and microphone locations are interchanged to allow easier testing of suitable loudspeaker positions. The loudspeaker takes the position of the listener's head. (Photos by Mendel Kleiner.)

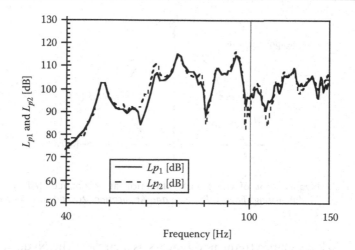

Figure 14.18 Two frequency response curves for a pair of points in a test chamber obtained by switching microphone and loudspeaker positions. (From Kleiner, M. and Lahti, H., Computer prediction of low frequency SPL variations in rooms as a function of loudspeaker placement, in *Proceedings of the 94th Audio Engineering Society Convention*, Berlin, Germany, 1993.)

14.5.3 Modal damping

High damping of low-frequency modes is important for low-frequency sound reproduction of modulated signals and can be measured in several different ways.

A simple way is to use a sine wave generator, find the resonance peak of the mode's frequency response, and switch off the generator and then measuring the decay rate using an oscilloscope or other graphic method. The decay of the mode will be fairly clear if the mode has low damping and there are few other modes in the same frequency range. If there are several modes excited, the decay rate will be difficult to read because of interference between the modes. The situation will be similar to that in reverberation time measurement discussed in a later section.

A more exact way is to measure the complex transfer function of the room. With this data, two approaches are possible. The simplest is to study the frequency response curve. If the resonance peaks of the modes are well separated in frequency, the bandwidth and Q-value of a peak can be determined as described in Chapter 2. The damping δ and the reverberation time T_{60} of each mode can then be obtained using Equations 2.41 and 2.44.

When the modes are close in frequency, a more involved method can be helpful. Plotting the real and imaginary parts of a measured room transfer function as a function of frequency, a curve similar to the one shown in Figure 14.19 is obtained [14]. Such data renderings are called Nyquist

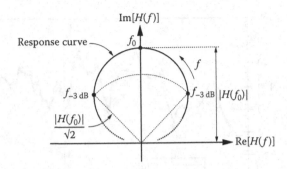

Figure 14.19 Nyquist plot of the transfer function at resonance. (After Ewin, D.J., *Modal Testing: Theory, Practice and Application*, 2nd edn., Wiley-Blackwell, 2000.)

plots. Each resonance frequency appears as a circle in the Nyquist plot and its damping calculated from the circle radius. While determination of the circle can be done with only a three complex transfer function values, a better circle fit is done with transfer functions measured for many frequencies. Special modal analysis software is usually used to extract the properties of transfer functions of multiresonant systems such as rooms. In contrast to many mechanical systems where the modes have about equal density in frequency, the room mode density increases with the square of the frequency. This makes modal analysis at higher frequencies difficult even using special software unless the transfer functions at many different measurement points are analyzed as done for multipoint experimental modal analysis described next.

14.5.4 Spatial response

It is sometimes important to study the spatial distribution of the sound pressure and particle field of a mode. This is done by a procedure called experimental modal analysis. Modal analysis is used to investigate structural vibration, but the method can be used in room acoustics as well to find the geometrical distribution of sound pressure and particle velocity or any other resonant multiple-degree-of-freedom system. The results can be used to optimize placement of sound absorbers such as Helmholtz resonators and tube traps, to find the damping of modes, and to optimize loudspeaker and listener positions. Experimental modal analysis is a combination of spatial sampling and spatial function finding [14].

To find the sound pressure distribution in the room, it is necessary to know the transfer functions between the sample points in the rooms. The number of measurement points must be high to avoid spatial aliasing. The points are usually spread out at about equal distance over a 3D grid with a spacing smaller than about ⅛ λ such as that shown in the car in

Figure 14.20 Microphone grid in a car compartment for experimental modal analysis. (Photo by Claes Fredö.)

Figure 14.20. This can be done by exciting the room using a constant volume velocity-type source. The excitation signal can be wideband sweeps or random or quasi-random noise (MLS or other) so that the transfer functions can be found conveniently using Fourier transforms.

The transfer functions must be measured with high-frequency resolution to allow circle curve-fitting procedures to extract the resonance frequencies and damping constants of the modes of interest. The curve-fitting procedure may be based on single or multiple-degree-of-freedom approaches.

Once the mode frequencies are known, the geometrical distribution of their relative pressure and particle amplitudes and phases may be visualized graphically using computer software. In room acoustics, it is primarily this type of data that is desired. The mode damping is usually dominated by sound absorption at the walls or by wall vibration that leaks energy away from the room system so the edge conditions are very important. The reverberation time of any mode is calculated from its damping constant. The shape of the mode may be much influenced by the placement of objects in the room that have flow resistance. If the mode shape is known, sound-absorbing material can be placed optimally for damping.

In systems that involve structure-borne sound, damping is caused by many mechanisms such as radiation, internal damping, air pumping, wave conversion, and energy leakage at the boundaries between the subsystems. The mode density and damping are usually smaller in structures than in rooms. In rooms, there is usually no internal damping (the losses in the air

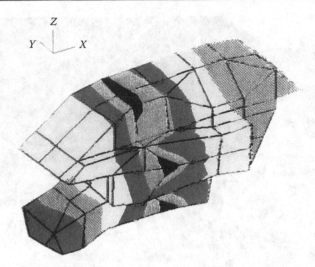

Figure 14.21 A 77 Hz half-wavelength lengthwise mode in a car compartment calculated from the complex sound pressure measured at more than 100 points. SPL distribution in this graph indicates the mode's rms pressure on the compartment's walls (black is high pressure). Only one-half of the compartment is shown.

being much smaller than any other losses at low frequencies). It is also much easier to measure sound pressure in rooms than it is to measure in-plane vibration in structures since sound pressure is a scalar quantity, whereas vibration velocity is vectorial and may not be dominated by vibration normal to the surface on which the accelerometer (vibration sensor) is fastened. Experimental modal analysis can be done investigating room and structure behavior simultaneously.

The sound pressure amplitude distribution in the room is typically presented in a 3D format such as that shown in Figure 14.21. This figure shows the sound pressure distribution of the first mode in the car compartment shown in Figure 14.20. The mode extends from the fire wall of the car to the rear shelf where the phase is 180° relative to the phase at the fire wall floor with a null at the headrest of the driver and front seat passenger.

Figure 14.21 shows that the sound pressure will be very different at the front and rear seats for this mode. The modal testing will usually show that the measured characteristic functions and resonance frequencies seldom coincide with those calculated by FEM models unless the volume of the compartment and the structure are coupled in the finite element analysis. It may be necessary to include the acoustic and mechanical properties of chairs, interior padding, doors, roof, and luggage compartment wall in the finite element model to obtain good agreement between calculated and measured values.

The mode data obtained by experimental modal analysis can then be used in finite element modeling of the room. In this way, the influence of furniture and other factors that have flow resistance and that affect the mode shape can be investigated as well as mode frequency shifts.

14.6 ROOM MEASUREMENT OVER THE SCHROEDER FREQUENCY

14.6.1 Impulse response measurement and analysis

As explained in Chapters 3 and 9, the impulse response is of considerable interest at medium and high frequencies because of the characteristics of human hearing and the signal properties of speech and music. One way of studying the influence of wall reflections and subsequent reverberation on the measurement of loudspeakers is to use impulse response measurement and apply time gating to the measurement. Figure 14.22 shows the principle of time gating. If the room is large, the reflection-free time available between the direct sound from the source and the arrival of the scattered sound from the walls will be long. Time gating corresponds to having a perfect anechoic chamber within the time slot under analysis.

Sometimes, it is necessary to limit the frequency range of the impulse response by filtering. Octave- and ⅓-octave-band filters are usually too

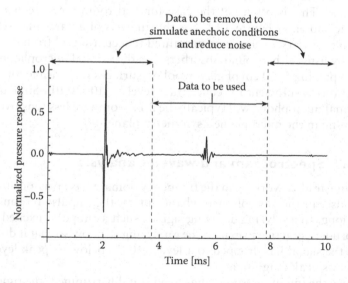

Figure 14.22 Application of time gating to isolate a reflection to be studied from the measurement of the impulse response of a room. (The reflection shown here arrives about 4 ms after the direct sound.)

narrowband so wider bandpass filters must be used. Such filters should be Bessel-type or similar-type filters that have constant delay so that the impulse response curve is distorted minimally. The amount of delay will depend on filter order and must be taken into account in calculating subsequent data. Butterworth and Chebyshev filters have frequency-dependent delay and will change the time history of the signal [15].

By application of proper window functions such as a raised cosine window, the values at the beginning and end of the resulting dataset can be made zero to minimize leakage in the Fourier transform [6]. Since the high-frequency components in the impulse response of rooms generally have a length of interest of about a few milliseconds, it is often advantageous to use time gating techniques [16].

The time gating approach works quite well at medium and high audio frequencies but gives problems in measuring the low-frequency properties of the loudspeaker unless one uses *zero padding* of the impulse response tail, extending its length. This approach, of course, causes errors in the measurement of the low-frequency response so it still makes sense to use a large room. One must remember that zero padding effectively assumes that the response of the system is negligible outside the open gate time, so very resonant systems characterized by strong, long time-extended oscillations will be incorrectly measured. Studying the SPL envelope decay characteristics of the impulse response may give valuable guidance to the appropriateness of time gating.

The type of analysis described is however mainly useful when the reflected sound component can be clearly isolated in the measured impulse response. This is often not the case for real room measurements such as the one shown in Figure 3.12. For the analysis of a particular reflection it is then necessary to eliminate unwanted reflected sound from the measurement by using either sound-absorbers or a directional microphone. In many cases applying 5–10 cm of glass wool on surfaces responsible for unwanted reflections is sufficient to reduce their level by 10–20 dB. Similarly a bidirectional microphone will typically be 20 dB or more less sensitive to sound incoming in the microphone's symmetry plane.

14.6.2 Spectrogram and wavelet analysis

Any linear filter working in the frequency domain has an associated impulse response. In the case of narrowband filters, the impulse response may be very long. In analyzing decaying signals such as impulse-excited modes, it is a prudent practice to extend the analyzing window so that it does not end until the signal has dropped to a level −40 dB below its peak level, that is, 1% of its peak magnitude.

Once the impulse response has been suitably trimmed, the time data can be analyzed to find interesting frequency response features of the system or room by the use of the Fourier transform.

Since hearing acts as a time/frequency analyzer, it is useful to have methods to study the information contained in the impulse response in a similar way as time–frequency histories. Two such methods are *short-time Fourier transform* (STFT) analysis for which the results are graphically presented as spectrograms and *wavelet* analysis for which the results are graphically presented as scalograms.

The basic principle of spectrogram analysis is to look at the frequency response magnitude in time snippets, that is, repeatedly offset local analyzing time windows, of the impulse response data. By using windowing techniques and moving the local analyzing time window along the time axis of the impulse response, one can graphically show how the frequency component develops over time in the response of the loudspeaker. Spectrogram analysis gives the same time and frequency resolution over the whole time/frequency region under study.

An alternative to the use of STFT-based spectrogram is to use wavelet analysis. While linear frequency response analysis is used in Fourier transforms, wavelet transforms are studied on a logarithmic frequency basis. The resulting scalograms offer different time and frequency resolutions in different parts of the joint time/frequency region under study and correspond to constant relative bandwidth frequency analysis. The difference in principle between the analysis methods is shown in Figure 14.23.

The scalogram graphs are typically slightly easier to study than conventional STFT spectrograms. Wavelet analysis involves convolution of time snippets of the impulse response by suitably chosen short impulses, or wavelets, that are scaled in length depending on their base frequency. The wavelets can be regarded as the impulse responses of narrowband filters.

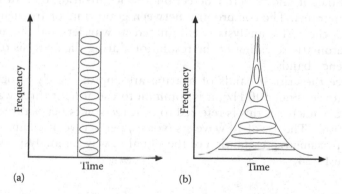

Figure 14.23 Time–frequency relationships in different octaves for (a) the STFT-based spectrogram and (b) the wavelet transform–based scalogram. The spectrogram consists of a number of Fourier transforms of a signal multiplied by a *local analyzing window*, gradually shifted over the signal. The ellipses indicate the relationship between time and frequency width.

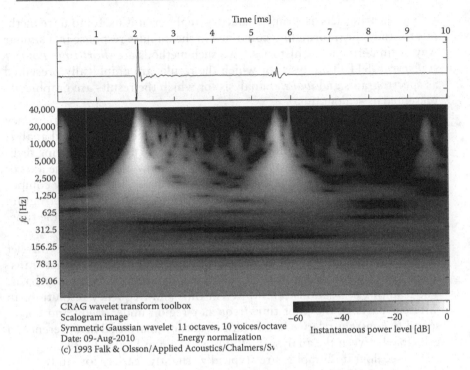

Figure 14.24 An example of the application of a wavelet transform–based scalogram to the impulse response shown at the top of the figure.

Wavelet analysis is advantageous when analyzing nonstationary signals, since it allows better detection and localization of transients than spectrograms. The compromise between good time or frequency resolution in the STFT analysis is eliminated in wavelet analysis, because of the automatic rescaling of the resolution during the analysis of different frequency bands.

Since the critical bands of hearing are approximately ⅓ octave band wide (over about 500 Hz), it is common to use at least three wavelets per octave (sometimes this is referred to as three voices per octave in wavelet literature). The use of 10 wavelets (voices) per octave often gives a graphically pleasing representation of the signal in wavelet analysis as shown in Figure 14.24.

14.6.3 Frequency response

While the frequency response can be derived from the impulse response, it is sometimes necessary to measure the approximate frequency response of the room directly. One way then is to use random noise filtered into

⅓-octave-bands or even ⅙-octave-bands. It is, however, better to drive the room with a signal that is less random than white noise such as an FM-modulated tone. A computer can be used for generating the signal. With the right choice of tone frequency, modulation frequency and bandwidth, and time averaging in the meter, very precise measurements can be made, which are quick and convenient if several places in a room need to be measured.

14.6.4 Metrics for room acoustics

It would be very advantageous if it were possible to specify the acoustic quality of a room by some simple numbers or indicators using impulse response data. However, a room may show large numerical differences between current quality metrics measured at various positions in the room while the subjective experience of listening in the room does not vary accordingly.

Metrics for room acoustics such as reverberation time, strength, speech transmission index, clarity, and lateral energy fraction can be derived from the measurement of the spatial impulse response from source to receiver.

The measured impulse response, h_{meas}, is a convolution (*) of the impulse responses of the loudspeaker, microphone, and other parts of the measurement system:

$$h(t) = h_{lsp}(t) * h_{room}(t) * h_{mic}(t) * h_{other}(t) \tag{14.15}$$

where h_{lsp}, h_{room}, h_{mic}, and h_{other} are the impulse responses of the loudspeaker, room, microphone, and other equipment (such as electronic filters and sound cards) involved in the measurement process and that would ideally be compensated for.

The influence of these other, undesired impulse responses is generally disregarded when considering the room acoustics metrics. For simplicity, the room impulse response is often set as the *raw* measured impulse response or filtered using a suitable bandpass filter such as an octave-band filter. It is important to note though that such filters will have group delay that must be compensated for. Using deconvolution, it is possible to eliminate some of the linear distortions generated by the electroacoustic equipment. Note, however, that most sound sources and microphones have directivity-dependent impulse responses that must either be acknowledged or taken into account as errors.

14.6.5 Reverberation time

The reverberation curve is central to the determination of the reverberation time. The curve can be determined either from the sound of recorded decays of interrupted noise in the room or from the measurement of room impulse response. Using the room impulse response, one avoids the uncertainty associated with narrowband-filtered noise signals discussed previously. If such signals are used, then several repeats or reverberation curves must be averaged to increase precision.

The reverberation curve is the curve that can be fitted to the decay of SPL versus time, shown in Figure 14.25b. In the case of a single, exponentially damped oscillation, the curve would be a straight line, the lower part of which would be obscured by room and microphone noise.

If the decaying signal is noise or otherwise contains a number of frequency components (such as an FM-modulated sine, as described previously), the reverberation curve becomes wiggly simply because of the summation of the different tones at each instant. In practice, the different frequencies excite different room resonances that typically have unequal damping and reverberation times. These effects make the reverberation curve appear as a noisy line, but the latter condition leads to the reverberation curve having double or more slopes. In the low-modal density and frequency region, the reverberation curves will be dependent on the number of modes excited as discussed in the previous section. Reverberation is normally measured using octave or ⅓ octave bandpass filters. These limit the number of modes available in the passband. For the sound field at the bandpass frequency measured to be considered diffuse enough for reverberation time measurement, there should ideally be at least 10 modes in the passband. For measurements where the number of modes per band is low, a special measurement technique can be used to measure the sound absorption of the room [17].

A better way to measure the reverberation time than by direct visual approximation to the wiggly decay curve is to evaluate the reverberation process by backward integration of the squared impulse response, *Schroeder's method*. The backward integration method is the common way to measure the reverberation curve. An example of such a curve is shown in Figure 14.25c. In any case, the determination of reverberation time is prone to vary depending on the method used. Measured reverberation time values will differ as a function of source and receiver locations in the room since the sound field is never ideally diffuse. Typically, at least two source and five receiver positions are measured. since microphones are easier to move than loudspeakers.

The sound absorption coefficient for a material depends on sound field diffusivity and will differ between rooms. For direct sound and early reflections, the angle of incidence is important as explained in Chapter 8, so the diffusivity of the sound field is important for the results. Comparison of results between laboratories shows large differences in measured sound absorption values, sometimes about as large as ±10%. The differences between chambers were investigated in a Scandinavian round-robin test. The same test sample was placed on the floor of the

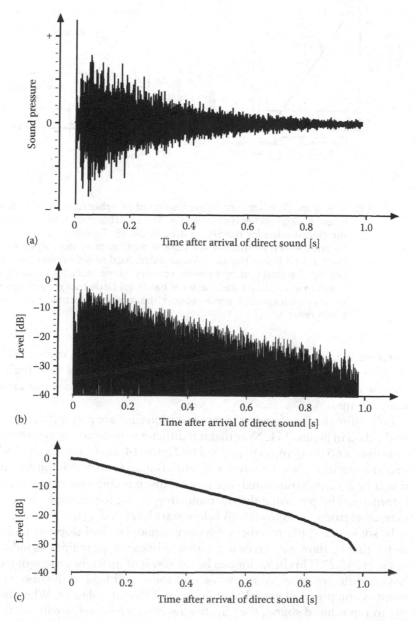

Figure 14.25 Three ways of representing the same measured sound decay characteristics: (a) impulse response, (b) SPL drop, and (c) reverberation curve obtained by reverse summation of the energy of the impulse response in the top curve.

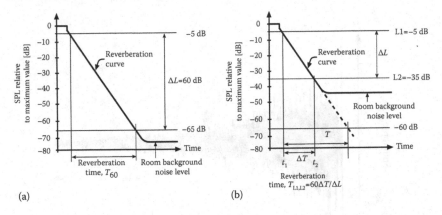

Figure 14.26 This graph illustrates the determination of reverberation times under different background noise conditions. The reverberation time is defined as the time it takes for the SPL to drop by 60 dB. To ensure that the reverberant sound is diffuse, the start of the measurement (for large rooms) is usually 5 dB below the steady-state value. Real reverberation curves will not be characterized by smooth, constant slope, particularly in narrow, low-frequency octave and narrower bands. (a) Little background noise and (b) much background noise reduces the available dynamic range of the measurement.

different measurement chambers of the laboratories. All rooms fulfilled the ISO 354 standard criteria [32], but a room that was very high and narrow gave higher sound-absorption coefficients than the ones that were less extreme.

Even quite small differences in reverberation time are perceptible as shown by the data in Figure 9.11. Note that it is difficult to measure the reverberation time over a 60 dB span, as illustrated by Figure 14.26. Generally, one is limited to estimating the decay over a 30–40 dB span from which the reverberation time is then extrapolated. Because of this, it is common to give the span information by providing the start and stop levels, for example, for a span extending from −5 dB to −35 dB below start level as $T_{-5,-35}$.

In some cases, the reverberation curves may be dual-slope or multiple-slope, that is, showing two or more almost linear slope regions as indicated in Figure 14.27. This behavior can be the result of different groups of modes having different decay constants or that the sound field at the observation point is composed of sound coming from different volumes. When listening to reproduced sound, the effective reverberation curves will usually be dual-slope because of the presence of listening room reverberation along with the recorded reverberation.

Typically, such multislope reverberation decay curves can be found for rooms that have one large surface that is more highly absorptive than the

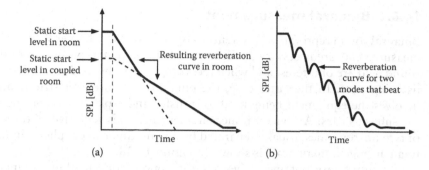

Figure 14.27 (a) Example of a dual-slope reverberation curve for a room that is acoustically somewhat coupled to another room having a longer reverberation time. (b) Example of reverberation curve resulting from the decay of two approximately equal excited modes with slightly different resonance frequencies but similar reverberation times.

others or when two rooms are coupled to one another through an opening such as an open door. Dual-slope reverberation processes can also be found as a result of structural resonance.

According to the ISO 3382 standard—mainly concerned with the acoustics of large rooms—the reverberation time is obtained by extrapolation to 60 dB of a line fitted to the measured reverberation curve [33]. Unless otherwise noted, the room impulse responses are to be measured using an omnidirectional microphone. The standard also requires the measurements to be done using at least two source positions and five receiver positions, spaced in a specified way on the stage and in the audience seating area. The standard also defines many of the quality metrics used for large auditoria. These are also summarized in Reference 9.

To start the reverberation time measurement from the point of –5 dB limit may not be optimal from the perspective of listening. In large rooms, the fine structure of reverberation may be audible so late as 400 ms after the arrival of direct sound [16], but the forward masking of hearing effectively masks at least 30 ms of the initial sound.

Research on the perceived reverberation time of large rooms for running music shows that the subjectively determined reverberation time can be composed of at least two parts: an initial part that is determined by the fall of SPL during the first 150 ms of the impulse response (and that determines the subjectively perceived reverberation time during most of the music performance) and a later part that should take its start at this point (but that is only heard at the end of a piece of music and determines the subjectively perceived late reverberation time that is likely to be fixed in memory after the concert).

14.6.6 Binaural measurement

Binaural sound reproduction systems were discussed in Chapter 9. Binaural measurements are necessary particularly when the spatial properties of the sound field are of interest and when recordings are to be made for objective listening tests. In the latter case, the binaural measurement strictly also involves the equipment being used to transfer and replay the recording to the subject's ears. All binaural measurement methods will involve the use of two microphones, usually separated by a manikin head or sphere similar to an average human head as shown in Figure 14.28.

Best results are obtained when an anthropometric manikin, equipped with binaurally mounted microphones, is used such as the one shown in the earlier figure. The acoustic properties of the manikin such as the HRTFs are determined by the torso, the head, and the ear replicas as discussed in Chapters 6 through 9. The HRTFs describe the sound pressure spectrum at the head center position, without the head present, and the sound pressure spectrum at some point in the respective listener's ears with the head present, for a plane wave at various angles of incidence.

To be useful, binaural recordings have to be made at typical listening levels and replayed at the correct sound levels. This requires little distortion and large dynamic range. A common difficulty is to reproduce the low-frequency information adequately. In the original sound field, strong very low-frequency sound is also felt by the body. It is easy to make

Figure 14.28 The anthropometric head and torso simulator KEMAR manikin for binaural sound recording and measurement is based on worldwide average human male and female head and torso dimensions. (Photo courtesy of G.R.A.S., Copenhagen, Denmark.)

mistakes in judgment of audio quality by only listening in the laboratory via a headphone system.

The most important property of binaural sound reproduction systems is that sound fields can be simulated reasonably well without cognitive factors influencing the subjective judgment of sound quality. However, good sound quality is sometimes defined as *adequacy of sound relative to expectations*, which means that optimum sound quality as judged by a human listener in the car will usually not coincide with objectively measured sound field properties.

Binaural measurements are used to document sound fields and to record acoustic events. Binaural measurement systems are best suited to determine factors that are different between left and right ears such as phase differences, time delay, and intensity differences. This makes them, in principle, suitable, for example, for the determination of sound stage and spaciousness. The sound stage, the feeling of presence, the feeling of immersion, diffusivity, and reverberance are all examples of properties that can be characterized to some extent by binaural information.

The IACC metric was introduced in Section 8.3.2 and gives a value of the similarity between the left and right ear binaural room impulse response channels. One is usually interested in the reverberation-sounding diffuse. Since reverberation diffuseness is closely connected to correlation, it is reasonable to assume that it is advantageous that the late reverberation part of the spatial binaural impulse response has a small value of IACC.

14.7 MEASUREMENTS USING LISTENERS

Measurement is also possible using less precise tools such as listeners in *listening tests*. While the physical properties of sound can be described quite precisely by, for example, sound pressure, phase, frequency, and duration, it is not possible to describe the auditory experience by similarly well-defined dimensions. The verbal descriptors usually used by untrained listeners to describe sounds are interrelated and not suitable to be used as metrics. In addition, the auditory experience of listeners is individual and different. Also, any listener's experience may change with time as a result of aging, for example.

Visual and other sensory cues and cognitive factors are always present while listening and difficult to remove. In the laboratory, these cues and factors may be neutralized. This, however, does not render the measurements more true necessarily since the neutralization process, for example, darkened room and blindfolds, may themselves detract from the true listening experience or add bias by their use. Of course the interaction between aural and visual cues is in itself interesting, but such investigations can only be systematically done under limited laboratory conditions as shown in the example in Figure 14.29.

Figure 14.29 Listening tests in an anechoic chamber, combining aural and visual cues. (In a listening test the lights will be switched off and only the screen visible.) (Photo by Pontus Larsson.)

14.7.1 Simple listening tests

In most listening tests, the test organizer has prepared a task for the listener that is presented with fewer degrees of freedom. Such listening tests are a special type of subjective measurement used both to compare and to give *absolute* judgment of different products, situations, and/or listening conditions and are similar to the taste testing of food and drink [18,19].

In audio and room acoustics listening tests, one or more persons are typically presented with audio samples—either in situ or by binaural sound reproduction—and asked to evaluate the room or loudspeakers. Listening may be either objective (what do they hear) or subjective (what do they prefer or dislike). Typically, a group of experienced listeners form a jury or expert panel. The composition of the group is likely to affect the outcome of the test. The test jury is typically asked to judge not only the sound character but also the acceptability of the sound and to give their preference among sounds.

Because of time and economic constraints, trained *expert listeners*, sometimes also called *golden ears*, that have a high degree of sensitivity and experience in listening and who are able to make consistent and repeatable assessments of various sounds are often chosen to perform the listening test. A central problem in any listening tests is that some audio properties are only discernible after very prolonged listening (and training).

14.7.2 Verbal descriptors

Much more freedom is given to the listener when verbal elicitation is used. As shown in Figure 7.1, the listener first does an internal evaluation and then communicates that evaluation, usually by using language, which by verbal descriptors conveys meaning. These descriptors are likely to be adapted to the receiver of the communication, a lay person requiring different descriptors than a colleague or expert listener.

Brainstorming in focus groups can be very useful to elicit verbal descriptors. The desired outcome is a list of *orthogonal verbal descriptors*, words that do not have overlapping meaning. Usually, the process of finding these words starts by eliciting words from the various individuals involved during listening sessions. These words are then analyzed qualitatively and a lexicon is assembled particularly to each individual. The lexicons can then be discussed by the group. This is done to harmonize the lexicon and reduce the number of words. Depending on the choice of words, unipolar or bipolar scales may be assembled, and finally, the ends of these scales are decided on by the group.

Communication involves two people so the language must be commonly understood by both. This common language is mostly based on longtime informal learning by example by identification, recognition, and verbalization as outlined in Figure 14.30. The language of critical listening is however seldom learned in a rigorous manner. Most persons will have difficulty in describing what they hear. By attending appropriate training, many are able to enhance their listening skills and learn proper verbalization [20,21]. The situation is somewhat analogous to our ability to describe taste of food and drink. A person has enjoyed a good dinner but the person's descriptions of the taste and texture of the food are seldom sufficient for a cook to recreate the dish from those sensory descriptions alone.

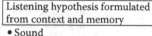

> Listening hypothesis formulated from context and memory
> - Sound
> - Sensory transduction
> - Short-term memory activation
> - Recognition
>   - Without identification
>   - With identification
>   - Access to lexicon of names
>   - Verbalization

Figure 14.30 Schematic representation of auditory perception with an *ecological* point of view. Top-down organization. (After Gaillard, P., *Étude de la perception des transitoires d'attaque des sons de steeldrums: particularités acoustiques, transformation par synthèse et catégorisation*, Université de Toulouse II - Le Mirail, Toulouse, France, 2000.)

The words chosen to represent what is heard are most often carried over from other modalities such as kinesthetics, vision, and taste. This is called synesthesia and may be, for example, direct from vision to hearing (*white noise* is such an example) or partially by the extension of meaning, for example, from dimensions and shapes to hearing (*a sharp sound* is such an example) as indicated in Figure 14.31. Touch, taste, and smell are the sensory modalities that are the first to be experienced and thus learned [23].

Table 14.2 shows some examples of real and partial synesthetic terms that apply to hearing.

In an experiment on descriptive analysis of low-frequency audio in small rooms, the following words were found: bassy, clean, clear, distant, dry, flat, fuzzy, imprecise, muddy, muffled, precise, punchy, rattles, resonant, reverberant, strong, tight, undefined, weak, and wooly [26]. Out of these, three words were devised for the test, namely, articulation, resonance, and bass content.

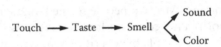

Touch ⟶ Taste ⟶ Smell 〈 Sound / Color

Figure 14.31 A scheme by which synesthetic words develop from one sensory field to another. (Adapted from Rioux, V., Sound quality of flue organ pipes, Dissertation, Chalmers University of Technology, Gothenburg, Sweden, 2001.)

Table 14.2 Real and partial synesthetic terms applicable to the hearing sensory mode

Vision	Dark, light, bright, brilliant, dusky, colorful, colorless, clear, translucent
Touch	Sharp, dull, cutting, rough, smooth, dry, fluid, even warm, cold, cool, hot, soft, hard, massive, muddled
Surface structure temperature consistency	
Physical description	High, low, deep, small, full, empty, light, heavy, solid
Shape	Thick, thin, cracked, compact, scattered, distorted
Bodily property	Strong, weak, fat, lean

Sources: Adapted from Rioux, V., Sound quality of flue organ pipes, Dissertation, Chalmers University of Technology, Gothenburg, Sweden, 2001; Abelin, Å., Studies in sound symbolism, Dissertation, Göteborgs Universitet, Gothenburg, Sweden, 1988.

In a study on sound reproduction in small rooms, the following words were found: echo, fluttering, rattle, spectral coloration, temporal coloration, timbral difference, wavering, and whooshing [27]. The reader is left to decide on whether these words correspond to his or her experience in listening.

Word elicitation by listening as discussed in the previous paragraph is a top-down approach that is extremely time consuming. In the university environment, the bottom-up approach is usually used instead since it comparatively quickly gives results that have good statistical validity. The research is then concentrated on finding fundamental psychoacoustic dimensions by which the auditory experience of any sound may be described, as a starting point.

REFERENCES

1. ISO Standard handbooks "Acoustics" (1995) Volume 1: General aspects of acoustics; methods of noise measurement in general; noise with respect to human beings, Vol. 2: Noise emitted by vehicles; noise emitted by specific machines and equipment; acoustics in building. ISO, International Organisation for Standardisation, Geneva, Switzerland.
2. G.R.A.S, Holte, Denmark, www.gras.dk (accessed May 2013).
3. Brüel & Kjær. (1996) *Microphone Handbook*, Vol. 1 (BE1447-11). Nærum, Denmark.
4. Wong, G.S.K. and Embleton, T.S.W. (eds.) (1995) *AIP Handbook of Condenser Microphones: Theory, Calibration and Measurements*. American Institute of Physics, New York.
5. Farina, A. (2000) Simultaneous measurement of impulse response and distortion with a swept-sine technique, in *Proceedings of the 108th Audio Engineering Society Convention*, Paris, France, Paper Number: 5093.
6. Broch, J.T. (1990) *Principles of Experimental Frequency Analysis*. Elsevier, London, U.K.
7. Slaney, M. (1993) An efficient implementation of the Patterson-Holdsworth auditory filterbank. Technical Report 35, Perception Group, Advanced Technology Group, Apple Computer Co. Cupertino, CA.
8. Tocci, G.C. (2000) Room Noise Criteria—The state of art in the year 2000. *Noise News Int.* 8(3), 106–119.
9. Beranek, L.L. (1996) *Concert Halls and Opera Houses: Music, Acoustics, and Architecture*. Acoustical Society of America, New York.
10. Harris, C.M. (ed.) (1991) *Handbook of Acoustical Measurements and Noise Control*. McGraw-Hill, New York (reprinted by Acoustical Society of America).
11. Lyon, R.H. (2003) Product sound quality—From perception to design, sound and vibration, March 2003. Available at http://www.sandv.com/downloads/0303lyon.pdf (accessed May 2013).
12. Kleiner, M. (2011) *Acoustics and Audio Technology*, 3rd edn., J. Ross Publishing, Ft. Lauderdale, FL.

13. Kleiner, M. and Lahti, H. (1993) Computer prediction of low frequency SPL variations in rooms as a function of loudspeaker placement, in *Proceedings of the 94th Audio Engineering Society Convention*, Berlin, Germany.

14. Ewin, D.J. (2000) *Modal Testing: Theory, Practice and Application*, 2nd edn. Wiley-Blackwell, New York.

15. Zverev, A.I. (2005) *Handbook of Filter Synthesis*, Revised edition. Wiley-Interscience, New York.

16. Kleiner, M., Svensson, P., Dalenbäck, B.-I., and Linusson, P. (1994) Audibility of individual characteristics in reverberation tails of a concert hall, in *Proceedings of the 127th Meeting of the Acoustical Society of America*, Cambridge, MA.

17. Zha, X., Fuchs, H.V., Nocke, C., and Han, X. (1999) Measurement of an effective absorption coefficient below 100 Hz, *Acoust. Bull.*, January/February 1999. Leaflet published by the Fraunhofer Institute, Stuttgart, Germany.

18. Bech, S. and Zacharov, N. (2006) *Perceptual Audio Evaluation—Theory, Method and Application*. Wiley, London, U.K.

19. Meilgaard, M., Carr, B.T., and Civille, G.V. (2006) *Sensory Evaluation Techniques*, 4th edn. CRC Press, Boca Raton, FL.

20. Berg, J. and Rumsey, F. (2003) Systematic evaluation of perceived spatial quality, *AES 24th International Conference on Multichannel Audio*, Banff, Alberta, Canada.

21. Bech, S. (1992) Selection and training of subjects for listening tests on sound-reproducing equipment. *J. Audio Eng. Soc.*, 40(7/8), 590–610.

22. Gaillard, P. (2000) *Étude de la perception des transitoires d'attaque des sons de steeldrums: particularités acoustiques, transformation par synthèse et catégorisation*. Université de Toulouse II - Le Mirail, Toulouse, France.

23. Williams, J. (1976) Synaesthetic adjectives: A possible law of semantic change, *Language*, 52(2), 461–478.

24. Rioux, V. (2001) Sound quality of flue organ pipes. Dissertation, Chalmers University of Technology, Gothenburg, Sweden.

25. Abelin, Å. (1988) Studies in sound symbolism. Dissertation, Göteborgs Universitet, Gothenburg, Sweden.

26. Wankling, M. et al. (2010) Improving the assessment of low frequency room acoustics using descriptive analysis, in *Proceedings of the 129th Audio Engineering Society Convention*, San Francisco, CA, Paper 8311.

27. Rubak, P. et al. (2003) Coloration in natural and artificial room impulse responses, in *Proceedings of the Audio Engineering Society 23rd International Conference on Signal Processing in Audio Recording and Reproduction*, Helsingør, Denmark.

28. Kleiner, M. (2013) *Electroacoustics*. CRC Press, Boca Raton, FL.

29. MLSSA Software by DRA Laboratories, Sarasota, FL www.mlssa.com (accessed May 2013).

30. ISO 9614-1:1993 Acoustics - Determination of sound power levels of noise sources using sound intensity - Part 1: Measurement at discrete points. ISO, International Organisation for Standardisation, Geneva, Switzerland.

31. ISO 532:1975 Acoustics - Method for calculating loudness level. ISO, International Organisation for Standardisation, Geneva, Switzerland.

32. ISO 354:2003, Acoustics – Measurement of sound absorption in a rever-
    beration room. ISO, International Organisation for Standardisation, Geneva,
    Switzerland.
33. ISO 3382 Acoustics – Measurement of room acoustic parameters. Part 1:
    Performance rooms; Part 2: Ordinary rooms. ISO, International Organisation
    for Standardisation, Geneva, Switzerland.
34. IEC 651:1979 - Sound Level Meters. IEC, International Electrotechnical
    Commission, Geneva, Switzerland.

Index

room acoustics, 163–164
and scattering, *see* Scattering
sound absorption, 178–180
sound reflection, 163
volume diffusers, 182–185
Direct and reverberant sound field
draw-away curve, 88–90
energy density, 86
sound power, 87
steady-state sound pressure level, 86
Distortion
audio and acoustics, 416
linear, 416
nonlinear, 416
swept sine waves, 416

E

Ear
canal, eardrum and middle ear,
196–197
cochlea, 197–198
hair cells, 198–200
head-related coordinate system,
189–190
HRTFs, 190–193
ILD and ITD, 193
pinnae, 194–196
signal processing blocks,
189–190
transducer system, 189
Equivalent rectangular bandwidth
(ERB), 219–220
ERB, *see* Equivalent rectangular
bandwidth (ERB)
Euler's equation
continuity, 6–7
3D gradient operator, 4
1D sound field, 3
Newton's law, 4
particle velocity, 9, 23
sound pressure, 3
sound propagation direction, 4
Experimental modal analysis
car compartment
microphone grid, 432–433
sound pressure amplitude
distribution, 434
description, 432
finite element modeling, room, 435
mode damping, 433–434
spatial aliasing, 432

F

FEM, *see* Finite element method (FEM)
Fermat's principle, 101
Finite element method (FEM)
collocation method, 396
description, 395
Helmholtz equation, 395
limitations, 400–401
meshing, 396–397
room acoustics, 395
Fitzroy's formula, 85
Flow resistance
definition, 127
micromanometer, 127
rayl [N s/m], 127
Flutter echo
basses and cellos, 299
coloration, 297
description, 295–296
floor and ceiling reflections, 297
high-frequency sound, 298
repeated reflections, 296
room impulse response, 297–298
vertical array loudspeakers, 298
Forced waves, enclosures
analysis and application, 93
analytical evaluation, 92
bandwidth of modes, 53–54
damping effects, 52–53
equation of continuity, 90
formulation, 51–52
reverberation and modal
bandwidth, 54
room excitation and frequency,
54–58
room response, 58–60
sound pressure, 93
thermodynamic law, 90–91
Fourier transform
aperiodic functions, 19–20
autocorrelation theorem, 18
complex line spectrum, 14
convolution function, 19
correlation function, 17
cosine function, 15
signal $f(t)$, 15
sin and cos functions, 16
Free sound waves
cavity, vibrating wall, 38
compression and rarefaction, 37
description, 38

Printed in the United States
by Baker & Taylor Publisher Services

Printed in the United States
by Baker & Taylor Publisher Services